A Primer of Ecological Aquaculture

Endorsements

"This important new volume on aquaculture presents a remarkable synthesis of all the key issues involved in the commercial culturing of freshwater and marine species. It provides a comprehensive overview of the basic biological elements of aquaculture, including new information on the use of genomic and genetic tools, and incorporates these analyses into a broader ecological, environmental, and sociological context. The emphasis on strategies for protecting water quality—a vital part of the aquaculture enterprise—is especially well-presented. The volume deserves wide readership among students and scientists who not only have interests in aquaculture, but also in sustaining robust aquatic ecosystems in freshwater and marine environments."

George N. Somero, David and Lucile Packard Emeritus Professor of Marine Science, Hopkins Marine Station, Stanford University, USA

"Dietmar Kültz has distilled many years of teaching an undergraduate course in aquaculture into an engaging and accessible text. The book is a huge endeavour and presents an honest and rigorous assessment of aquaculture practices with clear, detailed descriptions of invertebrate and vertebrate aquatic culture, diseases, and water management. The 'Key Conclusions' at the end of every chapter are particularly useful for students."

Patricia A. Wright, Professor, Department of Integrative Biology, University of Guelph, Canada

"A must-read primer for anyone considering aquaculture either as undergraduate and postgraduate students or aquaculture farmers. Dietmar Kültz takes readers on a fascinating journey into the world of ecological aquaculture to stimulate awareness of the need to reach a more sustainable future."

Silvia Gomez-Jimenez, Director of Accredited Aquaculture Diagnosis Laboratory, Research Centre for Food and Development, Mexico

"*A Primer of Ecological Aquaculture* by Dietmar Kültz is the latest addition to the growing list of entry-level books on the topic of aquaculture. It differs from other books, however, on its emphasis on ecological aspects of aquaculture. As such, it is like a new tool that seems impossible to have lived without before it was acquired. The book succeeds by providing a holistic assessment of various aquaculture production systems including inputs, outputs and how they contrast and compare among various species groups. Most importantly, the book presents both problems and opportunities associated with aquaculture in a concise and balanced manner. Aquaculture production must increase to meet future seafood demand and this book provides solid information and guidance on how society can meet seafood demand by employing sustainable aquaculture practices."

Ronald Hardy, Professor Emeritus, Department of Animal, Veterinary, and Food Sciences, University of Idaho, USA

"As the demand for wild caught seafood is continuing to increase while stock volumes are rapidly decreasing, our future food security will rely more heavily on ecologically sustainable aquaculture practices. Kültz presents a detailed exploration of its limitations and novel paths to diminish its ecological footprint, enriched with comprehensive analysis and clear graphics. This accessible primer systematically dissects the rationales behind current practices, from monocultures and polycultures to microcosms and mesocosms. It also provides clear signposts to sustainable practices aimed at minimizing the exploitation of natural ecosystems, while maximizing production efficiency and food security."

Gillian Renshaw, Professor, School of Allied Health Sciences, Griffith University, Australia

A Primer of Ecological Aquaculture

Dietmar Kültz

Professor of Physiological Genomics, College of Agricultural and Environmental Sciences, University of California Davis, USA

OXFORD
UNIVERSITY PRESS

Great Clarendon Street, Oxford, OX2 6DP,
United Kingdom

Oxford University Press is a department of the University of Oxford.
It furthers the University's objective of excellence in research, scholarship,
and education by publishing worldwide. Oxford is a registered trade mark of
Oxford University Press in the UK and in certain other countries

Impression: 1

Published in the United States of America by Oxford University Press
198 Madison Avenue, New York, NY 10016, United States of America

British Library Cataloguing in Publication Data
Data available

Library of Congress Control Number: 2022930150

ISBN 978–0–19–885022–9 (hbk)
ISBN 978–0–19–885023–6 (pbk)

DOI: 10.1093/oso/9780198850229.001.0001

Printed and bound by
CPI Group (UK) Ltd, Croydon, CR0 4YY

Cover image: High-density aquaculture of Bigeye scad (Selar crumenophthalmus)
is supported by their natural schooling behaviour. Photo by author.

About the Author

Dietmar Kültz is a Professor of Physiological Genomics at the College of Agricultural and Environmental Sciences, University of California Davis, USA. His laboratory focuses on investigating the mechanisms of stress-induced evolution in fish and marine invertebrates. His research spans molecular to organism levels of biological complexity and utilizes reductionist synthetic biology, biochemical, and holistic systems level approaches to dissect causality between environmental effects on cells and organisms, physiological responses, and complex adaptive phenotypes. He teaches a molecular genetics laboratory course, an introductory aquaculture course, and a stress physiology course at UC Davis. Professor Kültz received his BSc/MS and doctoral degrees from the University of Rostock in Germany. He was a DAAD postdoctoral fellow at Oregon State University, a Fogarty Visiting Fellow at the National Institutes of Health (Bethesda), and Assistant Professor at the University of Florida before joining the faculty at UC Davis.

Preface

Aquaculture exemplifies the ongoing global struggle to strike a sustainable balance between the conflicting needs of a large human world population, human health, ecosystem health, the welfare of wild and domesticated animals, and the economic principles of globalized economies. On the one hand, aquaculture has great potential for providing us with a healthy and nutritious food supply while alleviating pressure on capture fisheries and reducing fisheries-induced habitat destruction, overfishing, genetic modification of wild populations, and wholesale waste of bycatch. On the other hand, aquaculture relies heavily on clean water, an increasingly precious (and dwindling) resource that is subject to intense pressure through being used for many competing purposes. The multifaceted nature of aquaculture is most evident in the division of approaches between those that emphasize minimal reliance on ecosystem services (e.g. recirculating aquaculture systems, aquaponics) versus those aimed at maximizing profitability (e.g. open pond, cage, and raceway systems). Ecological considerations have received increasing attention in aquaculture since the turn of the millennium and their proper implementation promises to greatly accelerate the rate at which the field moves forward.

This book aims to introduce students to the basic concepts, opportunities, and challenges of aquaculture with an emphasis on ecological considerations. I contrast the many values and promises of aquaculture (including specific approaches and commodities) with their pitfalls, drawbacks, and challenges. My goal is to provide students with a broad understanding of the general state of aquaculture to equip them with the knowledge they need to contribute to future advances in the field. To achieve this goal, a critical assessment of current aquaculture practices from a broad, interdisciplinary perspective is necessary. The book carries out such an assessment from the standpoint of how best to align the two major (and often conflicting) goals of future aquaculture development: minimizing reliance on ecosystem services while maximizing productivity.

This book represents a broad snapshot assessment of the current state, past developments, and prospects of applying knowledge in aquatic animal biology and engineering to the captive culture of aquatic animals. My intent was to transcend traditional disciplinary boundaries whenever possible and present contrasting perspectives whenever applicable to stimulate critical assessment of aquaculture practices by students. There are many different opinions about the best way to further develop aquaculture. The perspective presented here should not be taken dogmatically, but rather should stimulate critical thinking in readers. To make the book more accessible, e.g. to students taking a semester- or quarter-long course on this subject, it presents the material concisely and provides select examples to illustrate pertinent points and key concepts, and it explains scientific and aquaculture-specific terminology when first used. It is not my intent to comprehensively review specific commodities or go into undue detail regarding specific aspects of aquaculture if such detail is unnecessary to support salient facts and arguments.

The motivation for writing this book arose from feedback provided by my students and colleagues. I have been teaching a general education (GE) class on aquaculture (ANS 18) at the University of California, Davis for more than a decade but a suitable text accompanying the class material was missing. Most students I have taught prefer single-author texts that require little prior knowledge of a subject and are written in a consistent style, rather than more specialized volumes comprised of chapters by different authors. Although there are many excellent books focusing on aquaculture, some of which are referenced in this volume, an integrated discussion of the ecological cornerstones of aquaculture in the form of a concise 'primer' was needed to help students learn pertinent concepts and principles. This book thus attempts to serve as such a 'primer' in a format that is accessible to readers from a broad range of academic backgrounds and experience. It is meant to consolidate and expand on my lecture notes to provide a study guide for students interested in both the natural biology and the culture of aquatic organisms, both marine and freshwater.

Although I majored in marine ecology when at university, my own research now focuses heavily on biochemical mechanisms of environmental tolerance and stress-induced evolution in fish and marine invertebrates using molecular, cellular, and systems biology approaches. It seems natural to investigate molecular phenomena at an increasing level of depth to explain how organisms function within their environment and how life works. However, at the same time it is important to integrate reductionist approaches with holistic approaches that explain the function of systems, e.g. ecosystems or aquaculture systems, that are more than just the sum of their parts. This book emphasizes the holistic view throughout.

This book could not have been completed without the support and encouragement from my family, friends, and colleagues. Many thanks go to my family, especially my wife, son, and parents for always supporting my pursuit of scientific research and enduring my spending long hours in the laboratory and at the desk pondering scientific questions. Their patience, encouragement, and feedback were invaluable for completing this book project, which took more effort and time than I had originally anticipated.

I am also particularly grateful to Professor George Somero (Stanford University), my post-doc mentor and long-term friend, who has read and commented on early drafts of all chapters. His knowledgeable feedback and constructive comments helped greatly in shaping the final draft of the book and pushing it over the finish line on time. I am also very grateful to other colleagues and friends who have read and made valuable suggestions for improving individual chapters of this book, including Professors Steve McCormick (University of Massachusetts), Ron Hardy (University of Idaho), Colin Brauner (University of British Columbia), Patricia Wright (University of Guelph), Brian Sardella (California State University Stanislaus), Wes Dowd (Washington State University), Jonathon Stillman (San Francisco State University), Silvia Gomez-Jimenez (CIAD and University of Sonora), Alison Gardell (University of Washington), Peter Allen (Mississippi State University), Frederick Silvestre (University of Namur), Jason Podrabsky (Portland State University), Katie Gilmour (University of Ottawa), Tyler Evans (California State University East Bay), Xiaodan Wang (East China Normal University Shanghai), and Esteban Soto and Jackson Gross (both University of California Davis). Many thanks also go to my graduate students Elizabeth Mojica, Meranda Corona, Larken Root, and Jens Hamar, who read chapter drafts and provided photos and valuable feedback that helped improve the book. I also thank my colleagues including Professors Avner Cnaani (Agricultural Research Organization, Beth Dagan), Delin Duan (Institute of Oceanology Chinese Academy of Sciences), Steve McCormick (University of Massachusetts), Olivier DeClerck (Ghent University) and Esteban Soto (University of California, Davis), and Miguel Sepulveda (Rio de Janeiro, Brazil), Christian Veterlaus (marinecultures.org), and Mark Zivojnovich (Hydromentia Technologies) who have kindly shared illustrations for the book. Their contributions are acknowledged in the corresponding figure legends.

Finally, I thank my editorial contacts at Oxford University Press, Ian Sherman, Charlie Bath and Sharmila Radha, who kept my writing on track with the planned schedule for publication and compatible with the projected size of this book. Ian and Charlie helped with all aspects of the book production process and in keeping the page number from increasing beyond a limit that we felt would diminish accessibility. The collective feedback and great advice I have received from many people helped substantially improve this book although, of course, I am solely responsible for any omission or errors that may remain.

Dietmar Kültz,
Davis, CA, November 10, 2021

Contents

PART 1

General Principles of Ecological Aquaculture

CHAPTER 1

Aquaculture terminology and basic concepts

'Science and engineering is meant to be questioned, tested and re-tested, but the road of true progress can be long, convoluted and tiresome.'

Adey and Loveland, *Dynamic aquaria: building living ecosystems* (2007)

This book introduces the basic concepts, opportunities, and challenges of aquaculture with an emphasis on ecological considerations. The road to progress in aquaculture has been followed for several millennia leading to a dramatic increase in its global productivity. We have arrived at a major, challenging junction on this road and future advancement of aquaculture depends on a thorough understanding of the ecological value of clean water as its main resource.

Therefore, this book outlines links between aquaculture, trophic cascades, nutrient cycles, and ecosystem services that determine water quality and ecosystem stability. A basic understanding of such links reveals opportunities for increasing the ecological sustainability of aquaculture and makes the diverse but interwoven facets of ecological aquaculture accessible to students entering this field.

I assess the prospects of various aquaculture practices for key species groups based on the premise of aligning two major (and often conflicting) development goals: minimizing reliance on ecosystem services while maximizing productivity.

The first part of the book builds a general understanding of terminology, the history, objectives, approaches, and ecological implications of aquaculture. It also emphasizes prospects for further development of ecologically sustainable aquaculture approaches. The discussion of aquaculture approaches focuses on their potential for meeting future demand in an economically feasible manner while minimizing the exploitation of common property resources and reliance on ecosystem services.

Although socio-economic aspects are critical for practical implementation of aquaculture solutions, they are not considered here in the interest of brevity and to maintain focus. They have been discussed extensively elsewhere and the reader is referred to recent works that highlight this important topic (Bunting, 2013; FAO, 2020).

Specific aquaculture practices are illustrated for a selection of economically important aquatic species in Part 2 of this book. These examples illustrate key concepts and contrast the values and promises of aquaculture with their pitfalls, drawbacks, and challenges to instil a broad understanding of the current state of aquaculture and prospects for increasing its ecological sustainability.

It was not my intent to comprehensively review specific aquaculture commodities or practices and omissions in this regard are unavoidable in this introductory overview. The reader is referred to excellent recent volumes that focus on particular aspects of aquaculture more comprehensively (e.g. Pillay and Kutty 2005; Bert, 2007; Davenport *et al.*, 2008; Costa-Pierce, 2014; Boyd and McNevin, 2015; Stickney, 2017; Lucas, Southgate, and Tucker, 2019).

A Primer of Ecological Aquaculture. Dietmar Kültz, Oxford University Press. © Dietmar Kültz (2022). DOI: 10.1093/oso/9780198850229.003.0001

In addition, there are many volumes that cover aquaculture practices for a particular species or species group in great detail, e.g. oysters (Hanes, 2019), shrimp (Leung and Engle, 2010), and tilapia (El-Sayed, 2019).

Part 3 provides an overview of abiotic parameters and biotic factors that are critical for water quality management in aquaculture. Clean water is a common property resource that affects all. Aquaculture competes for this resource with other uses, including as drinking water, for sanitary purposes, for ecosystem health and species conservation, for agricultural irrigation, and for human recreational and cultural activities.

Most importantly, clean freshwater is limited and its value as a resource is increasing in parallel to the rising world population. Nevertheless, freshwater (FW, also referred to as land-based) aquaculture has greater prospects for future sustainable development than mariculture. Throughout this book I provide examples supporting this argument. Clean water, whether fresh or marine, can no longer be considered a free natural resource but must be valued by properly incentivizing its conservation.

1.1 What is aquaculture?

According to the Food and Agriculture Organization (FAO) of the United Nations (UN), 'Aquaculture is the farming of aquatic organisms in both coastal and inland areas involving interventions in the rearing process to enhance production.' (FAO, 2019). Farming is the rearing of organisms under (semi-)controlled conditions and implies ownership (by an individual or corporation) over the organisms that are being reared. In contrast to aquaculture, capture fisheries collect seafood from wild populations, i.e. from natural, unmanaged resources.

The organisms that are subject to capture fisheries are considered common property resources that can be exploited and are not owned by anyone, although fishing rights often apply. Capture fisheries yield has plateaued at approximately 90 million metric tons annually since the turn of the millennium while aquaculture yield continues to increase steadily (Figure 1.1). In fact, 52% of world marine fish stocks are already fully exploited, 17% are overexploited, and 8% are depleted (with 1% recovering), while 20% are moderately exploited and only 3% are underexploited by capture fisheries (FAO, 2011).

Aquaculture and capture fisheries play an essential role in ensuring food supply and security in the context of climate change, in particular in developing countries. These tasks are recognized in the 2030 UN Agenda for Sustainable Development that comprises seventeen sustainable development goals (SDGs) (United Nations, 2015). One of these goals (SDG 14) emphasizes the conservation and responsible use of the oceans and marine resources for sustainable development, which includes aquaculture development governed by proper ecological stewardship.

Aquaculture exemplifies the ongoing global struggle to strike a sustainable balance between the many (often conflicting) needs of a rapidly increasing world population. Such needs pertain to human health and nutrition, ecosystem health, the welfare of wild and domesticated animals, the cultural history of peoples, and the economic principles of globalized economies.

On the one hand, aquaculture has great potential for securing a healthy and nutritious food supply while alleviating pressure on capture fisheries and reducing fisheries-induced habitat destruction, overfishing, genetic modification of wild populations by fisheries-induced evolution, and wholesale waste of bycatch.

On the other hand, aquaculture relies heavily on clean water, an increasingly precious (and dwindling) resource that is subject to intense pressure of being used for many competing purposes. Although transitions between solid and liquid state water occur more rapidly because of climate change, the total amount of water on earth has not changed much.

The amount of available clean water is decreasing because water pollution has increased greatly over the past few centuries, which affects the utility of available water for many human needs. The rate of pollution of the world's inland water bodies mirrors the exponential growth of the human population, although environmental protection efforts have improved this trend in recent decades. Even the seemingly vast reservoir of ocean

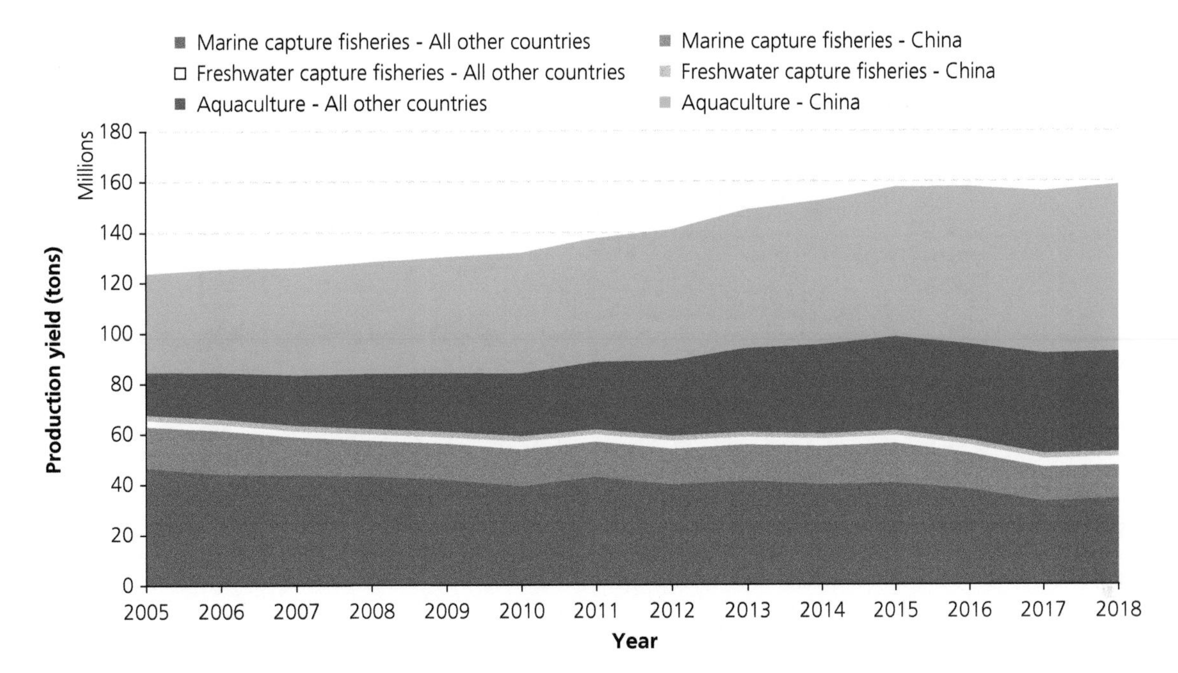

Figure 1.1 Yields of seafood production by aquaculture and capture fisheries from 2005 to 2018. Global yields excluding China and yields produced in China are shown separately. Data source: OECD.stat.

water is increasingly impacted by anthropogenic pollution.

A rising world population requires more FW for drinking, sanitary, agricultural, and other purposes. FW is also essential for maintaining the balance of inland ecosystems. Clean seawater (SW) is required to maintain ecosystem health and species diversity in the oceans, which is particularly critical in near-shore areas exploited by aquaculture, including coral reefs, kelp forests, and mangrove lagoons.

1.2 Sustainable aquaculture development

Ecological considerations have received increasing attention in aquaculture during the past few decades and their proper implementation promises to greatly accelerate the rate at which the field moves forward. As a result of the industrial revolution, clean water, in particular FW, became progressively precious as a resource and competition over water rights and access have resulted in fiercer rivalries as this resource became scarcer and its usage more prolific.

Although crucial for assessment of the ecological impact of aquaculture practices, 'comparatively little effort has been expended on documenting and analysing the range of methods used in exploitation of aquatic environments' (Beveridge and Little, 2014). Nonetheless, consumer awareness and interest in the sources of seafood is increasing.

Educational programmes, such as the SeaFood Watch programme of the Monterey Bay Aquarium, offer relevant knowledge and many markets now provide detailed information about seafood sources and methods of harvest such that this knowledge can be factored into consumer choices (Figure 1.2).

There is a growing public resentment against semi-intensive farming systems that rely heavily on common property resources and ecosystem service capacity (open systems). Another main concern with semi-intensive farming is animal welfare. To meet high seafood demand and address such concerns, aquaculture development goals should prioritize practices that maximize yields while also

Figure 1.2 Consumer information at a seafood market in Davis, California. Each seafood item is tagged with information about its origin and the method of catch if obtained by capture fisheries. This information educates consumers about their choice of seafood products and gives them the option to consider factors other than price (blacked out), appearance, and familiarity with the product for their selection. Tag information such as aquaculture farm-raised locally, aquaculture farm-raised imported, wild-caught by hook-and-line locally, wild-caught by bottom trawling imported, etc., expands the information available for educated consumer choices. Photo by author.

minimizing exploitation of ecosystem services and maximizing animal welfare.

Development of better wastewater treatment strategies should be accompanied by utilization of species that occupy low levels in trophic (nutrient) networks and are most amenable to high-density culture in captivity under high animal welfare standards.

Sustainable development was defined by the United Nations General Assembly in 1983 as 'development that meets the needs of the present without compromising the ability of future generations to meet their own needs' (Brundtland, 1987). This vague definition requires an assessment of present and future needs of humanity, which may be very subjective.

Prioritizing needs for political decision makers is the domain of lobbying efforts, which are highly polarized towards achieving specific objectives. For example, the need for access to clean water is prioritized differently by aquaculture producers, agricultural producers that depend on irrigation, metropolitan cities relying on water for sanitizing and drinking purposes, conservationists, or individuals requiring water for recreational and health activities.

Each of these needs has its own merits, but which is more important if the cumulative demand is higher than the supply? The answer to this question often depends on local elites with the strongest influence and their elitist priorities: 'the tendency for local elites to capture projects and programmes and use them for their own benefit should indeed be recognized as a fact of life' (Chambers, 2005).

Addressing increased seafood demand in a sustainable manner is not limited to increasing the sustainability and output of production. A key aspect for achieving this goal is to reduce food loss and waste, which has great potential to improve the efficiency of the whole food chain. Approximately one-third of food produced for human consumption (about 1.3 billion tons annually) is lost (Gustavsson, Cederberg, and Sonesson 2011). Reduction of this massive food waste must be a priority to sustain earth's limited natural resources.

Not only are we losing large amounts of the much-needed supply of safe and nutritious food, but a large part of the carbon footprint associated with food production is also generated in vain. Food loss occurs along all parts of the food chain, including the consumer level. It has been estimated that per capita food waste by consumers is highest (95–115 kg annually) in Europe and North America and lowest (6–11 kg annually) in sub-Saharan Africa (Gustavsson, Cederberg, and Sonesson, 2011).

Consumer behaviour, high food safety standards, and lack of coordination between producers and other commercial entities in the food supply chain are major causes of food waste in industrialized countries. In lower-income countries of the developing world the major causes of food loss are managerial and technical limitations in production, food preservation, and marketing systems.

Food loss can be reduced by raising awareness and coordination among food producers, retailers, and consumers in industrialized countries; by improving infrastructure, preservation, and management of the food supply chain in developing countries; and by globally increasing the ratio of food safety versus food waste. Aquaculture producers, commercial entities involved in the seafood supply chain, and seafood consumers need to enhance their efforts for reducing food waste as an integral and important part of sustainable development.

1.3 Cell-based seafood

Cell-based seafood is an emerging alternative to capture fisheries and conventional aquaculture. In this approach the seafood is produced from animal cells in tissue culture that yields tissue resembling muscles. Cells that make up muscle tissue in animals are used but these cells are grown in bioreactors and incubators rather than in the animal.

Ideally, no animals are required for cell-based seafood production, other than for taking initial tissue biopsies that are used to derive and immortalize cells. The immortalized cells grow and multiply in growth media inside bioreactors or cell culture incubators (in theory indefinitely). Part of the grown tissue is regularly harvested while the remainder is allowed to continue to grow to support regular harvests. This area of food technology represents undoubtedly a promising approach for ecologically sustainable food production systems.

The concept of cell-based foods is already well established for mammals (e.g. to produce burgers) or birds (e.g. to produce chicken breast substitutes). Many start-up companies are now exploring a similar strategy for a variety of different fish and invertebrates (e.g. shrimp) that are key aquaculture species.

There are advantages of cell-based seafood over cell-based foods derived from mammalian or avian cells. Fish cells and aquatic invertebrate cells can be grown at ambient CO_2 while mammalian and avian cells require CO_2 supplementation. Cells from cold-water fish and invertebrates do not require heated incubators or minimal energy for heating. Moreover, they naturally produce a higher level of healthy, polyunsaturated fatty acids to maintain an optimal fluidity of their plasma membranes and intracellular membranes at lower temperature (to counteract slower Brownian motion).

In theory, cell-based seafood could dramatically improve the ratio of seafood produced relative to the amount of nutrients used to grow it, while minimizing the amount of waste that is generated. Possibilities for optimizing and controlling a cell culture system in terms of inputs and outputs of matter and energy are much greater than for growing complex multicellular animals. Moreover, 100% of the cells or tissue produced would be utilized while generally less than 50% of the biomass of aquaculture animals is used for seafood production. In addition, given proper knowledge of the species, it is possible to direct the differentiation of cells in culture towards specific desirable cell types (e.g. myocytes or adipocytes).

In praxis, many challenges remain for reliable economical production of cell-based seafood. Generating cell lines from fish and aquatic invertebrates is anything but trivial, especially if differentiated adult tissues (e.g. muscle tissue) are used as the source of biopsies for establishing primary cultures. Immortalization is the process of preventing cellular senescence in culture to achieve indefinite mitotic cell division and growth in culture. This process is easier to achieve when embryonic, relatively undifferentiated cells are used as the source of primary cultures.

Although it is easier to create immortalized cell lines from embryonic tissues, the resulting cell lines will likely have a fibroblast or neuroepithelial cellular phenotype and need to be further manipulated to steer cell differentiation towards the desired phenotype of myocytes plus adipocytes. Manipulation of cell differentiation can be achieved via synthetic biology approaches, but such approaches may be unattractive to consumers because they are equivalent at the cellular level to producing genetically modified organisms (GMOs) at the animal level.

The alternative to synthetic biology approaches for directing cell differentiation are methods that employ stress-induced evolution and various supplements in cell culture media to steer cell differentiation towards the desired phenotype. Such media supplements may be tissue-specific growth factors, proteins, metabolites, or abiotic factors that control cell differentiation and pathways.

Another important consideration is that cellular phenotype may change with population doubling time while cell lines are being propagated continuously in incubators or bioreactors. For example, even if muscle biopsies are obtained and primary cultures are successfully immortalized, the resulting cell lines may dedifferentiate as they grow and multiply. Therefore, it is necessary to monitor the molecular phenotype (protein composition) of cells during the culture process. Such monitoring is possible using modern proteomics technologies, but costs are still relatively high.

There are very large differences in the propensity of cells from different species of fish and invertebrates to immortalize spontaneously in primary cell cultures. Obtaining immortalized cell lines from some species (few) is relatively easy while for (most) others this process is notoriously difficult. This difficulty applies especially when spontaneous cell immortalization approaches are pursued. Using very large amounts of starting biopsy material represents a possible remedy but is often impractical. An example of a spontaneously immortalized fish cell line from a major aquaculture species used for seafood production is the tilapia OmB cell line generated in the author's laboratory (Figure 1.3).

Spontaneous immortalization refers to a process during which primary cell cultures overcome cellular senescence and acquire the ability to grow and divide indefinitely without any invasive genetic manipulation. This process is based on natural selection of cells that can bypass senescence under culture conditions *in vitro*. It is comparable to traditional selective breeding or domestication at the level of whole animals. Selection of spontaneously generated genetic variants is generally preferred by consumers over synthetically creating specific genetic variants to promote cell immortalization. The latter approach is comparable, in principle, to producing GMOs (Chapter 6).

Another challenge for cell-based seafood is the development of appropriate edible matrices for cell attachment that mimic the three-dimensional structure of meat (muscle tissue), maximize the surface area for cell growth while ensuring proper media circulation, promote formation of an extracellular matrix (ECM consisting of secreted proteins and other components) that resemble ECM in muscle tissue, and minimize intercellular space to achieve a desirable texture.

Growing cell lines from fish and aquatic invertebrates in affordable, defined media is another task that requires careful consideration for commercial development of cell-based seafood. Animal serum such as foetal bovine serum (FBS) or fish serum is commonly used as a supplement in concentrations

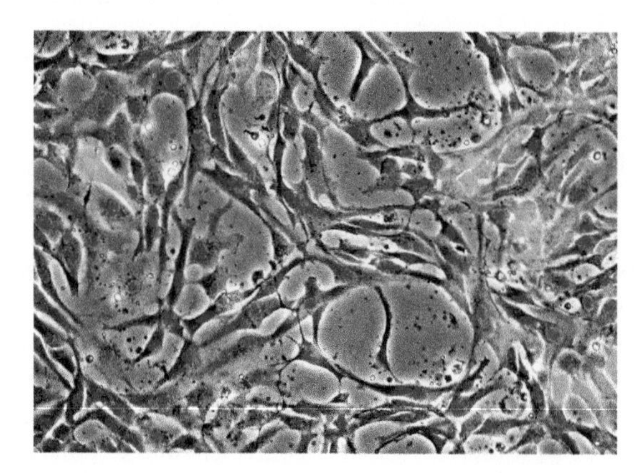

Figure 1.3 Immortalized tilapia OmB cell line that can be grown indefinitely *in vitro*. The cell line was produced in the author's laboratory (Gardell *et al.*, 2014). The image was taken by a combination of phase contrast and fluorescence microscopy. For better visualization, cell nuclei in this batch of cells have been stained with a blue dye (NucBlue) and some cells express green fluorescent protein (GFP). The original OmB cell line was derived by stress-induced immortalization of primary cell cultures obtained from a tilapia biopsy without any genetic manipulation. Photo by Jens Hamar.

of up to 10% for cell culture media. If this is the case, then the argument that no animals are required to produce cell-based seafood no longer holds up.

Vertebrate serum contains many growth factors and metabolites that promote cell growth and differentiation. Finding specific combinations of these factors that can be used as defined media supplements in lieu of animal serum represents a highly active area of research. Associated with developing defined serum-free media is the development of methods for media recycling and for minimizing the retention of media hormones and growth factors in the final cell-based seafood product.

The technical challenges pointed out here are tricky but not unsurmountable, given the level of knowledge and technology that is currently available. However, other major obstacles to the adoption of cell-based seafood are consumer acceptance and consumer preference. Ecologically sustainable production of high quality, healthy food for seven billion or more people cannot be achieved for another century using current approaches. Novel and innovative solutions need to be implemented. Part of those solutions is education to influence consumer preference.

If, in fact, more or equally healthy and nutritious food can be produced by ecologically more sustainable cell-based systems, then it will be important that the look and feel of these new products mimics that of traditional food favourites as closely as possible. Research efforts for achieving such mimicry must consider regional differences in culinary cultures. They should be paralleled by education efforts for incentivizing a shift in consumer preference. The look and feel of current beef and chicken cell-based foods is already so much better than that of dietary supplement pills, but further improvements are necessary.

Although cell-based seafood represents a promising approach for sustainable food production it is still in its infancy and is not considered further in this book. Readers interested in cell-based seafood are directed to recent publications on this topic, including Rubio *et al.*, 2019 and Potter *et al.*, 2020.

1.4 Aquaculture concepts and terminology

In the remainder of this chapter, I briefly define key concepts and terms used often throughout the book.

Seafood is an umbrella term that describes edible products derived from aquatic organisms including FW, brackish, and marine organisms that are produced for the purpose of human consumption. Seafood is produced by aquaculture, capture fisheries, or most recently also by cell-based approaches. Methods of seafood production differ greatly in the extent to which they rely on common property ecosystem resources.

Ecosystems are 'diverse communities of organisms, supported and constrained by a given physical-chemical environment, interacting to capture and process energy and nutrients in food webs' (Adey and Loveland, 2007). Ecosystem services are defined as benefits provided by ecosystems to humankind for all purposes. Ecosystem services are critical for the provisioning of clean drinking water, the decomposition of organic wastes, and providing many other benefits. Because the capacity for ecosystem services is limited their provisioning must be monitored and, if necessary, regulated wisely.

Ecosystem services return the matter and energy exiting from aquaculture systems to the global nutrient cycle. This recycling capacity has limits that are different for each locality. Moreover, these limits are not static but are subject to dynamic changes over time. Spatiotemporal dynamics of ecosystem service capacity is impacted by anthropogenic pollution and acceleration of climate change, seasons, weather, and many other factors.

The dynamic nature of capacity limits and ecosystem succession must be considered in models aiming to predict the impacts of aquaculture systems that are open to the natural environment (flow-through pond, raceway, or cage cultures). The area of wild wetland habitat required to recycle wastewater produced by aquaculture depends on these dynamic limits of ecosystem service capacity.

Extensive aquaculture refers to the culture of aquatic species with a minimal level of control that requires little technology, no fertilization or feeding, and depends to a large extent on natural resources and ecosystem services.

In contrast, *intensive aquaculture* is defined as utilizing exclusively supplementary (including live) feeds and no feeds provided by surrounding ecosystems. In addition, intensive aquaculture accommodates the highest density of culture organisms. In this book, I refer to intensive systems only if, in addition to the points mentioned earlier, they also afford a high level of operator control, require elaborate management, and are highly dependent on technology such as aeration, pumps, wastewater treatment, and water quality monitoring systems.

Semi-intensive aquaculture is characterized by an intermediate level of human control and technology utilization. The reliance of semi-intensive aquaculture systems on natural resources such as clean water is very high when open systems such as flow-through earthen ponds, flow-through raceways, or cage culture systems are used that permit high stocking densities but are not very technology intensive. These systems rely to a great extent on ecosystem services.

Such reliance is considerably reduced when closed systems such as *recirculating aquaculture systems (RAS)* are being used. RAS are intensive systems that recycle wastewater produced by aquaculture using filters that remove solid waste and promote assimilation of soluble waste into bacterial and algal biomass, removal by chemical filters, or conversion to volatile gases (e.g. N_2).

The multifaceted nature of aquaculture is most evident by contrasting approaches that emphasize minimal reliance on ecosystem services versus those aimed at maximizing yield and profitability by fully exploiting common property resources. The former category is exemplified by indoor RAS and outdoor water reuse systems that include designated wastewater treatment ponds. The latter category is exemplified by open raceways and cage culture systems that are fully reliant on ecosystem services for providing clean water and treating disposed wastewater. Currently, the extent of using these different aquaculture systems is inversely proportional to their long-term ecological sustainability.

Mariculture is a special form of aquaculture that refers to the culture of organisms in SW or brackish water. Brackish water encompasses a wide range of salinity ranging from 0.5 g/kg (greater than FW) to 30 g/kg (less than SW). The majority of aquaculture utilizes FW or near-shore coastal systems. A minor fraction is contributed by animal mariculture.

The opposite applies to capture fisheries, which is dominated by marine offshore fisheries. Some modelling approaches have suggested that vast areas of the ocean can be used for mariculture (Gentry*et al.*, 2017). However, it is unclear how fragile marine ecosystems will be impacted by extensive aquaculture practices and irreversible, unintended consequences are possible and likely in the longer term.

An example for unintended consequences of once saluted aquaculture innovations are irreversible invasive species problems that arose from introducing aquaculture species to non-native habitat during the first half of the twentieth century. A recent review outlines links between mariculture and ecosystem services and emphasizes the importance of proper regulatory incentivization to improve the ecological sustainability of mariculture (*Alleway et al.*, 2019).

Sea ranching is a method of stocking aquatic animals in a natural aquatic ecosystem for grow out over extended periods of time without human interference before recapture (Bunting, 2013). It applies mostly to migratory species like salmonids, which use homing behaviour to return to the site of their release. This method is increasingly being adopted, although concerns about genetic dilution of wild stocks prevail (Section 6.5).

Such concerns also apply to *stock enhancement*, which refers to the culture of early developmental stages of a target species and the release of cultured juveniles to replenish wild populations. Stock enhancement has been practiced since the late nineteenth century, e.g. Atlantic cod (*Gadus morhua*) hatcheries have been operated for restocking since 1882 in Norway (Bunting, 2013).

The success of stock enhancement programmes is highly variable and depends on species, developmental state and size at release, time of release, and location of release. Released juveniles can be tagged with genetic markers that increasingly replace physical tags to afford more accurate assessment of stock enhancement impacts on wild populations. Sea ranching and stock enhancement are best practised with species and local populations that are native in the affected area.

Monoculture is the culture of a single species, which is more commonly practised but less ecologically sustainable than its counterpart, *polyculture*, which is the culture of multiple species. It is less commonly practised but more ecologically sustainable than monoculture. Historically, polyculture of different carp species has been practised for many centuries in China. However, most polyculture combinations of species have been explored only within the past fifty years.

Integrated MultiTrophic Aquaculture (IMTA) is a special form of polyculture that combines the farming of aquatic organisms belonging to different trophic levels. The goal of IMTA is to recycle waste from one trophic level by utilizing it as nutrients for another trophic level. IMTA can be extensive or intensive aquaculture.

Open IMTA systems often combine coastal cage mariculture of fish with the culture of filter feeders such as oysters, detritivores such as sea cucumbers, and primary producers such as seaweed. Some semi-intensive forms of IMTA are also referred to as integrated aquaculture-agriculture (IAA) when they combine waste from terrestrial animals (chickens, mammals) to fertilize aquaculture ponds and stimulate aquatic plant growth as a food source for herbivorous or omnivorous fishes (e.g. tilapia or certain carps).

Aquaponics is a form of IMTA that combines fish (e.g. tilapia) aquaculture and hydroponics, which is a method of growing macrophyte plants (e.g. lettuce, basil) without soil by direct submersion of roots in nutrient-containing water.

An overview of key aquaculture terminology would be incomplete without mentioning domestication. *Domestication* is the artificial (agent-driven as opposed to natural) selection of organisms for life in an environment that is controlled (to various extent) by humans. Artificial selection implies the human sorting out of genotypes according to their suitability for performing well in a human-controlled environment. This process can only be done over many generations and thus its duration depends on the generation time of animals being domesticated.

Domestication of terrestrial vertebrates (mammals, birds) has a longer history than domestication of aquatic animals. Carp domestication has been practised for centuries in Asia and Europe to yield koi (*Cyprinus carpio*) and goldfish (*Carassius auratus*), which are closely related to the Prussian carp (*Carassius gibelio*). These carps are arguably the longest domesticated aquatic species. Domestication of thousands of ornamental aquatic species started at the beginning of the twentieth century, including taxonomically diverse plants and animals.

Key conclusions

- Advancement of aquaculture depends on a thorough understanding of the ecological value of clean water as its main resource; Ecologically sustainable aquaculture must be properly incentivized.
- Aquaculture development goals are: 1) maximal production yield, 2) minimal reliance on ecosystem services, 3) maximal standards of animal welfare, and 4) maximal ratio of food safety vs. food waste.
- Seafood is produced by capture fisheries and aquaculture, but only aquaculture continues to increase to keep step with a growing human population and increased per capita consumption.
- Animal aquaculture development is now paralleled by the development of methods for growing cell-based seafood, which do not require animal husbandry but rely on aquatic animal cell lines.
- Most aquaculture is practised in FW while most capture fisheries take place in the oceans.
- Mariculture represents aquaculture in SW and brackish water.
- Aquaculture systems are diverse, including extensive, semi-intensive, intensive, open, closed, sea ranching, restocking, mono- and polyculture, and multi-trophic integrated approaches.
- Currently, the most common aquaculture systems are open semi-intensive ponds, cage, and flow-through systems, which rely heavily on common property ecosystem services.
- Domestication of many aquatic species has started at the beginning of the twentieth century although some carps (koi, goldfish) have been domesticated for much longer.

References

Adey, W. H. and Loveland, K. (2007). *Dynamic aquaria: building living ecosystems*, 3rd edn. Boston: Academic Press.

Alleway, H. K., Gillies, C. L., Bishop, M. J., Gentry, R. R., Theuerkauf, S. J., and Jones, R. (2019). 'The ecosystem services of marine aquaculture: valuing benefits to people and nature', *BioScience*, 69, pp. 59–68.

Bert, T. M. (ed.) (2007). *Ecological and genetic implications of aquaculture activities*. Dordrecht: Springer.

Beveridge, M. C. M. and Little, D. C. (2002). 'The history of aquaculture in traditional societies', in Costa-Pierce, B. (ed.) *Ecological aquaculture: the evolution of the blue revolution*. Oxford: Wiley-Blackwell, pp. 3–29.

Boyd, C. and McNevin, A. (2015). *Aquaculture, resource use, and the environment*. Hoboken: Wiley-Blackwell.

Brundtland, G. H. (1987). *Our common future*. Oxford: Oxford University Press.

Bunting, S. W. (2013). *Principles of sustainable aquaculture: promoting social, economic and environmental resilience*. Abingdon: Routledge.

Chambers, R. (2005). *Ideas for development*. London: Routledge.

Costa-Pierce, B. (2014). *Ecological aquaculture: the evolution of the blue revolution*, 2nd edn. Philadelphia: Lippincott Williams & Wilkins.

Davenport, J., Black, K. D., Burnell, G., Cross, T., Culloty, S., Ekaratne, S., Furness, B., Mulcahy, M., and Thetmeyer, H. (2008). *Aquaculture: the ecological issues*. Oxford: Wiley-Blackwell.

El-Sayed, A-F. M. (2019). *Tilapia culture*, 2nd edn. Cambridge: Academic Press.

Food and Agriculture Organization of the United Nations (FAO). (2011). *Review of the state of world marine fishery resources*. FAO Fisheries and Aquaculture Technical Paper 569. Rome: Food and Agriculture Organization of the United Nations.

Food and Agriculture Organization of the United Nations (FAO). (2019). *Aquaculture*. Rome: Food and Agriculture Organization of the United Nations.

Food and Agriculture Organization of the United Nations (FAO). (2020). *The state of world fisheries and aquaculture 2020*. Rome: Food and Agriculture Organization of the United Nations.

Gardell, A. M., Qin, Q., Rice, R. H., Li, J., and Kültz, D. (2014). 'Derivation and osmotolerance characterization of three immortalized tilapia (*Oreochromis mossambicus*) cell lines', *PLoS ONE*, 9, p. e95919.

Gentry, R. R., Froehlich, H. E., Grimm, D., Kareiva, P., Parke, M., Rust, M., Gaines, S. D., & Halpern, B. S. (2017). 'Mapping the global potential for marine aquaculture', *Nature Ecology and Evolution*, 1, p. 1317–1324.

Gustavsson, J., Cederberg, C., and Sonesson, U. (2011). *Global food losses and food waste: extent, causes and prevention*. Rome: United Nations.

Hanes, S. (2019). *The aquatic frontier: oysters and aquaculture in the progressive era*. Amherst: University of Massachusetts Press.

Leung, P. and Engle, C. R. (eds.) (2010). *Shrimp culture: economics, market, and trade*. Hoboken: Wiley-Blackwell.

Lucas, J. S., Southgate, P. C., and Tucker, C. S. (2019). *Aquaculture: farming aquatic animals and plants*, 3rd edn. Hoboken: Wiley-Blackwell.

Pillay, T. V. R. and Kutty, M. N. (2005). *Aquaculture: principles and practices*, 2nd edn. Oxford: Wiley-Blackwell.

Potter, G., Smith, A. S. T., Vo, N. T. K., Muster, J., Weston, W., Bertero, A., Maves, L., Mack, D. L., and Rostain, A. (2020). 'A more open approach is needed to develop cell-based fish technology: it starts with zebrafish', *One Earth*, 3, 54–64.

Rubio, N., Datar, I., Stachura, D., Kaplan, D., and Krueger, K. (2019). 'Cell-based fish: a novel approach to seafood production and an opportunity for cellular agriculture', *Frontiers in Sustainable Food Systems* [online]. https://doi.org/10.3389/fsufs.2019.00043.

Stickney, R. R. (2017). *Aquaculture: an introductory text*, 3rd edn. Wallingford: CABI.

United Nations. (2015). *Transforming our world: the 2030 Agenda for Sustainable Development*. New York: United Nations.

CHAPTER 2

The historical origins of aquaculture

'The origins of aquaculture are lost in history and little evidence remains . . . '
Beveridge and Little, in *Ecological aquaculture: the evolution of the blue revolution* (2014)

Unambiguous archaeological evidence and written accounts that communicate clear evidence about the beginnings of aquaculture are scarce. Often, existing evidence is vague regarding whether it supports prehistoric capture fisheries (hunting) or aquaculture (farming) practices. However, the lack of records or archaeological evidence does not rule out the existence of ancient aquaculture practices. Consequently, the beginnings of aquaculture are hazy and can only be estimated.

Regardless of these uncertainties, there is general agreement that the historical time frame during which aquaculture activities have emerged can be traced back at least four millennia (Beveridge and Little, 2014; Nash, 2011). Despite this long history, the impact of aquaculture on the development of human society overall and its global ecological footprint remained very low until the dawn of the twentieth century.

2.1 The earliest records of aquaculture practices

Aquaculture is predated by agriculture, which originated after the last ice age at least eleven to twelve millennia ago (Larson *et al.*, 2014). Moreover, the domestication of aquatic animals started about 10,000 years later than that of terrestrial mammals (Balon, 2004). The marginal historical role of aquaculture contrasts starkly with the steady intensification and the major impact that terrestrial agriculture has had on the evolution of human society.

Multiple reasons for this discrepancy in the development and impact of food resource cultivation in terrestrial (agriculture) and aquatic (aquaculture) environments have been suggested. First, until the twentieth century the degree of resource-limitation was much greater for terrestrial than for aquatic resources, and second, aquatic habitats posed much greater challenges than terrestrial habitats regarding access and management and they instilled greater fear into ancestral peoples. Third, ancient civilizations had better opportunities to observe, and greater knowledge of, terrestrial than aquatic organisms (Beveridge and Little, 2014).

Notwithstanding the surmised overall miniscule impact of aquaculture on the development of ancient human societies, there are notable exceptions. For instance, ancient coastal societies may have been inspired by sustainable fisheries practices to manipulate crop plants and enter the 'era of agricultural domestication' (Malindine, 2019).

In particular the development of some island societies, such as the Polynesian cultures of Hawai'i (Section 2.7), was strongly influenced by aquaculture and integrated agriculture practices that reached a high level of sophistication almost 1000 years ago (Costa-Pierce, 2014). The unique mariculture and integrated agriculture practices and systems produced by the Hawai'ian Polynesians support the hypothesis that they have originated independently.

Archaeological evidence on early aquaculture practices in Africa, Asia, and Europe also supports multiple independent origins of aquaculture

A Primer of Ecological Aquaculture. Dietmar Kültz, Oxford University Press. © Dietmar Kültz (2022). DOI: 10.1093/oso/9780198850229.003.0002

in these different geographic regions, although some geographic and chronological parallels are also apparent. Moreover, recent archaeological discoveries suggest an independent invention of floodplain aquaculture in the Amazon basin of South America long before the arrival of Spanish conquistadors (Nash, 2011).

Furthermore, in the Victoria region of Southeast Australia, the aboriginal Gunditjmara people sustainably managed wetland ecosystems primarily for Shortfin eel (*Anguilla australis*) production as early as 4000 years ago (Builth et al., 2008; Malindine, 2019). These ancient Australian eel farming systems utilized the same conceptual management innovation as the ancient mariculture systems of Hawai'i.

Despite their independent origins, both systems are based on the key observation that catadromous fishes (e.g. river eels, milkfish, and mullets) migrate from the ocean to nutrient-rich brackish or freshwater ecosystems where they grow and mature before returning to reproduce in the ocean. This observation prompted the construction of gated channels to control access from grow-out ponds and marshlands to the ocean. As a result of such management, these aquaculture systems facilitated a very convenient harvest of catadromous fishes when they arrived at the gates during their seaward migration.

Intriguingly, aquaculture of catadromous fishes has a much older and richer history than that of anadromous species. Nevertheless, the present aquaculture production of anadromous salmonids exceeds that of catadromous species. The main reason for this historical development is that technology-intensive freshwater hatcheries can be used for producing anadromous species while reproduction of catadromous fishes in captivity is economically more challenging and has only been implemented on a commercial scale for milkfish (*Chanos chanos*).

Although aquaculture originated independently in many parts of the world, the transition from capture fisheries (hunting) to aquaculture (farming) practices has been gradual in each case. It included intermediate stages that defy clear-cut classification as either hunting or farming. Primitive management of fisheries resources such as transplantation of fertilized eggs between different environments, entrapment of fish in suitable areas, and environmental enhancement to support reproduction and growth likely led to the development of management strategies that eventually culminated in more sophisticated aquaculture systems (Beveridge and Little, 2014).

Four theories—oxbow, catch-and-hold, concentration, and trap-and-crop—exist on how the transition from hunting to farming fish may have occurred (Rabanal, 1988).

The oxbow theory suggests that some oxbows formed by rivers winding through the landscape sometimes periodically became separated from the rivers (e.g. during floods followed by droughts or as result of the river changing its course). People discovered that it is much easier to catch more fish than immediately needed and keep them in these natural entrapments to let them grow larger until needed. Eventually, 'enterprising individuals' may have reinforced the embankments that enclose oxbow areas and started to manage fish in separated oxbows by stocking fry.

The catch-and-hold theory postulates that ponds and moats with a primary function that is unrelated to aquaculture (defending medieval castles in Europe or being a mill reservoir) have later been adopted in a multi-purpose capacity, that is, to hold and grow fish as a secondary function. Perhaps, during prime hunting season when fish supply was plentiful hunters disposed of excess fish in such artificial bodies of water, which turned out advantageous as some species grew well in these man-made habitats and were later available for harvest when hunting yields were low.

This theory would be applicable to the multipurposing of dammed storage reservoirs for aquaculture, secondary to their primary purpose of controlling water flow for powering flour mills in medieval European monasteries (Nash, 2011). Similar multi-purposing of rice field floodplains may have sparked the origin of aquaculture in Asia (Beveridge and Little, 2014).

The concentration theory proposes that natural seasonal flooding of large plains and subsequent recession of floodwater from the plains left water holes of varying size in natural depressions on these plains. As the dry season progressed these water holes became progressively smaller, thus, concentrating fish they harboured. People inhabiting such floodplains may have taken advantage of this phenomenon as it greatly simplified the capture of fish. Eventually, this practice is thought to have led to active modification of water holes and construction of artificial ponds along with stocking fry and supplemental feeding.

The trap-and-crop theory differs from the other three theories by focusing on brackish and marine habitats. Coastal peoples of Southeast Asia and Oceania (e.g. Hawai'i) learned that fish, crustaceans, and molluscs regularly enter tide pools, lagoons, estuarine marshland, and other coastal ponds to feed in these nutrient-rich nursing grounds. These animals were much easier to catch in these shallow, protected bodies of water than on the open ocean. Over time, gradually increasing management of such coastal ponds and pools led to the development of sophisticated mariculture systems. One of the earliest forms of such management may have been the installation of gates to control the flow of water and prevent the escape of fish and aquatic invertebrates once they entered.

2.2 Religious drivers of aquaculture development in Africa and Assyria

The need for a stable food supply represented only one of the various incentives that promoted the development of aquaculture practices. Other important drivers included religious and cultural motives.

The high standing of fish as religious symbols in ancient Egypt is exemplified by naming entire large metropolitan areas after specific species of sacred fish. For instance, the city of Oxyrhynchus was named after the sharp-nosed Nile elephantfish (*Mormyrus kannume*) (Wallis Budge, 1904a).

During the time of pharaohs, Egyptians believed that Nile fish (e.g. tilapia) symbolize the concept of rebirth and virtues of the goddess of the West, who receives the deceased and can take on different forms (e.g. as Nut or Hathor) (Desroches-Noblecourt, 1954). *The Book of the Dead* mentions two sacred fishes of the river Nile—Abtu and Ant—that guide and protect the boat of the sun god Ra carrying the deceased to the underworld (Wallis Budge, 1904b, 1904a).

Another ancient Egyptian deity associated with fish is the goddess Hatmehit, who bears a fish head and represents the female counterpart to Ba-neb-Tettu (the personification of the souls of Ra, Osiris, and other gods) (Wallis Budge, 1904a). In ancient Egyptian mythology (e.g. in *The Book of the Dead*) the fish god Remi is mentioned as the personification of Ra's tears (Wallis Budge, 1904b).

The eel was worshipped in Upper Egypt and mummied eels were buried during religious rituals in sepulchral boxes:

> That the crocodile, ibis, dog-headed ape, and fish of various kinds were venerated in Egypt is true enough; they were not, however, venerated in dynastic times as animals, but as the abodes of gods . . . the Egyptians began to keep . . . fishes in sanctuaries, and to worship them as deities incarnate' (Wallis Budge, 1904b).

Such practices of keeping fish for worshipping purposes may have resulted in reproduction of tilapia and other readily reproducing species in artificial, man-made, and managed sanctuaries. The opportunity to closely observe life cycles of these fish in such sanctuaries likely improved their management and promoted aquaculture development.

Frescos in Egyptian Tombs illustrate that ecologically sustainable concepts of aquaculture, which are currently revelled by technocratic language such as integrated multi-trophic aquaculture (IMTA) and aquaponics, were embraced in ancient Egypt more than three millennia ago (Figure 2.1).

Interestingly, artificial (human-driven) selection for most efficient resource utilization and pollution resilience in human-made artificial ponds yielded herbivorous and omnivorous fish such as tilapia, goldfish, common carp, and milkfish as historically preferred aquaculture species. Aquaculture of herbivores and omnivores is generally more ecologically sustainable than that of carnivorous fish. Predatory carnivores such as salmonids, bass, or seabream occupy top positions in trophic networks and require more ecological resources per kg of body mass produced than herbivores or omnivores. In that regard, the history of aquaculture teaches us an important lesson about trophic ecology.

The roots of aquaculture in Northern Africa (Egypt) may be intertwined with those in Western Asia (the Middle East). Assyrians ruled much of the Middle East from about 5000 to 2500 years ago. During this time, they constructed many fish ponds that surrounded temples and other sacred places—not unlike the sacred ponds build in ancient

Figure 2.1 Part of a fresco from the Tomb of Nebamun at Thebes (1550–1229 BC), showing an artificial fishpond containing what looks like Nile tilapia among aquatic plants and waterfowl. This fresco suggests that ecological aquaculture has been practiced for more than 3000 years (perhaps unintentionally) by integrating fertilization from waterfowl manure with the production of water lilies and tilapia in a polyculture system. Photo reproduced under CC0 1.0 licence from Wikimedia Commons.

Egypt. Moreover, wealthy citizens of many Assyrian towns owned fish ponds within their estates that were used for religious/ornamental purposes. Evidence of such sacred Assyrian fish ponds appears in written documents that date back to 422 BC (Nash, 2011).

Moreover, these people, including the Sumerians and Babylonians, created freshwater reservoirs by damming the Tigris and Euphrates to manage crop irrigation. More than 2500 years ago, such reservoirs existed for grow-out of large numbers of fish, presumably for producing food (Nash, 2011). The Assyrians probably used different species of fish for food production and religious/ornamental use, but reliable documentation hasn't yet been found.

2.3 Origins of freshwater aquaculture and fish domestication in Asia

Some of the earliest known/documented origins of aquaculture are rooted in China and date back approximately four millennia. Fish rearing operations in China appear in 4000-year-old records. Around 2070 BC the emperor Si Wen Ming (Da Yu the Great) wrote about the practice of stocking fish ponds with larvae and the regulation of specific periods when harvests of fish were permitted (Nash, 2011).

The first written record devoted entirely to aquaculture was *The Chinese Fish Culture Classic*, a short book by Fan Lai (Fan Lee, Fan Li) from between 475 and 473 BC (Rabanal, 1988).

Slightly later (300 BC) in India, the philosopher Kautilya comments on the management of fish in freshwater reservoirs (Nash, 2011). The polyculture of fish and rice (and some other plant species) is documented in written publications that are at least 2000 years old. These early origins of aquaculture in China and India are all focused on the culture of freshwater fish (mostly carp species).

Although freshwater carp aquaculture is undoubtedly the earliest form of aquaculture documented in China, the particular carp species that was originally cultured is somewhat controversial. Many authors believe that common carp (*C. carpio*) was native to China and that it was the species originally cultured (Tamura, 1961). Other researchers argue that early Chinese aquaculture used riverine grass carp species, and it was only later when domesticated forms of common carp were (during the Dark or Middle ages) imported from Europe (Balon, 2004).

There are also records of domestication in China of the goldfish (*Carassius auratus*), a close relative of the common carp. Orange-golden colour variants (xanthic aberrations) and stunted growth forms occur at low frequency in wild populations of fish in this genus (including *C. gibelio*, *C. carassius*) and they formed the basis for early attempts of artificial selection (Chen, 1956).

A major driver of aquaculture and eventual domestication of goldfish in Asia was the belief that these rare xanthic variants had supernatural powers, illustrating that religious motives were important for aquaculture development in Asia, just as they were in ancient Egypt and medieval Europe (Section 2.5).

Around 970 AD, 'ponds of mercy' were established to cultivate sacred goldfish. These ponds represented a symbol for worshipping goldfish, to honour Buddhism's decrees of abstinence from taking the life of any creature, and to perform at least one good deed per day (Chen, 1956).

By the middle of the thirteenth century Chinese rulers had established a special fish breeding guild that was responsible for managing 'ponds of mercy', including fish feeding with optimized natural feeds, fish reproduction, and artificial selection of desirable phenotypes. The transition from keeping goldfish in 'ponds of mercy' to holding them in small aquarium-like bowls made from jade and earthen materials supposedly happened during the sixteenth century as a result of artificially selected genetic variants that performed well in small spaces (Chen, 1956).

Containment in small vessels facilitated the transport and distribution of goldfish leading to their introduction to Japan and other parts of Asia during the sixteenth century. Domesticated forms of *C. auratus* have a bright yellow to orange body colour, stunted growth form, and greatly altered proportions of the body and fins. Such 'domesticated goldfish monstrosities' (Balon, 2004) have been exported from China all across the world. Goldfish were the earliest ornamental aquatic species, and they are still among the most common ornamental fish world-wide.

The tradition of breeding common carp colour variants in Japan may go back as far as the beginning of the nineteenth century, although the earliest known record that documents koi colour variants is less than a century old (Ishikawa *et al.*, 1931). The emergence of modern koi (nishikigoi) has happened only during the first half of the twentieth century as a result of rearing common carp in polyculture with rice in Japan (Balon, 2004). It is possible, however, that koi-like colour variants of *C. carpio* were produced even earlier, incentivized by traditions of Buddhism favouring development of masterful gardens that include ornamental ponds stocked with fish (Figure 2.2).

A recent discovery shows that the Oujiang colour carp, a *C. carpio* colour variant known from China for almost 200 years, is genetically very closely related, and possibly ancestral, to Japanese koi (Mabuchi and Song, 2014). Today, ornamental fish such as goldfish and koi remain popular inhabitants of ponds that have been masterfully integrated within garden landscapes surrounding temples and prestigious estates in Japan, China, and other countries in Southeast Asia.

2.4 Freshwater aquaculture origins in Europe

Following the collapse of the Roman Empire until the Middle Ages, aquaculture in Central Europe was thought to have been practised mostly in ponds managed by monasteries. During this time, religious motivations fostered the culture of fish. Fish had an important religious aspect for Christianity, which rapidly spread throughout Europe after the collapse of the Roman Empire. For Christians, fish symbolized human souls and early Byzantine churches are decorated with frescos that depict artificial fish ponds serving as sanctuaries (Drewer, 1981), similar to those depicted in Egyptian tombs.

In addition, several Christian orders introduced the habit of eating fish on Fridays and during Lent, which reinforced the construction and management of artificial fish ponds in medieval monasteries. During this period, more than 100 days of fasting

Figure 2.2 This figure shows some examples of Koi, domesticated common carp (*Cyprinus carpio*), that display highly variable patterns of scale colours. Photos by author.

Figure 2.3 A medieval castle near Münster (Germany) surrounded by a moat. The primary use of moats as defence structures was gradually multipurposed to support carp aquaculture. Photo by author.

per year were established, which prohibited the consumption of homoeothermic (warm-blooded) animals (mammals and birds) and promoted the consumption of poikilothermic (cold-blooded) fish and invertebrates (Beveridge and Little, 2014).

Besides religious motives, aquaculture of fish in Europe also originated from practical considerations. For instance, medieval rulers often defended their castles by surrounding them with an artificially constructed moat to limit access to the castle via a drawbridge. The multipurpose use of such ponds promoted the development of carp aquaculture (Figure 2.3).

Other examples of multipurpose use fish ponds are millponds and reservoirs that resulted from damming a stream in a narrow gorge. Millponds and moats were stocked with fish and fertilized with village or castle waste, which promoted plant growth that served as food for herbivores and omnivores.

Some species, in particular the common carp (*C. carpio*), were able to reproduce and thrive under these conditions, although more demanding species such as the Brown trout (*Salmo trutta*) have also been cultured since at least the fourteenth century in France (Pillay and Kutty, 2005).

Eventually, fish produced in moats became a convenient source of food that was readily accessible even during times of war and besiegement of the castle. Ecological management of such ponds became necessary if the amount of introduced waste and fertilizer and the produced fish biomass exceeded the recycling capacity of ecosystem services. Ecosystem services represent the sum of activities of a variety of decomposers (e.g. nitrifying, denitrifying, and organic phosphorus metabolizing bacteria) and primary producers (photosynthetic plants) that maintain a balanced nutrient cycle in the pond ecosystem (Section 4.6).

Medieval fish farmers realized (by trial and error) that fallow periods are necessary for maintaining pond productivity and preventing the fouling of pond sediment. Draining ponds during fallow periods and drying them allows removal of solid waste that builds up over time. Moreover, draining ponds rapidly oxidizes sediment as air, which contains a 20,000-fold greater amount of oxygen than water, enters the pores between drying sediment particles.

The necessary duration of fallow periods depends on sediment porosity. Soils of higher porosity are reconditioned much more rapidly than those with very fine porosity, which have smaller and less accessible spaces for aeration. During reconditioning the redox potential of pond sediment is restored to support the establishment of a healthy community of aerobic decomposing bacteria upon re-flooding of the ponds. Such bacterial communities are critical for organic waste assimilation. The ancient practice of allowing for fallow periods has been rediscovered in recent decades for pond aquaculture and cage mariculture.

2.5 Carp domestication in Central Europe

The common carp was one of the earliest freshwater fish species exploited by the Celts, cultured by the Romans, and managed in artificially constructed ponds and moats by medieval Europeans (Balon, 1958). Keeping fish in man-made and human-controlled environments promoted (initially unintentional) artificial selection, which supported their domestication.

The common carp represents the second-oldest domesticated aquatic species, goldfish being the oldest. Genetically and morphologically distinct lines of *C. carpio* started to originate in Europe between the twelfth to fourteenth centuries (Balon, 2004). Fish ponds, constructed by humans in medieval Europe, were stocked with a variety of naturally occurring species, which were then subject to selection pressure imposed by the unique environmental conditions in these man-made artificial habitats.

Besides the direct benefit of transplantation of animals into new habitats by humans, common carp benefited indirectly from urbanization. This species is superior to other fish regarding the tolerance of wastes and poor water quality, and it thrived near the growing medieval hubs of human population.

Artificial fish ponds can be viewed as a novel ecological niche that was available for colonization by species best equipped to quickly adapt to its environmental conditions and outcompete other species in that regard. It is no coincidence that freshwater fish, which have been used in aquaculture for many millennia, including tilapia, carp, and goldfish, are among the most resilient (and highly invasive) species of fish.

For instance, common carp tolerate a very wide range of environmental stress, including highly turbid water, extreme and long-lasting fluctuations in food availability, anthropogenic pollution, and stress associated with climate change (Pietsch and Hirsch, 2015). Common carp and goldfish naturally overwinter under ice during the winter months.

Primary production and oxygen concentration in lakes are very low in winter, especially if snow cover prevents the penetration of light. During this period, which can last months, carp greatly reduce oxygen consumption, activity, and feeding by adjusting their metabolism. For example, common carp produce ethanol, which can readily diffuse from the body, unlike lactate, which accumulates. This fermentative strategy prolongs carp tolerance of long periods of low oxygen availability (Sorvachev, 1957, 1959).

During the summer, these fish tolerate high temperature, saturated oxygen levels, a large pH range, salinities ranging from freshwater to 6 g/kg, and high levels of common environmental toxins and pollutants (Pietsch and Hirsch, 2015). This wide environmental tolerance range promoted the establishment of common carp in man-made ponds that were subject to variable (often poor) water quality.

Along with rapid growth and large size, the wide environmental tolerance range represents a selective advantage that renders carp, tilapia, and similar species superior for aquaculture but, at the same time, underlies their success as invasive species.

The common carp may be the first example of an invasive aquatic species that has been intentionally introduced to a wide range of non-native habitats by humans. The historical epicentre of the *C. carpio* native habitat is in Pannonia, the area of the Roman Empire bordered to the North and East by the Danube River, roughly corresponding to modern Hungary (Balon, 2004).

From this area of Central Europe, *C. carpio* was exported to Italy more than 2000 years ago and to

Asia many centuries ago (Balon, 2004). More than 1000 years ago the distribution range of common carp is thought to have expanded towards the North and West of Europe, perhaps as a result of transplantation by humans (Hoffmann, 1994).

A second phase of range expansion has followed during the Middle ages (twelfth to fourteenth centuries), during which common carp radiated across most of medieval Central and Western Europe (Hoffmann, 2005). Another range expansion of this species took place during the late fourteenth century, when common carp were introduced to the British Isles and Scandinavia (Balon, 2004).

It is uncertain how (much) the invasion of common carp has changed the affected aquatic ecosystems and whether it resulted in displacement or extinction of other species in large parts of Europe and Asia. However, it is clear that the range expansion of this species was greatly facilitated (if not entirely mediated) by human management (Balon, 2004).

Nevertheless, today common carp is not considered an invasive species in large parts of Europe and Asia outside of its historical centre of distribution. This example illustrates that consideration of invasive species from a practical conservation and management perspective depends on when the invasion of new habitat has occurred, in addition to whether it had been facilitated by humans and how much potential it bears to alter ecosystems.

2.6 The development of mariculture

Mariculture of marine species in brackish water or seawater appears to have originated around the same time as freshwater aquaculture but early mariculture attempts have not been as well preserved or discovered in the historic record.

In Europe, the Romans started keeping fish in special artificial fish ponds or reservoirs called *piscinae* more than 2000 years ago (Friedländer, 1934). *Piscinae* were initially used to store fish and other aquatic animals to ensure year-round access to exotic foods imported from different geographic regions of the Roman Empire or to circumvent seasonality in the availability of certain food species. Soon these *piscinae* became a symbol of Roman luxury and power, and they were also used for entertainment and ornamental purposes (Higginbotham, 1997).

Marine *piscinae* were considered by the Roman patricians more prestigious than their freshwater counterparts, which is why the latter more commonly retained their original purpose as food reservoirs. They were common throughout the Mediterranean provinces of the Roman Empire (Hoffmann, 2005).

Ancient invertebrate mariculture has been practised by the Etruscans and Romans who managed oysters in coastal ponds and lagoons at the Adriatic and Tyrrhenian coasts of the eastern Mediterranean Sea as early as 2500 years ago. It has been suggested that oysters were also cultured in Japan as early as three to four millennia ago (Iversen, 1976; Stickney, 2017).

Moreover, oyster aquaculture may have been practised on mainland Asia across the Sea of Japan for more than four millennia, for example, on the coasts of Peter the Great Bay. In Papua New Guinea mussel aquaculture may be as old as 2000 years and oysters have been consumed in large quantity in North America since 500 to 1000 years ago. However, whether the widespread consumption of oysters and mussels indicates the existence of aquaculture or merely capture fishery practices in North America is uncertain (Malindine, 2019).

In Java (Indonesia), regulatory policy on true aquaculture dates back at least 600 years, when laws aimed at the protection of milkfish farmers from poachers were established (Iversen, 1976). Documents by Hindu record keepers describe milkfish (*Chanos chanos*) mariculture in artificial brackish ponds as early as the twelfth to fourteenth centuries, perhaps as a result of observing and eventually managing the seasonal flooding of tide pools and salt ponds along the coasts of Java and Indonesia (Schuster, 1952).

In this part of Southeast Asia, the shallow brackish/marine fish ponds were known as *tambaks*, while freshwater ponds were referred to as *siwakan* (Schuster, 1952). The timeframe when *tambaks* were first mentioned in historical records coincides with, or slightly precedes, the development of milkfish and mullet mariculture systems in Hawai'i (Section 2.7) and other Polynesian islands (Nash, 2011).

2.7 Origins of ecologically sustainable mariculture in Hawai'i

Hawai'ian Polynesians started to construct and manage coastal fish ponds (loko i'a) for mariculture from at least the thirteenth century (Burney, 2002; Kikuchi, 1976). Loko i'a were developed by altering and managing naturally occurring anchialine pools (from the Greek *ankhialos*, 'by the sea'). The water levels and salinity of anchialine pools are subject to large fluctuations depending on tidal cycle, temperature, wind speed, and rain fall.

Anchialine pools represent unique and highly productive ecosystems that serve as nurseries and refugia for many species of coastal fish. They represent preferred sites for early Hawai'ian settlements. For instance, at the Big Island's Anahoomalou Bay, the archaeological remains of an early Hawai'ian settlement are located adjacent to many natural anchialine pools and a pair of coastal loko i'a.

The Hawai'ian name Anahoomalou Bay merges the words anae (mullet) and ho'omalu (protected), which indicates that this area was a protected site for providing the Hawai'ian people with mullet and other coastal fish. Unlike in other parts of the world, fish ponds were not fertilized with human or animal waste, as such practice was prohibited by the Hawai'ian rulers (Kikuchi, 1976). Rather, the naturally high productivity of loko i'a was utilized, resulting in ecologically highly sustainable management.

The Anahoomalou Bay site contains a larger (Ku'uali'i) and a smaller (Kahapapa) loko i'a, both of which harbour brackish water of varying salinity. The salinity depends on the amount of freshwater runoff and underground springs fed by rain accumulating in the mountains versus the extent of ocean water entering through canals and during storms. The smaller Kahapapa pond is connected to the ocean via a gated (makaha) channel ('auwai) that permits the passage of small fish larvae and juveniles (Figure 2.4).

Makaha' auwai were made from lava rocks and coral rubble cemented together with secretions from coralline algae. The makaha gates were originally made from porous lava rock and coral rubble or wood poles, which have been replaced in recent times by metal gates. The larger Ku'uali'i pond is connected to the smaller Kahapapa pond via another channel through which small fish can enter. This larger pond is separated from the ocean by a sand bar that can be breached during storms to occasionally spill in large amounts of seawater and marine larvae.

Figure 2.4 A channel that connects a fishpond with the open ocean in Anaehoomalu Bay, Big Island of Hawai'i. Gates facilitate the trapping of small catadromous fish larvae and juveniles while preventing their escape after grow-out. Channels were built from lava rocks cemented together using secretions from coralline algae. Photo by author.

Loko i'a harbour elaborate food webs that consist of many plants, bacterial decomposers, herbivorous zooplankton, small fish, and top-level predatory fishes such as barracuda, jacks, and moray eels. Annual pond productivity has been estimated as high as 17,000 kg/acre for plants and 170 kg/acre for fish and, on average, about 2000–3000 kg of fish were harvested from a single pond (Costa-Pierce, 1987).

Larval and juvenile fish develop and grow rapidly in the nutrient-rich brackish ponds, and they quickly exceed the size for fitting through the Makaha gate, which prevents their escape. Several of the fish species entering coastal ponds as juveniles could be considered catadromous, that is, they develop in brackish or freshwater and return to seawater for spawning.

This group of coastal fishes is represented by many euryhaline species from diverse taxa, including milkfish, mullets, parrotfishes, estuarine moray eels, jacks and trevallies, goatfishes, scads, barracudas, surgeonfishes, bonefish, gobies, and nehu. Many of these fishes may be facultatively catadromous, meaning that a brackish or freshwater phase during their life history is optional.

Milkfish and mullet (e.g. *Mugil cephalus*) were the most commonly cultured species that produced the largest biomass of fish in loko i'a. These catadromous species even survive abrupt transfer from seawater into freshwater ponds and grow very well in freshwater, brackish, and marine ponds (Costa-Pierce, 2014).

Already many hundreds of years ago, the Hawai'ian people were keen observers of the ecology and behaviour of catadromous fish. They utilized the resulting knowledge to manage the culture and harvest of these fish in the loko i'a. From observing the behaviour of catadromous fish, they knew that these species migrate from brackish and freshwater ponds to the ocean to spawn. This migration occurs at night and is synchronized seasonally with the lunar cycle. During spring full moons, large numbers of mullet and milkfish attempt to leave the loko i'a to start their seaward migration.

The Hawai'ians utilized this migratory fish behaviour to conveniently harvest a large number of fish with relatively little effort in a short time. Fish were easy to net out as they piled up at the makaha gates that closed off the canals connecting the loko i'a to the ocean. Harvesting was further facilitated by installing a second gate towards the end that connected the canal to the ponds to trap fish before netting in the narrow canal between the two gates.

Clever use of these two gates, in combination with the kinetic energy provided by the tides, allowed periodic cleaning of the ponds and canals by removing accumulating silt and debris, which kept those ecosystems functioning well. These examples illustrate that Hawai'ian aquaculture has a long tradition of ecological sustainability and technological innovation.

2.8 Diversification of aquaculture following the industrial revolution

During the age of the industrial revolution in the nineteenth century, many technological advances occurred that greatly increased the efficiency of capture fishery practices. In particular, steam engines and the mechanization of fishing aids resulted in the construction of fishing vessels and fleets capable of rapidly harvesting from large and remote areas far offshore in relatively short time (Nash, 2011).

Exploiting the resources of the oceans, which must have seemed plentiful and inexhaustible during industrialization's early years, was far more attractive and profitable than inland fisheries, which were well established in Europe at that time. In addition, industrialization greatly accelerated the rate of anthropogenic pollution, which became evident in rapidly deteriorating water quality of many European freshwater ecosystems.

Because of these negative side effects, the industrial revolution led to a collapse of freshwater fisheries and fish farming in many European countries (Nash, 2011). Ironically, the term aquaculture originated during this time, even though the concept existed under different names for many centuries. In Germany, for example, the term *Teichwirtschaft* (pond management) has long been (and still is) used in lieu of freshwater aquaculture (Schäperclaus, 1933).

Despite the initially pervasive negative side effects of the industrial revolution on aquaculture, the technological advancements resulting from it also provided new opportunities for aquaculture. During the second half of the nineteenth century, in parallel with the age of industrialization, technological advancements developed and introduced by naturalists and other innovators in Europe and North America represented the first scientific contributions to the field of aquaculture (Stickney, 2017).

For instance, paddle wheel aerators were inspired by paddle wheel steam boats during the nineteenth century. Such aeration devices facilitated an increase in fish pond stocking density and, thus, contributed greatly to aquaculture intensification.

The first marine laboratories were founded during this time, starting with the Stazione Zoologica in Naples, Italy in 1872, followed by the Woods Hole Marine Biological Laboratory in the US in 1888, the Marine Biological Laboratory in Plymouth, England in 1888, and the Oceanographic Museum and Aquarium in Monaco in France in 1906 (Nash, 2011).

These institutions fostered biological research on aquatic organisms and contributed to aquaculture innovations. The first artificially manufactured pellets for feeding fish were developed because of biological research and contributed to the intensification of aquaculture. Pellets simplified the feeding process and streamlined access to defined feeds throughout the year. The main ingredients used for the first fish pellets were similar to those that are commonly used today. They consisted, among other ingredients, of fish meal, soy or sunflower oil, and wood sawdust as binder (Schäperclaus, 1933).

In addition, breeding techniques were optimized for key aquaculture species during this time. For instance, in 1870, Polish fish farmer Thomas Dublisch developed a system for breeding carp in natural ponds, transferring fry to small nursery ponds, and eventually to larger grow-out ponds—an approach that later became known as the Dublisch method of fish aquaculture.

The production of fish larvae and fry in hatcheries was incentivized by government laws and regulations that promoted restoration of freshwater and anadromous fish populations, which were decimated because of industrial pollution and damming of major inland waterways.

Moreover, ground-breaking scientific progress in emerging fields such as developmental biology, evolutionary ecology, genetics, and environmental physiology provided a sound theoretical basis for artificial fertilization and selective breeding efforts.

Throughout the nineteenth century government hatcheries arose throughout Europe and North America to promote the production of salmon, trout, and bass fry for stocking aquatic ecosystems with those species primarily to promote recreational fishing, which was negatively impacted by the industrial revolution (Pillay and Kutty, 2005).

One of the first fish hatcheries was in Huningue, France, built in 1852 after French researchers had worked out a method for artificial fertilization of trout eggs based on observations and experiments published by German Naturalist Ludwig Jacobi in 1763 and refined by Italian Monk and Università di Pavia professor Lazare Spallanzani (Nash, 2011).

In 1856, the first hatcheries for producing salmonid fish were constructed in Canada's eastern provinces, followed in 1870 by the first US Atlantic salmon hatchery in Maine, and in 1884 by a Pacific salmon hatchery in British Columbia. In California, the first hatchery for Pacific salmon originated from an 'egg-taking station' established in 1872 on the McCloud River, while in Oregon the first Pacific salmon hatchery began operating in 1877 (Nash, 2011).

The practice of 'sea ranching' was first introduced in the 1930s by scientists at the Swedish Salmon Research Institute. Sea ranching represents a special method of salmon aquaculture that takes advantage of the homing behaviour of salmon returning from the ocean to their freshwater spawning grounds. In particular, tagging experiments by Börje Carlin contributed greatly to the development of this aquaculture approach and the salmon tags he developed are now referred to as Carlin tags (Carlin, 1969; Isaksson, 1988).

In 1879, the first farm for culturing Japanese eel (*Anguilla rostrata*) was built in Japan and within twenty years many others followed, supplying over 80% of Japan's eel harvest. Eel farming is unique to the present day and differs from other forms of aquaculture by its reliance on wild-caught juvenile developmental stages of this catadromous species (glass eels).

Glass eels are caught in coastal areas just before they start their upstream migration in rivers to enter the freshwater phase in their life cycle. Thus, eel farming relies on healthy wild populations and setting aside a portion of adult eels produced in aquaculture farms for restocking rivers represents an integral part of this form of aquaculture. Such practice ensures that a sufficiently large number of eels can complete their life cycle by returning to their marine spawning grounds for reproduction and maintaining stable populations (Section 15.1).

In Eastern Asia, seaweed farming was developed during the first half of the twentieth century by expanding on the practice of harvesting wild seaweeds. Although seaweed was harvested and already used for medicinal purposes 5000 years ago in China, it was only about 400 years ago in Japan that the management and farming of seaweed for human consumption originated (Nash, 2011).

Seaweed farming represents the oldest form of plant aquaculture (besides the culture of ornamental plants in the sacred ponds and *piscinae* of ancient societies around the Mediterranean and in Asia). However, managed propagation of the entire life cycle of several commercially important species of seaweed (*Porphyra spec.*, laver, red macroalgae) only became possible in 1949 after British scientist Kathleen Drew discovered an intermittent conchocelis phase, which produces asexual spores and alternates with the sexually reproducing thallus phase (Drew, 1949).

A multitude of additional innovations resulted in improvements and novel approaches for aquaculture during the first half of the twentieth century. Between about 1920 and 1930, oyster aquaculture production in Japan tripled after the introduction of a new culture technique—growing oysters freely in the water column by attaching larvae (spat) on ropes suspended from rafts. Since oysters are filter feeders, this technique provided large numbers of oysters with optimal access to food, resulting in faster growth on a much smaller footprint compared to bottom culture. In addition, harvesting oysters was simplified with this new technique. This technique and variations thereof are now commonly used for culturing pearl oysters and other invertebrates (e.g. sponges).

By the end of the nineteenth century, while it was possible to produce in very large numbers fertilized eggs and larvae of commercially important fish and invertebrates, it was not yet conceivable to raise the larvae of most species under controlled conditions in hatcheries. Therefore, billions of eggs and larvae of fish and aquatic invertebrates were transplanted and restocked for grow-out in non-native habitats all over the world (Stickney, 2017).

2.9 Aquaculture intensification and the emergence of invasive species

The second half of the nineteenth century and the first half of the twentieth century witnessed an enormous boom in the development of techniques for mass production, transport, and stocking of fish and mussel larvae. While it was possible to produce large numbers of larvae under controlled culture conditions, the lack of appropriate feeds, technologies, and hatchery systems hindered their grow-out in managed environments. Instead, it was common practice at the time to release large numbers of larvae into natural ecosystems for stocking purposes.

As early as 1904, the suggestion was that fishermen should strip milt and eggs from captured marine fish, mix them, and release them back into the ocean to sustain future populations (Nash, 2011). Unfortunately, such restocking efforts were not limited to native habitats. Moreover, the release of fertilized eggs and small larvae relied to a great extent on natural ecosystem services to eventually yield harvestable organisms.

The original intent of stocking natural ecosystems with larvae was to foster the production of fish, molluscs, and crustaceans for providing high-quality seafood to the growing human population and for supporting recreational fisheries. At the time, experience with and scientific information about invasive species was very limited and, therefore, there were few concerns about disturbing the balance of natural ecosystems that had evolved over millions of years.

Fortunately, many of the species that were transplanted all across the world to non-native habitats were not able to survive and reproduce in foreign environments (Stickney, 2017). However, several of the transplanted species were highly invasive and have firmly established themselves in the new habitats, often by out-competing native species that occupied a similar ecological niche. In that sense, the aquatic species transplantation flurry that occurred around the late nineteenth to early twentieth centuries represented an unintended, gigantic artificial selection experiment, which identified the most highly invasive aquatic species.

Such invasive species include certain salmonids, striped bass (*Morone saxatilis*), common carp, and tilapia (*Oreochromis spp.*). The transport of aquaculture species deemed of high potential for future seafood production to non-native habitats throughout the world included not only fish, but also invertebrates. For instance, millions of Eastern oysters (*Crassostrea virginica*) were transplanted from the Atlantic coast to the Pacific coast of North America and millions of Japanese oysters (*Crassostrea gigas*) were introduced from Asia to the North-American Pacific coast in the 1920s (Blakeslee *et al.*, 2013).

Fertilized rainbow trout eggs were exported by the US Fish Commission to France in 1879, to Germany in 1882, and to England in 1884. In hindsight, it is difficult to grasp the underlying logic that led to the introduction of brown trout (*Salmo trutta*) from Europe to North America while, at the same time, the introduction of North American rainbow trout (*Oncorhynchus mykiss*) to Europe.

Likewise, Atlantic salmon were transported from Europe to North America and Pacific salmon were transplanted from the west coast to the east coast of North America. Atlantic salmon was not (at least not yet) able to outcompete Pacific salmon and establish itself on the west coast of North America. However, invasive trout species are firmly established in Europe and North America and a Pacific salmon population thrives in the Great Lakes of North America.

Rainbow trout were also introduced from the US to New Zealand in 1868, followed by Chinook (Quinnat) salmon (*Oncorhynchus tshawytscha*), which originated at California's McCloud River hatchery (Nash, 2011). Both species established viable populations in New Zealand. Moreover, trout were introduced to the British colonies in Africa, primarily to foster the development of recreational fisheries (Pillay and Kutty, 2005) and for controlling disease-transmitting mosquitos, particularly in Southeast Africa (Nash, 2011).

In 1877, fertilized eggs of rainbow trout and brown trout were also introduced to Japan and established populations in Lakes Inawashiro and Chuzenji (Nash, 2011). The goldfish is native to East and Central Asia but during the second half of the nineteenth century was transplanted to and rapidly populated Europe, South Africa, Madagascar, Mauritius, North America, Oceania, South and Central America, and the Caribbean (Balon, 2004).

Tilapia were first farmed in their native habitat in Kenya in 1924 and they quickly established themselves as prime aquaculture species throughout most of Africa. The prime suitability of tilapia for aquaculture rapidly made news throughout the world and during the late 1930s tilapia (*Oreochromis spp.*) were transported from their native African habitats to the Middle East and Southeast Asia and later also to Hawai'i, Florida, and the California Salton Sea.

In each case, tilapia established themselves quickly as a dominant invasive species. Striped bass (*M. saxaitils*) was introduced to California from the east coast of the US and common carp had established itself after introduction throughout much of North America. For instance, common carp was imported to California from Germany in 1872 and has since flourished not only in freshwater, but also in slightly brackish habitats like Suisun Marsh (Nash, 2011).

American shad (*Alosa sapidissima*) was introduced to California's Sacramento River from its native habitat on the North American east coast in 1871. From California, the shad population rapidly expanded along the Pacific coast as far north as the Columbia River (Nash, 2011).

Retroactively, the 'gigantic artificial selection experiment' conducted by transplanting many aquaculture species at the turn of the nineteenth to twentieth century resulted in serious ecological consequences by introducing invasive species and associated pathogens that permanently altered ecosystems in many parts of the world.

For example, during this time crayfish plague was introduced via transplantation of aquaculture crustaceans from North America to Europe.

North American crayfish species are resistant to *Aphanomyces astaci*, the fungal pathogen causing this disease, but European crayfish species suffer high mortality following infection (Alderman, 1996).

Thus, the transplantation of invasive species with the intent to foster aquaculture had (and continues to have) significant ecological impact. However, such adverse effects are not unique to invasive aquatic species. They have resulted on a much larger scale from the transplantation of domesticated terrestrial animals and plants and their associated pathogens. Because transplantations of many terrestrial organisms and associated invasions are historically much older, they are often not scrutinized to the same extent as those of aquatic species.

2.10 Blue revolution: an upsurge of aquaculture research and development

According to Beveridge and Little (2014), 'farmed aquatic production prior to 1950 was insignificant' from a global economic perspective. However, the global impact of aquaculture greatly increased during the second half of the twentieth century due to several key technical breakthroughs and innovations.

Although many informative and influential monographs on various aspects of aquaculture existed for centuries, the term aquaculture was not used routinely in its current meaning prior to the second half of the twentieth century. Articles reporting original aquaculture research in peer-reviewed scientific journals started to appear regularly from 1950. Over the next five decades, the number of such articles rose at an exponential rate before gradually levelling off (Figure 2.5). This period, in particular the years between 1960 and 1980, have been dubbed the 'blue revolution'.

During these two decades, the commercialization of aquaculture expanded on a grand scale and saw established many extensive scientific research programmes that focused on aquatic species' basic biology and applied culture. Aquaculture technology and engineering advanced rapidly and globally.

Selective breeding and hybridization of many aquatic species intensified and allowed systematic promotion of specific traits of interest. Major progress in reproducing and maintaining numerous aquatic species was made in the 1960s–1980s due to research programmes at universities and other scientific institutions as well as the efforts of devoted hobby aquarists. The aquarium trade and market for ornamental aquatic species started to boom and aquaculture production gained global economic importance.

In 1939, for the first time it was possible to raise fish larvae and keep them alive in captivity during critical developmental stages by hatchery feeding of brine shrimp (*Artemia spp.*) nauplii (Rollefson, 1939). This convenient source of live food revolutionized aquaculture as it permitted the propagation of early fish larvae and fry from species for which these early developmental stages had to be previously released into the wild.

Previous propagation of these species by releasing them into wild habitats was problematic because survival rates of released larvae and fry were generally very small due to predation and other environmental challenges. Even for highly invasive species such as tilapia and carp the initial survival rates of first-generation fish released into novel habitats were relatively low.

In contrast, survival of larvae and fry fed with brine shrimp nauplii in hatcheries was dramatically improved and much larger numbers of fish were produced. In addition, it became possible to control and manage the extent to which ecosystem services were needed and to reduce the potential for invasion of aquaculture species into non-native environments.

Brine shrimp inhabit hypersaline environments, for example, the Great Salt Lake in Utah and they form a dormant cyst-like life cycle stage that can persist in a dehydrated state for many years (Clegg, 1964, 1976). These dormant brine shrimp cysts float on the surface of salt lakes and can be easily harvested, dried, canned, and shelved until needed.

Rehydration (i.e. placing the cysts into water) represents the signal for the termination of dormancy and triggers resumption of development, which results in nauplii emerging within minutes. Brine shrimp and their nauplii are extremely euryhaline (i.e. they can survive for long periods in brackish water, seawater, hypersaline water, and even for approximately an hour in freshwater). This high salinity tolerance and convenient storage as canned

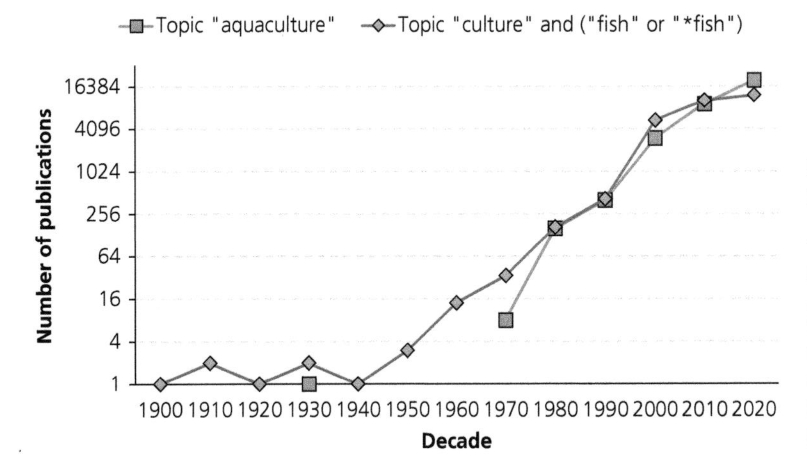

Figure 2.5 Exponential growth of published aquaculture research between 1950 and 2018 based on searching the 'Web of Science' literature database. Topic searches were performed on 12 February 2019 using the terms indicated in the figure legend. Additional publications using alternative terminology (mariculture, etc.) increases the overall number of publications but does not change the general trend.

cysts renders brine shrimp an ideal live food source for raising early developmental stages of marine and freshwater fish and invertebrates in aquaculture hatcheries.

Another key technical innovation benefitting aquaculture were plastics like polyvinyl chloride (PVC), fiberglass, and various plastic foams (polycarbonate, polyphenylene oxide, polyurethane). The new lightweight plastic materials were highly durable and, thus, were ideal replacements for rapidly corroding metal pipes and containers used up until the mid-twentieth century in hatcheries (Nash, 2011).

Unfortunately, the advantages of plastics materials also represent their biggest shortcomings, that is, their high durability renders plastics materials very difficult to degrade and dispose of. Today, the accumulation of disposable plastic waste in the oceans threatens many pelagic species of fish and invertebrates and, in turn, negatively affects seafood production by capture fisheries (Haward, 2018).

Therefore, significant scientific research efforts are now focusing on the generation of plastics materials that are not only highly durable, but also readily biodegradable under specific conditions that exist in waste processing plants but not in environments where the plastics are needed.

The invention of the *'aqua-lung'* by Émile Gagnan and Jacques-Yves Cousteau (i.e. self-contained underwater breathing apparatus, or SCUBA) diving technology, promoted the diversification of aquaculture approaches (Cousteau and Dumas, 1953). For instance, SCUBA diving facilitates regular monitoring and maintenance of extensive offshore aquaculture systems such as sponge and oyster rafts, horizontal line systems, and cage culture systems.

Other technical innovations that propelled aquaculture development during the second half of the twentieth century included new tests and associated equipment for monitoring and optimizing water quality parameters. Such monitoring aids included portable oxygen meters and probes, conductivity meters, pH meters, osmometers, test kits for ammonia, nitrite, and nitrate, and others.

Moreover, protein skimmers, rotary and air pumps, reverse osmosis systems, bioreactors, temperature and lighting regulators, automatic feeders, and other technological innovations made it possible to maintain a high level of control over artificial aquatic environments, which propelled the adoption of freshwater and marine aquaria for keeping ornamental species on a large scale.

These technological innovations have also facilitated the design and development of intensive recirculating aquaculture systems (RAS) that maximize the efficiency of resource usage while minimally depleting the capacity of natural ecosystem services.

Towards the end of the twentieth century and the beginning of the twenty-first century, computers and cell phone towers enabled the installation of automatic alarm systems. These alert aquaculture farmers when water quality parameters reach

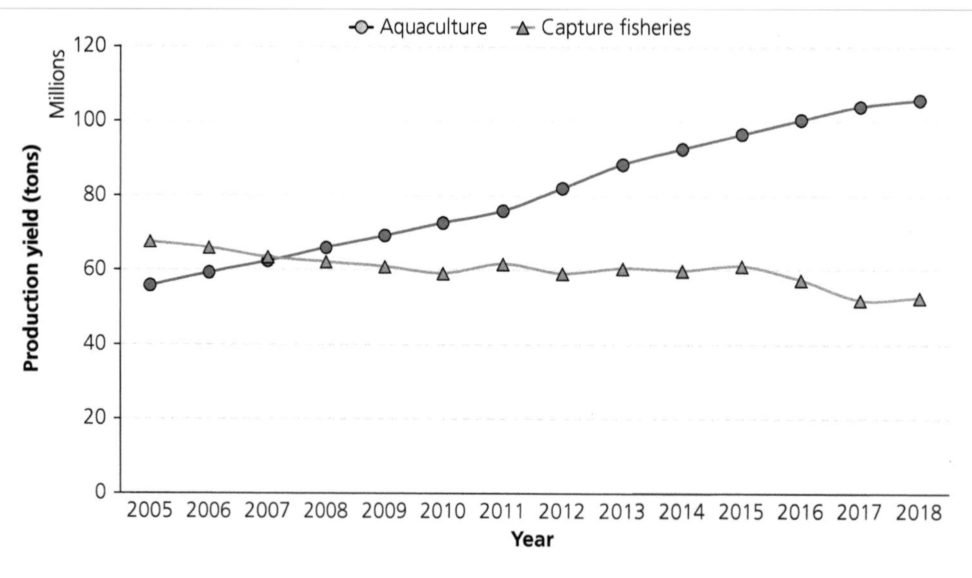

Figure 2.6 Annual aquaculture production compared to capture fisheries production (both in million tons/year) from 2005 to 2018. Note that estimates for capture fisheries vary somewhat and are higher when FAO estimates are used (FAO, 2020). Regardless, aquaculture production now tops that of capture fisheries. Data source: OECD.stat.

critical levels such that rapid correction of water quality is possible, reducing adverse effects on aquaculture organisms and improving animal welfare.

These key innovations boosted aquaculture into the focus of significant economic development and made it a lucrative business. As a result, aquaculture production overtook the lead from capture fisheries seafood production shortly after the turn of the millennium (Figure 2.6).

Still, technological aids were often not utilized to their fullest extent due to high costs and decreased profitability of aquaculture businesses to such an extent that economical operation was compromised or impossible. Instead of recycling resources, many aquaculture operations heavily rely on large amounts of high-quality water and ecosystem services for processing the waste that is generated.

The regulatory and economic incentives for utilizing costly technological aids that reduce the depletion of natural resources and exhaustion of ecosystem service capacity were insufficient to prevent increasing damage of natural ecosystems by large aquaculture businesses. Consequently, at the start of the twenty-first century focus shifted to ecological considerations and concerns about the sustainability of aquaculture.

2.11 The rising ecological impact of aquaculture and sustainability concerns

Aquaculture regulation intensified during the nineteenth century, especially during the second half, where keen observers, naturalists, scientists, aristocrats, and public servants realized that industrialization not only provided benefits, but also generated significant new challenges (overfishing, poaching, pollution of rivers, uncontrolled transplantation of invasive species, etc.).

In response to such problems, in many countries the second half of the nineteenth century gave rise to agencies and laws that regulate aquaculture and fisheries. For instance, the US government formed the Commission of Fish and Fisheries, which later became the US Fish and Wildlife Service. Almost simultaneously, in December 1870, the American Fish Culture Association was founded to represent the interests of fish farmers and ichthyologists on a national scale. In 1884 it was renamed as the American Fisheries Society (AFS).

In 1870, the Deutscher Fischereiverein, which was later renamed to Deutscher Fischerei-Verband (German Fisheries Society) was founded to regulate and promote fisheries and aquaculture in Germany. Shortly thereafter, in 1884, the Marine Biological Association of England was formed.

Founded in Rome, Italy, in 1905, the International Institute for Agriculture (IAA) performed statistical analyses of fisheries and aquaculture trends. The IAA was reorganized and became the United Nations Food and Agriculture Organization (FAO), which now monitors the state of world fisheries and aquaculture, lobbies on behalf of those industries, and develops recommendations for their future development, including recommendations on ecologically sustainable practices, regulatory recommendations, and funding for improving aquaculture.

Since 1994, the FAO Fisheries and Aquaculture Department has published the *State of World Fisheries and Aquaculture* (SOFIA) in two-year intervals. The intended purpose of this document is advocacy on behalf of the fisheries and aquaculture industries to inform 'policy-makers, civil society and those whose livelihoods depend on the sector' about developments in capture fisheries and aquaculture economics, associated policies, research, and challenges.

The construction of dams to harness hydropower and create freshwater reservoirs for urban and agricultural use had dramatic consequences for anadromous salmonids and associated fisheries. A striking example is the Columbia river system in the US Pacific Northwest, where salmon runs have decreased from about 15 million adults per year at the beginning of the twentieth century to fewer than 2.5 million adults per year a century later (Nash, 2011).

The US government enacted the Federal Power Act in 1920, the Fish & Wildlife Conservation Act in 1934, and the Mitchell Tax Allocation Act in 1938 to alleviate the effect of dam construction on diminishing salmon runs by mandating the funding of hatcheries, restocking programmes, fish ladders, and other mitigations. As a result, this regulation helped to establish many government-funded salmonid hatcheries that still operate today.

The rapid rise of semi-intensive monocultures of Atlantic salmon, marine shrimp, and other aquaculture species during the 1980s led to dramatic increases in production, vigorous market development and expansion, and mitigated the stagnation of capture fisheries, which became imminent during this time.

However, the extensive growth of these monoculture systems also raised renewed awareness of ecologically more sustainable aquaculture practices such as polyculture, integrated multi-trophic, and more closed, water-reuse systems.

Rapid development of salmon farming in Europe and shrimp farming in Southeast Asia resulted in a noticeable and significant environmental impact. For example, mangrove ecosystems were converted into semi-intensive shrimp ponds by enterprises that were sprouting up during the second half of the twentieth century. The rapid rise of coastal shrimp aquaculture enterprises caused great concerns about massive environmental destruction of mangrove habitats. As a result, a 1995 Supreme Court order restricted aquaculture development in India to specific coastal zones. Similar legislation was passed in other countries of Southeast Asia (Nash, 2011).

A build-up of pro and con aquaculture arguments by lobbyists on both sides and political opposition developed during the last quarter of the twentieth century, and the debate continues. The many facets of aquaculture technology and business practices were initially lost in the lopsided polemics, but over time, arguments both for and against were evaluated from perspectives that give more consideration to the diversity of aquaculture approaches.

While much of the debate surrounding modern aquaculture has been fruitful and has led to significant improvements, it is also true that political lobbying intended to influence regulatory policy and consumer acceptance, both pro and con, has used up much precious energy and resources.

The debate surrounding aquaculture is symptomatic for the competitive nature of human society (influence, power, and even the emotional support of our fellow beings). Properly incentivizing and harnessing this competitive nature for future progress and promoting better environmental stewardship in a humanistic spirit represents the biggest challenge of the twenty-first century.

The current trend still shows a growing number of human interventions in natural cycles and increasing competition for exploitation of aquatic and terrestrial habitats with short-term gain at the expense of long-term destruction. Alarming projections for severe consequences of such practices on global climate, species diversity, and human society underline the need to correct them (UN Environment, 2019; Brondizio *et al.*, 2019).

Aquaculture faces its share of challenges, which reflect legitimate concerns regarding its ecological impact and long-term sustainability while also fulfilling the high expectations and hopes that it carries for supporting instant and future needs of the human population, such as the demand for seafood.

The main challenges future aquaculture operations must overcome are to minimize global water usage, pollution, reliance on ecosystem service capacity, habitat fragmentation, escape and invasiveness of aquaculture species (especially GMOs), genetic dilution of wild populations, the spread of pathogens and diseases, development of antibiotic resistance in pathogens, extinction of wild species, and visual pollution (e.g. floating cages in coastal areas). At the same time, productivity, food safety, and the welfare of aquaculture animals must be maximized. The following chapters discuss examples and potential solutions for these challenges.

Key conclusions

- Aquaculture developed independently several millennia ago in multiple regions of the world.
- Aquaculture of catadromous fishes has a much longer history than that of anadromous species, though the latter are more commonly cultured today.
- There are four scenarios, referred to as Oxbow, Catch-and-Hold, Concentration, and Trap-and-Crop theories, that attempt to explain the transition from hunting to farming fish.
- Aquaculture development was driven by demand for seafood and by ornamental and religious motives.
- Good historical records of ancient aquaculture development exist for the Eastern Mediterranean regions of Egypt and Southern Europe, Assyria, China, Southeast Asia, and Oceania.
- In contrast to terrestrial species, only two aquatic species have been domesticated across the centuries: the goldfish (*Carassius auratus*) in China and the carp (*Cyprinus carpio*) in Central Europe and Eastern Asia.
- Development of freshwater aquaculture may have preceded that of mariculture.
- Evidence exists that ancient societies practised ecologically, highly sustainable, and integrated multi-trophic aquaculture, for example, in Hawai'i.
- The industrial revolution not only enabled key technological innovations that supported aquaculture development, but also greatly accelerated the unsustainable exploitation of aquatic resources.
- Massive transplantation of aquatic species at the dawn of the twentieth century led to the establishment of invasive species with high environmental resilience.
- The 'blue revolution' refers to a surge in aquaculture research and development during the second half of the twentieth century, which resulted in major economic growth of aquaculture businesses accompanied by growing concerns about the sustainability and ecological impacts of this new industry.

References

Alderman, D. J. (1996). 'Geographical spread of bacterial and fungal diseases of crustaceans', *Revue scientifique et technique*, 15, pp. 603–632.

Balon, E. K. (1958). 'Die Entwicklung der Beschuppung des Donau-Wildkarpfens', *Zoologischer Anzeiger*, 160, pp. 68–73.

Balon, E.K. (2004). 'About the oldest domesticates among fishes', *Journal of Fish Biology*, 65, pp. 1–27. https://doi.org/10.1111/j.0022-1112.2004.00563.x

Beveridge, M. C. M. and Little, D. C. (2014). 'The history of aquaculture in traditional societies', in Costa-Pierce, B. (ed.) *Ecological aquaculture: the evolution of the blue revolution*. Oxford: Wiley-Blackwell, pp. 3–29.

Blakeslee, A. M. H., Fowler, A. E., Keogh, C.L., 2013. 'Marine invasions and parasite escape: updates and new perspectives', in: Lesser, M. (ed.) *Advances in marine biology*. Oxford: Academic Press, pp. 87–169. https://doi.org/10.1016/B978-0-12-408096-6.00002-X

Brondizio, E., Settele, J., Díaz, S., and Ngo, H. T. (eds.) (2019). *Global assessment report on biodiversity and ecosystem services*. Intergovernmental Science-Policy Platform on Biodiversity and Ecosystem Services Report 7. Bonn: IPBES Secretariat.

Builth, H., Kershaw, A. P., White, C., Roach, A., Hartney, L., McKenzie, M., Lewis, T., and Jacobsen, G. (2008). 'Environmental and cultural change on the Mt Eccles lava-flow landscapes of southwest Victoria, Australia', *The Holocene*, 18, pp. 413–424. https://doi.org/10.1177/0959683607087931

Burney, D. A. (2002). 'Late quaternary chronology and stratigraphy of twelve sites on Kaua'i', *Radiocarbon*, 44, pp. 13–44.

Carlin, B. (1969). 'Salmon tagging experiments', *Swedish Salmon Research Institute Report*, 3, pp. 8–13.

Chen, S. C. (1956). 'A history of the domestication and the factors of the varietal formation of the common goldfish, *Carassius auratus*', *Scientia Sinica*, 5, pp. 287–321.

Clegg, J. S. (1964). 'Control emergence and metabolism by external osmotic pressure and role of free glycerol in developing cysts of *Artemia salina*', *Journal of Experimental Biology*, 41, pp. 879–892.

Clegg, J. S. (1976). 'Hydration measurements on individual Artemia cysts', *Journal of Experimental Zoology*, 198, pp. 267–272. https://doi.org/10.1002/jez.1401980217

Costa-Pierce, B. A. (1987). 'Aquaculture in ancient Hawaii', *BioScience*, 37, pp. 320–331. https://doi.org/10.2307/1310688

Costa-Pierce, B. A. (2014). 'The Ahuapua'a aquaculture ecosystem in Hawaii', in: *Ecological aquaculture: the evolution of the blue revolution*. Oxford: Wiley-Blackwell, pp. 30–43.

Cousteau, J.-Y. and Dumas, F. (1953). *The silent world*. London: Harper & Brothers.

Desroches-Noblecourt, C. (1954). 'Poissons, tabous et transformations du mort. Nouvelles considérations sur les pélérinages aux villes saintes', *Kêmi*, 13, pp. 33–42.

Drew, K. M. (1949). 'Conchocelis-phase in the life-history of *Porphyra umbilicalis* (L.) Kütz', *Nature*, 164, pp. 748–749. https://doi.org/10.1038/164748a0

Drewer, L. (1981). 'Fisherman and fish pond: from the sea of sin to the living waters', *The Art Bulletin*, 63, pp. 533–547. https://doi.org/10.1080/00043079.1981.10787920

FAO (2020). *The state of world fisheries and aquaculture 2020*. Rome: Food and Agriculture Organization of the United Nations (FAO).

Friedländer, L. (1934). *Sittengeschichte Roms*. Vienna: Im Phaidon.

Haward, M. (2018). 'Plastic pollution of the world's seas and oceans as a contemporary challenge in ocean governance', *Nature Communications*, 9, 667. https://doi.org/10.1038/s41467-018-03104-3

Higginbotham, J. (1997). *Piscinae: artificial fish ponds in Roman Italy*. Chapel Hill: University of North Carolina Press.

Hoffmann, R. C. (1994). 'Remains and verbal evidence of carp (*Cyprinus carpio*) in medieval Europe', in Van Neer, W. (ed.) *Fish exploitation in the past. Proceedings of the 7th Meeting of the ICAZ fish remains working group*. Tervuren: Musée Royal de l'Afrique Centrale, pp. 139–150.

Hoffmann, R. C. (2005). 'A brief history of aquatic resource use in medieval Europe', *Helgoland Marine Research*, 59, p. 22. https://doi.org/10.1007/s10152-004-0203-5

Isaksson, A. (1988). 'Salmon ranching: a world review', *Aquaculture*, 75, pp. 1–33. https://doi.org/10.1016/0044-8486(88)90018-X

Ishikawa, C., Okamura, K., Tanaka, S., Terao, A., Marukawa, H., Higurashi, T., and Seno, H. (1931). *Illustrations of Japanese aquatic plants and animals*. Tokyo: Fisheries Society of Japan.

Iversen, E. S. (1976). *Farming the edge of the sea*. Farnham: Fishing News Books, Ltd.

Kikuchi, W. K. (1976). 'Prehistoric Hawaiian fish ponds', *Science*, 193, pp. 295–299. https://doi.org/10.1126/science.193.4250.295

Larson, G., Piperno, D. R., Allaby, R. G., Purugganan, M. D., Andersson, L., Arroyo-Kalin, M., Barton, L., Climer Vigueira, C., *et al*. (2014). 'Current perspectives and the future of domestication studies', *Proceedings of the National Academy of Sciences of the United States of America*, 111, pp. 6139–6146. https://doi.org/10.1073/pnas.1323964111

Mabuchi, K. and Song, H. (2014). 'The complete mitochondrial genome of the Japanese ornamental koi carp (*Cyprinus carpio*) and its implication for the history of koi', *Mitochondrial DNA*, 25, pp. 35–36. https://doi.org/10.3109/19401736.2013.779261

Malindine, J. (2019). 'Prehistoric Aquaculture: Origins, Implications, and an Argument for Inclusion', *Culture, Agriculture, Food and Environment*, 41(1), pp. 66–70. https://doi.org/10.1111/cuag.12226

Nash, C. E. (2011). *The history of aquaculture*. Ames: Wiley-Blackwell.

Pietsch, C. and Hirsch, P. E. (2015). *Biology and ecology of carp*. Boca Raton: CRC Press.

Pillay, T. V. R. and Kutty, M. N. (2005). *Aquaculture: principles and practices*, 2nd edn. Oxford: Wiley-Blackwell.

Rabanal, H. R. (1988). 'History of aquaculture', in: *ASEAN/UNDP/FAO regional small-scale coastal fisheries development project*. Manila: Food and Agriculture Organization of the United Nations, pp. 1–17.

Rollefson, G. (1939). 'Artificial rearing of fry of seawater fish: preliminary communication', *Rapports et procès-verbaux des réunions*, 109, pp. 197–217.

Schäperclaus, W. (1933). *Lehrbuch der Teichwirtschaft*. Berlin: Paul Parey.

Schuster, W. H. (1952). *Fish culture in brackish water ponds in Java*. Rome: FAO Indo-Pacific Fisheries CouncilSpp.

Sorvachev, K. (1957). 'Changes in proteins of carp blood serum during hibernation', *Biokhimiy*, 22, pp. 822–827.
Sorvachev, K. (1959). 'Nitrogenous substances of the muscle of one-year-old carp during hibernation', *Biokhimiy*, 24, pp. 225–230.
Stickney, R. R. (2017). *Aquaculture: an introductory text*, 3rd edn. Boston: CABI.
Tamura, T. (1961). 'Carp cultivation in Japan', in Borgström, G.(ed.) *Fish as food*. New York: Academic Press, pp. 103–120.
UN Environment (2019). *Global Environment Outlook—GEO-6: healthy planet, healthy people*. Cambridge: Cambridge University Press.
Wallis Budge, E. A. (1904a). *The gods of the Egyptians: studies in Egyptian mythology*, vol. 1. London: Methuen & Co.
Wallis Budge, E. A. (1904b). *The gods of the Egyptians: studies in Egyptian mythology*, vol. 2. London: Methuen & Co.

CHAPTER 3

Seafood and beyond: key aquaculture objectives

'... most economic models call for ever-continuing growth, when this is clearly the root of our failure to meet environmental problems.'

Adey and Loveland, *Dynamic aquaria: building living ecosystems* (2007)

This chapter introduces seven major objectives of aquaculture, most of which have the common overarching benefit of relieving pressure on capture fisheries by allowing maintenance at more ecologically sustainable levels and reducing the potential for human-induced evolution as a result of overfishing (Kuparinen and Hutchings, 2012).

Incentives in contemporary human society are heavily dominated by economic interests. Any activity, including aquaculture, must be economically viable and is governed to large extent by economic drivers. Therefore, business and management aspects represent a critical part of aquaculture.

Nevertheless, the focus of this book is on the biology of aquaculture species, comparison of the conditions in their natural habitat with those under which they are produced in aquaculture, and the ecological implications of their production. For delving deeper into economic aspect of aquaculture the reader is referred to other sources (Lucas *et al.*, 2019; Pillay and Kutty, 2005; Stickney, 2017).

3.1 Economic benefits as drivers of aquaculture

Global harvest of animals from capture fisheries has plateaued at about 90 million tons annually during the dawn of the twenty-first century and has decreased since then. Although capture fisheries reached a plateau in the 1980s to 1990s, demand for seafood continued to increase, which sparked a rapid expansion of fish and invertebrate aquaculture enterprises.

In 2018, the estimated worldwide economic value of seafood produced by aquaculture was almost twice as high (US$250 billion for animals plus US$13 billion for plants) compared to seafood produced by capture fisheries (US$151 billion) (FAO, 2020).

Moreover, the total aquaculture production yield in 2018 exceeded that of capture fisheries. According to the United Nations Food and Agriculture Organization (FAO), global capture fisheries production was 96 million tons while total aquaculture production (including seaweeds) was 115 million tons in 2018 (Figure 3.1). Although these numbers are higher than production estimates given by the Organisation for Economic Co-operation and Development (OECD), the same trend applies (Figure 2.6).

This trend of rising aquaculture yields versus stable or declining capture fisheries yields is expected to continue in the foreseeable future as one-third of commercially harvested fish stocks are already over-exploited and much of the remainder is exploited at full capacity (Pauly and Zeller, 2016).

Expansion of aquaculture efforts is motivated and economically incentivized by growing demand. Two main reasons underlie the increased demand for seafood: human population growth and increased per capita consumption (Figures 3.2 and 3.3).

A Primer of Ecological Aquaculture. Dietmar Kültz, Oxford University Press. © Dietmar Kültz (2022). DOI: 10.1093/oso/9780198850229.003.0003

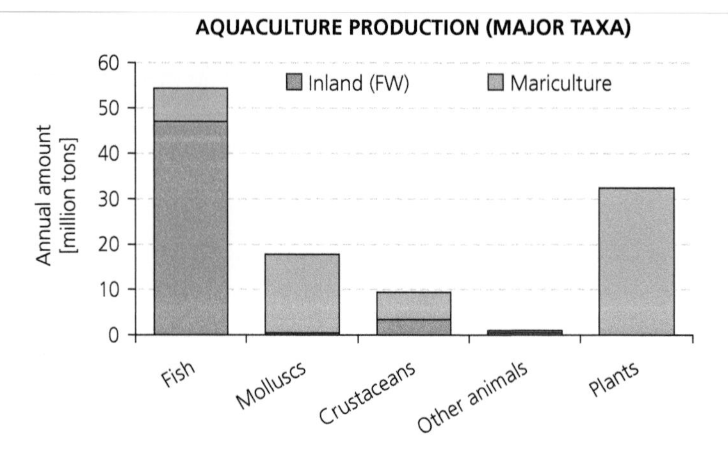

Figure 3.1 Annual production yields of inland freshwater (FW) aquaculture and mariculture for major taxa in 2018. Animals other than fish, molluscs and crustaceans do not contribute significantly to overall yield. Estimates for FW plant aquaculture (e.g. ornamentals, watercress, water hyacinth) are not available but might be negligible relative to marine seaweed. Data source: (FAO, 2020).

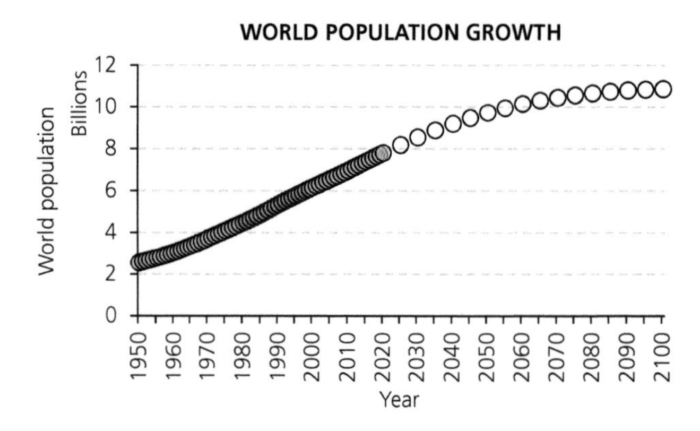

Figure 3.2 World population growth since 1950 as a driver for increased seafood demand. Values for 2020 and beyond are based on estimated projections. Data source: United Nations World Population Prospects 2019 (United Nations, 2019).

It is widely realized in the scientific community that 'the world population must be stabilized and, ideally, gradually reduced within a framework that ensures social integrity' (Ripple *et al.*, 2020). However, currently the human population still increases by more than 200,000 people every day with a predicted trajectory of exceeding 10 billion by 2050 (United Nations, 2019). Per capita consumption of seafood on a world-wide scale has also been projected to increase further in the foreseeable future.

High demand has rendered aquaculture produced seafood a global commodity with a 2018 annual estimated first-sale value of $US 263 billion. The first sale value refers to the price paid by the original vendor to the producer, which is significantly lower than the final retail value. The largest share of first-sale value for seafood produced by aquaculture was contributed by fish ($US 140 billion), followed by crustaceans ($US 69 billion), molluscs ($US 35 billion), seaweed ($US 13 billion), and other animals including reptiles, echinoderms, cnidarians, sponges, and other taxa ($US 7 billion).

China is by far the leading producer country for aquaculture products but many countries in the developing world are also contributing significantly to this growing sector, which represents an opportunity for their economic development. The largest

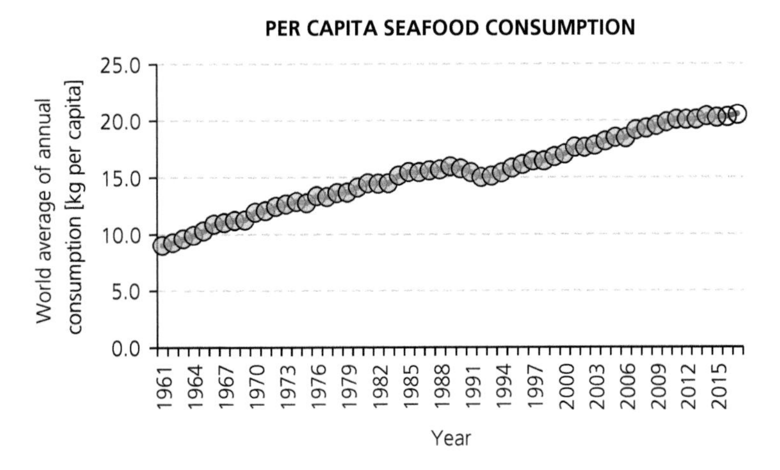

Figure 3.3 Annual per capita worldwide seafood consumption since 1961 as a driver for increased seafood demand. Values for 2016 and 2017 are based on estimated projections. Data source: Fishbase (FAO, 2018).

markets for aquaculture products are the European Union, the US, and Japan.

At present, the aquaculture industry employs approximately 20 million people worldwide, although fewer than 20% are woman (FAO, 2018). This large gender discrepancy provides opportunities for future development and innovation in aquaculture initiatives. Efforts directed at gender equality in aquaculture-related jobs are being promoted by multiple organizations. Examples include programs within the Cultivate Africa's Future initiative that are sponsored by WorldFish, the Consultative Group for International Agricultural Research (CGIAR), and the International Development Research Centre (IDRC) (Cole *et al.*, 2018).

3.2 Aquaculture relieves pressure on capture fisheries

As an overarching benefit that spans all objectives, sustainably practiced aquaculture provides a solution for reducing capture fisheries to more sustainable levels. Seafood produced by aquaculture can relieve increasing pressure on wild fish stocks resulting from overfishing. Such relief is necessary because of the heavy exploitation of wild fish stocks, as mentioned.

Seafood species of significant concern include many temperate species such as halibut, tuna, cod, Atlantic sea scallops, and others that are harvested in large numbers by commercial trawling and other net-based methods. The number of species vulnerable to overfishing in tropical areas may be even greater compared to temperate areas. Population sizes and fishing yields of tropical species are smaller compared to the large population sizes and yields of some temperate species. But the biodiversity of tropical species is much greater, putting more species at risk of overfishing in tropical areas (Figure 3.4).

A comprehensive assessment of island coral reef fisheries has been conducted for forty-nine island countries in the Pacific and Indian Oceans (Newton *et al.*, 2007). This study has shown that more than half (55%) of these countries exploited their coral reef fish stocks unsustainably and that fish harvests were on average 64% higher than can be sustained. To sustain such a high level of exploitation would require an additional area of coral reef that is four times the size of Australia's Great Barrier reef—the largest coral reef ecosystem of the world. By 2050 the additional area of coral reef ecosystem needed would more than double when current rates of exploitation are maintained.

Unfortunately, coral reef ecosystems are not predicted to increase in the foreseeable future. On the contrary, they have decreased dramatically over the past few decades and are predicted to further decrease. For instance, the cumulative population

Figure 3.4 Surgeonfish, including the yellow tang (*Zebrasoma flavescens*), are among tropical fish exploited by local island fisheries. These fish are critical for sustaining the ecological balance of coral reef ecosystems. Photo by author.

size of all corals in Australia's Great Barrier reef has decreased by 50% in just three decades since 1985. Aquaculture provides a promising alternative for managing the social and economic dependence of tropical island countries on seafood while maintaining a stable balance of their natural resources.

A crucial advantage of aquaculture over capture fisheries is the greater control for minimizing destructive side effects. Damaging fishing practices, habitat destruction, unwanted bycatch, losses due to perishability and food safety concerns, ocean pollution due to operation of fishing vessels, and disruption of breeding grounds and population dynamics of keystone species are common problems associated with capture fisheries. They can be avoided or minimized in well-management aquaculture systems.

Particularly destructive fishing practices include bottom trawling, cyanide—and explosives-based fishing in coral reefs for harvesting ornamental fish, and unintended fishing due to loss of fishing gear (ghost fishing). These practices can be avoided altogether if aquaculture methods for reproduction and culture of the species of interest in suitable captive systems can be developed.

In addition, transportation logistics including the associated carbon footprint, food waste, food safety and stochastic oversupply are concerns associated with capture fisheries. These critical aspects of seafood production are more manageable by providing well-designed aquaculture solutions.

Research and development efforts over the past half century have greatly expanded the number of species that are amenable to domestication and aquaculture. However, it remains very challenging to complete the life cycle of many commercially important seafood species, for example, tunas and eels, in captivity. More research is necessary to decrease our dependence on wild stocks and develop aquaculture solutions for more species.

3.3 Production of seafood for human consumption

Seafood production for human consumption is the first and most important aquaculture objective. Ninety per cent of the combined total yield supplied by capture fisheries and aquaculture is used as seafood for human consumption. The remainder is used for the other objectives outlined in this chapter.

A caveat in aquaculture and capture fisheries estimates is that the proportion of aquaculture products used for non-food products may be significantly underestimated. Only ornamental shells and pearls are typically quantified as non-food aquaculture products. Biomass contributed by aquaculture for other purposes is not well accounted for by conventional measurement and documentation metrics. Such alternative purposes of aquaculture include species conservation, restocking, ornamental species production, supplementing ecosystem services, and producing animal food or fertilizer.

Substantial health benefits strongly contribute to the increased popularity and per capita consumption of seafood. Seafood is rich in protein and animals grown in aquaculture are cold-blooded (poikilothermic). Like humans, other mammals (pigs, cows, goats, etc.) maintain a constant body temperature of 37 °C, while the body temperature of birds (chickens, turkeys, ducks, etc.) is kept even higher at 42 °C. At these high body temperatures, the ratio of polyunsaturated fatty acids (PUFAs) versus saturated fatty acids (SFAs) in biological tissues is relatively low.

The temperature of fish and aquatic invertebrates is much lower than 40°C as these cold-blooded animals do not have the ability to regulate their body temperature, which, therefore, is commonly equal to environmental temperature. An exception are large and highly active pelagic fish such as tuna that have a slightly increased body temperature at their inner core. Thus, even in surface waters of tropical areas the body temperature of aquaculture animals is at least 10 °C lower than that of humans and other mammals.

The temperature of colder (deeper or temperate) waters can be as low as 4 °C, or even −1.5 °C in polar oceans. The cold-bloodedness of aquaculture animals increases the ratio of PUFAs versus SFAs. Since SFAs are associated with heart disease and blood vessel plaque formation in humans a greater PUFA to SFA ratio in the diet is beneficial for health.

Why do cold-blooded aquatic animals have more PUFAs and fewer SFAs than mammals and birds? The answer lies in the temperature-dependent structural properties of biological membranes that encapsulate cells and cell organelles (Figure 3.5). Fatty acids comprise the hydrophobic (water-repelling) tails of lipids of which these biological membranes are made. Rising temperature causes an increase in the fluidity of cell and organelle membranes. However, the fluidity of these membranes must be within a proper range to ensure an optimal trade-off between stability and flexibility (elasticity) of these membranes.

Therefore, the effect of temperature on membrane fluidity must be counteracted. At reduced temperatures this is achieved by increasing the PUFA to

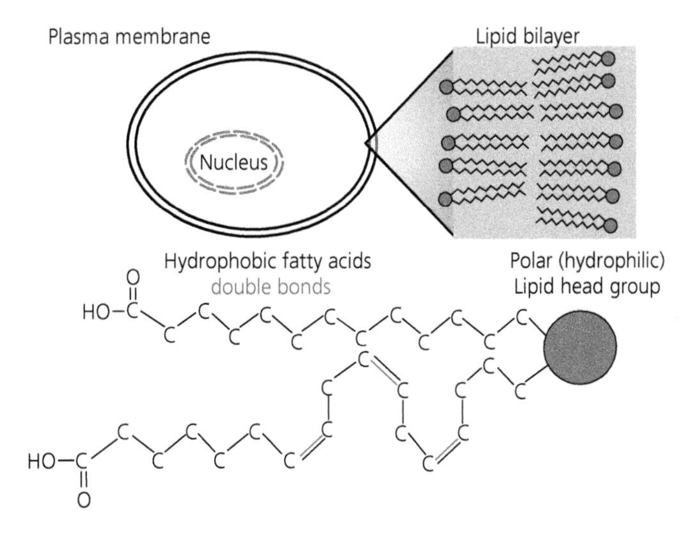

Figure 3.5 Cells that make up animal tissues and organs are enclosed by a plasma membrane as are their organelles, e.g. the nucleus (upper left panel). The plasma membrane and cell organelle membranes are lipid bilayers with hydrophilic head groups facing the aqueous outside and hydrophobic fatty acid tails buried in the interior (enlarged in upper right panel). Desaturation of fatty acids by incorporation of double bonds (red) between carbons of the fatty acids renders them unsaturated and the membrane lipids more fluid to compensate decreased Brownian motion at lower temperature. Hydrogens are not shown except at the termini.

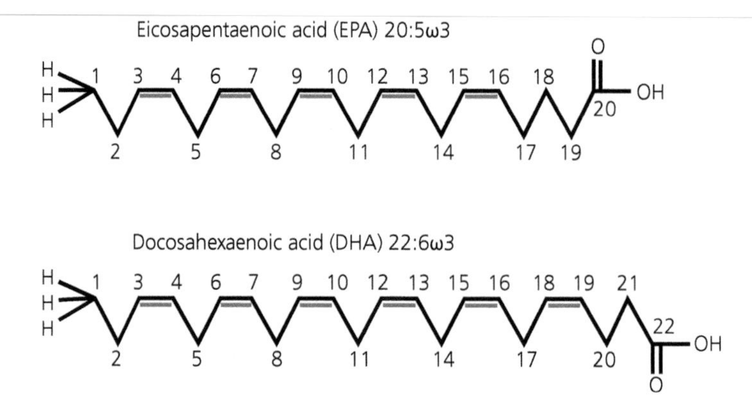

Figure 3.6 Structures of two polyunsaturated fatty acids (PUFAs), both of which are of the omega-3 type: eicosapentaenoic acid (top) and docosahexaenoic acid (bottom). Carbons in fatty acid chains are numbered starting at the methyl- and ending at the carboxy-terminus. Hydrogens are not shown except at the termini.

SFA ratio, that is, the mean degree of fatty acid desaturation, which is a function of the number of double-bonds in fatty acids (Figure 3.6, Box 3.1). Cold-blooded animals can adjust the amounts of double-bonds when the temperature in their environment changes (e.g. seasonally in temperate areas). This process of membrane fluidity adjustment is called homeoviscous adaptation (Cossins and Prosser, 1978).

Another health benefit of seafood derives from aquaculture animals being evolutionarily more distantly related to humans than the warm-blooded (homeothermic) terrestrial animals used for agriculture (mammals and birds). Therefore, animal welfare aspects are generally easier to address, especially for invertebrates, and the potential for zoonotic diseases is much lower than for domesticated mammals and birds. Zoonotic diseases are those that can be transmitted from animals to humans (Box 18.1). An example is bovine spongiform encephalopathy (BSE, or mad cow prion disease).

3.4 Aquaculture for restocking lakes, rivers, and the ocean

Aquaculture of fish and invertebrates for restocking aquatic habitats is classified as a non-food objective here because the primary goal is to produce juveniles that are released rather than adult fish that are harvested for seafood production. However, some of the fish released into the wild may later be captured by recreational fishing or by commercial capture fisheries. For instance, salmonids and bass are commonly produced in aquaculture hatcheries for this purpose. Restocking of anadromous salmonids is sometimes complemented by sea ranching (Box 3.2).

Stock enhancement is common for rainbow trout (*Oncorhynchus mykiss*) in mountain lakes and streams and has been practised for more than a century to support recreational fishing (Figure 3.8). In these cases, restocking practices blurry the lines between aquaculture and capture fisheries as some seafood harvest attributed to capture fisheries may have been bred and juveniles raised in aquaculture.

In other cases, restocking primarily serves a conservation purpose to prevent the decline or extinction of endangered species and populations. Examples include restocking of coral reefs with coral species and other invertebrates that are produced in aquaculture hatcheries and released as larvae or juvenile colonies into suitable ocean habitat.

Another purpose of restocking is vector-borne disease control. Mosquitofish (*Gambusia spec.*) are commonly used in this regard. Historically, restocking has often been practised in non-native habitats and the species introduced have become invasive. Their introduction has altered the balance of ecosystems and endangered other species. For example, California fairy shrimp populations have declined as a result of introducing mosquitofish

Box 3.1 Types of polyunsaturated fatty acids (PUFAS)

There are many different types of PUFAs. To classify PUFAs, the location of the first double-bond (C=C) is counted from the methyl (CH_3) end of the carbon chain, which is also known as the omega (ω) end, or the n terminus. For example, omega-3 fatty acids are long-chain PUFAs ranging from eighteen to twenty-two carbon atoms in length, with the first of many double bonds located after the third carbon atom counted from the methyl (CH_3) end of the carbon chain. The opposite end of the carbon chain is the carboxylic acid (COOH) end. The name of the fatty acid is based on total numbers of carbon atoms and double bonds. Omega-3 PUFAs include docosahexaenoic acid (DHA) with a chain length of twenty-two carbon atoms and six double bonds and eicosapentaenoic acid (EPA) having a chain length of twenty carbon atoms and five double bonds. DHA is commonly referred to as 22:6ω3 or 22:6n-3, that is, a PUFA having a chain length of twenty-two carbon atoms with six double bonds, the first of which occurs after carbon 3. EPA is commonly referred to as 20:5ω3 or 20:5n-3, that is, a PUFA having a chain length of twenty carbon atoms with five double bonds, the first of which occurs after carbon 3. The content of omega-3 PUFAs varies depending on species and the specific types of PUFAs present in aquaculture animals contribute to the unique, cherished taste of different types of seafood (Figure 3.7).

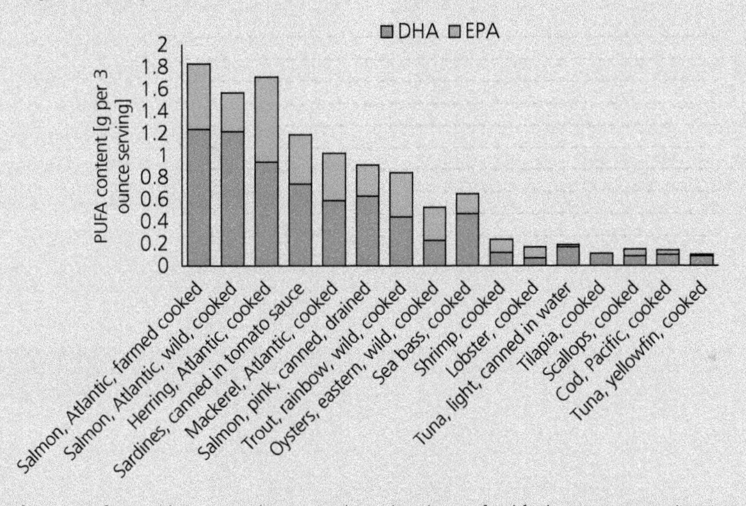

Figure 3.7 Content of omega-3-fatty acids in aquaculture animals produced as seafood for human consumption. Data source: USDA FoodData Central (https://fdc.nal.usda.gov/).

(Leyse *et al.*, 2004). Therefore, restocking practices are not without pitfalls. The other main concern about restocking is human-induced evolution (Section 6.5).

Stock enhancement of marine fishes is based on hatchery production of juveniles up to a developmental stage/size at which mortality rates in the wild decline substantially. At this stage they are released into the ocean. The primary purpose of marine stock enhancement is aiding conservation of declining populations and/or supporting stocks for capture fisheries. Therefore, many stock enhancement programmes are funded by government organizations and/or conservation funds.

More than 180 aquatic species in many countries have been tested for stock enhancement programmes with varying rates of success (Kitada, 2020). Large-scale marine stock enhancement has been established for fish (e.g. red drum, *Sciaenops ocellatus*) in the US, and invertebrates (e.g. Ezo abalone, *Haliotis discus hannai*, in Japan; Box 3.2).

Box 3.2 Sea ranching

A special form of aquaculture that is related to restocking but incorporates a harvest strategy after release of fish into the wild is sea ranching. Sea ranching is particularly attractive for anadromous species that spawn in freshwater but spend most of their lives growing in the ocean before returning to freshwater for reproduction (Chapter 14). This form of aquaculture is not uncommon for anadromous salmonids (e.g. Pacific salmon species, Atlantic salmon, and Steelhead trout) and practised in many countries, including Japan, the US, and Iceland. Salmonids use magnetic fields and their sense of smell for returning to the headwaters of rivers to reproduce at the same site where they were born. Their incredible homing behaviour is utilized during sea ranching. Broodfish are caught in streams in close vicinity of hatcheries, eggs are artificially fertilized, and juveniles raised in the hatcheries in large numbers. Offspring are reared and released into streams adjacent to the hatchery. Mature fish return to the site near the hatchery where they had been released after growing for several years in the ocean. They can then be harvested in a convenient and minimally invasive manner.

3.5 Culture of ornamental animals and plants

Aquaculture of ornamental species represents an objective for producing aquatic invertebrates, fish, and plants as pets, which is the focus of a rapidly growing industry. The associated multi-billion-dollar trade includes thousands of marine and freshwater species. Ornamental species are used as companion animals by hobby aquarists, in public aquaria and zoos, and in ponds and lakes of public or private gardens. Their culture and domestication represent the ultimate goals, rather than a means to an end.

Even though only a fraction of species traded as ornamentals is being produced by aquaculture, the number of ornamental species for which aquaculture methods have been developed as an alternative to capture fisheries is steadily increasing. The extraordinary diversity of evolutionary adaptations and ecological niches occupied by ornamental fishes and invertebrates requires thorough species-specific research to enable their domestication and

Figure 3.8 Restocking of a mountain lake with cutthroat trout transported from a hatchery in a tank carried by truck. Photo by US Fish and Wildlife Service reproduced under CC0 1.0 licence.

reproduction in captivity. Such research is aided by the efforts of many hobby aquarists and applying the comparative method provides opportunities to gain novel insight into fundamental aspects of biology.

Ornamental animal and plant production in aquaculture aids conservation efforts and contributes significantly to the preservation of freshwater and marine ecosystems from which ornamental species are taken by environmentally destructive methods of capture fisheries (e.g. cyanide fishing in coral reefs, see Chapter 12). Moreover, aquaculture preserves the balance of natural ecosystems by relieving pressure on wild populations of ornamental species.

3.6 Production of jewellery and souvenirs

The production of pearls and ornamental shells by aquaculture of marine and freshwater bivalves represents a sizeable non-food aquaculture objective for which both production volume and trade value have been well documented. In 2018 about 26,000 tons of these jewellery and souvenir items were produced in China, Japan, Australia, Indonesia, French Polynesia, Mexico, and several other countries.

China is the biggest producer of freshwater pearl mussels, while Japan produces most marine pearls. Since the average value of natural pearls is very high, ranging from hundreds to thousands of US dollars, the global pearl market has a significant economic value. The average annual trade value of all pearls produced in the world is approximately US$250 million.

However, unlike the steady increase of seafood aquaculture products, production volumes and prices of pearls have fluctuated greatly over the past decades. This trend indicates highly variable demand and the non-essential nature of jewellery and souvenirs when compared with seafood. Section 10.5 discusses pearl mollusc aquaculture and its ecological implications.

3.7 Aquaculture production of secondary metabolites and proteins

A specialized niche objective of aquaculture is the production of biologically active compounds, including small pharmaceutically active metabolites and peptides as well as macromolecules such as proteins. Two major approaches are being used.

The first of these approaches relies on harvesting secondary metabolites produced naturally by aquaculture organisms, mostly for the purpose of chemical defences against fouling and predation. These metabolites are often referred to as marine drugs because they have significant benefits for human medicine.

Examples include anti-proliferative compounds produced in sponges, which are used as tumour-suppressant drugs. The wide variety of secondary metabolites that are produced as natural marine products by a great diversity of marine species and their varied biological effects make them attractive as pharmaceuticals, for diagnostic purposes, and as nutritional supplements. Chapter 9 provides specific examples and information on their production in aquaculture systems.

The second approach used to achieve this aquaculture objective utilizes biotechnology to produce either small chemicals (metabolites) or proteins that have biomedical, nutritional, or other value. The organisms that are being cultured may be genetically engineered to produce these compounds.

Genetic engineering results in either increasing the amount of desirable endogenous compounds or, alternatively, in producing heterologous (foreign) compounds (for example, proteins from different species or metabolites that are not normally produced by the aquaculture species).

For example, ornamental zebrafish that emit fluorescence when illuminated with blue light express a heterologous (foreign) green fluorescent protein (GFP) that is encoded by a gene derived from a marine jellyfish (*Aequorea victoria*). This jellyfish gene has been instrumental for many advances in biomedical and basic research, which earned its discoverer, Osamu Shimomura, the 2008 Nobel Prize in Chemistry. Aquaculture of GloFish® harbouring a GFP gene (or modifications of this gene that result in different colours) fulfils a dual objective: production of heterologous protein and ornamental companion animal production.

Production of a heterologous protein has been combined with the objective of seafood production in AquAdvantage salmon, which grows faster and larger by expressing a foreign gene. In this case, Atlantic salmon (*Salmo salar*) has been genetically engineered to replace its endogenous growth hormone gene with a promoter sequence from ocean pout (*Zoarces americanus*) and the growth hormone gene from Chinook salmon (*Oncorhynchus tshawytscha*).

Although genetically engineered aquaculture organisms are invaluable models for basic research, their use for commercial aquaculture is complicated by consumer acceptance and food safety regulation debates, which are further discussed in Chapter 6.

3.8 Culture of aquatic organisms to aid ecosystem restoration

Another objective of aquaculture is ecosystem restoration. In this case, the culture organisms of interest fulfil a critical role in the global food web that is complementary to the ecological niche occupied by traditional aquaculture animals.

Global food webs have three components: primary producers, consumers, and decomposers. All aquaculture animals, even filter feeders, are heterotrophic consumers from a trophic level perspective. In contrast, most plants are primary producers, and many bacteria are decomposers.

Food webs and trophic levels are covered in more detail throughout this book, especially in Chapters 4 and 5. Here, it suffices to point out that heterotrophic consumers require energy-rich organic compounds as a food source while generating organic waste. The vast number of organic compounds differs from most inorganic compounds by covalent linkage of carbon atoms to nitrogen, oxygen, hydrogen, sulfur, and/or phosphorus.

Energy-rich organic compounds originate initially from photosynthesis, which is performed by primary producers that convert inorganic molecules (water and carbon dioxide) to an organic sugar (glucose). Green plants perform photosynthesis, which produces glucose by converting energy from sun light into chemical energy.

Glucose represents the substrate for the synthesis of many other organic compounds that are consumed by animals. Animals metabolize these organic nutrients and excrete organic waste, which is broken down into inorganic compounds by decomposers (bacteria). These inorganic compounds, many of which are rich in nitrogen and phosphorus, are then taken up as nutrients by plants, which maintains trophic cycles that constitute the food web.

A proper equilibrium of producers, consumers, and decomposers is essential to balance material and energy flows in ecosystems. If there is an excess of consumers, including humans and aquaculture animals, that generate organic waste then ecosystem services are required to recycle this waste. In other words, a large fraction of the primary producers and decomposers present in natural ecosystems receiving untreated animal and human waste would be required to assimilate the organic waste into their biomass.

Depending on the extent to which food web balance is impacted by the waste from aquaculture animals, ecosystems will inevitably change. Ecosystem change is natural and referred to as ecosystem succession. However, to attain a new equilibrium, succession of natural ecosystems receiving waste from food production systems will be forced into emphasizing decomposer communities at the expense of consumer (animal) and higher plant biodiversity.

Aquaculture of suitable plants and bacteria represents a promising solution to shift the burden of wastewater treatment from natural ecosystems to managed aquaculture systems. This aspect of aquaculture is often neglected because the culture organisms are not animals or edible plants. Instead, suitable microalgae and bacteria (primary producers and decomposers) are the organisms that are being cultured.

The culture medium is wastewater generated by animals and humans (heterotrophic consumers). Organic waste present in the water is extracted by the bacteria and algae and assimilated into their biomass, which results in bioremediation. Water treatment by bioremediation removes organic and inorganic waste and reduces the burden on ecosystem services.

If the wastewater is high in toxic compounds, then additional pre-treatment may be necessary. The biomass produced in the process of waste assimilation has added value in addition to the recycling of clean water. Microbial biomass grown in such aquaculture can be used as fertilizer for

Figure 3.9 Algal Turf Scrubber® technology combines the culture of aquatic bacteria (decomposers) and microalgae (primary producers) for wastewater treatment and production of fertilizer and biofuel. Photo reproduced with kind permission from Hydromentia Technologies.

forestation (silviculture), horticulture, and crops; as food for herbivorous aquaculture animals; and for producing biofuel.

The Algal Turf Scrubber® technology represents an excellent example for this innovative form of aquaculture, which yields mats composed of algal and bacterial biomass that can be regularly harvested (Figure 3.9). Aquaculture of microalgae and bacteria has outstanding potential for development of minimal discharge, low footprint, integrated multi-trophic aquaculture (IMTA) systems that are not limited to different trophic levels of consumers but integrate all trophic levels at proper ratios.

3.9 Bait production

Aquaculture of bait species has the objective to produce live fish and invertebrates as bait for recreational fishers and anglers. In addition, live bait species are used as a highly valued food source for ornamental aquatic animals. Since the production of ornamental species is the fastest growing branch of aquaculture, the demand and production of live food for this sector is also increasing.

Baitfish production has significant economic value, comprising a US$1 billion industry in Canada and the US alone (Litvak and Mandrak, 1993). Both freshwater and marine animals are being produced for the purpose of providing live food for baiting sport fish or feeding ornamental fish. Fish species comprise the largest fraction of bait produced in aquaculture but crustaceans, molluscs, polychaete worms, and other invertebrates are also produced as live bait.

Key conclusions

- Aquaculture is economically incentivized by growing demand due to 1) human population growth and 2) increased per capita consumption.
- Aquaculture relieves pressure on capture fisheries and has high potential to be ecologically more sustainable than fishery exploitation of natural ecosystems.
- The most important aquaculture objective is the production of seafood for human consumption.
- Seafood has substantial health benefits compared to other sources of nutrition.
- Other significant aquaculture objectives include:
    - Restocking of lakes, rivers, and the oceans with species of interest and/or concern;
    - Culture of ornamental fishes and aquatic invertebrates as companion animals;
    - Production of jewellery and souvenirs;
    - Production of compounds (metabolites, proteins) with health or other benefits;
    - Culture of aquatic organisms (bacteria, plants) to aid ecosystem restoration; and
    - Bait production.

References

Adey, W. H. and Loveland, K. (2007). *Dynamic aquaria: building living ecosystems*. Amsterdam: Academic Press.

Cole, S., McDougall, C., Kaminski, A., Kefi, A., Chilala, A., and Chisule, G. (2018). 'Postharvest fish losses and unequal gender relations: drivers of the social-ecological trap in the Barotse Floodplain fishery, Zambia', *Ecology and Society*, 23, pp. 18.

Cossins, A. R. and Prosser, C. L. (1978). 'Evolutionary adaptation of membranes to temperature', *Proceedings of the National Academy of Sciences*, 75, pp. 2040–2043.

Food and Agriculture Organization of the United Nations (FAO). (2018). *The state of world fisheries and aquaculture 2018: meeting the sustainable development goals*. Rome: Food and Agriculture Organization of the United Nations.

Food and Agriculture Organization of the United Nations (FAO). (2020). *The state of world fisheries and aquaculture 2020*. Rome: Food and Agriculture Organization of the United Nations.

Kitada, S. (2020). 'Lessons from Japan marine stock enhancement and sea ranching programmes over 100 years', *Reviews in Aquaculture*, 12(3), pp. 1944–1969. https://onlinelibrary.wiley.com/doi/abs/10.1111/raq.12418

Kuparinen, A. and Hutchings, J. A. (2012). 'Consequences of fisheries-induced evolution for population productivity and recovery potential', *Proceedings of the Royal Society B: Biological Sciences*, 279, pp. 2571–2579.

Leyse, K., Lawler, S., and Strange, T. (2004). 'Effects of an alien fish, *Gambusia affinis*, on an endemic California fairy shrimp, *Linderiella occidentalis*: implications for conservation of diversity in fishless waters', *Biological Conservation*, 118, pp. 57–65.

Litvak, M. K. and Mandrak, N. E. (1993). 'Ecology of freshwater baitfish use in Canada and the United States', *Fisheries*, 18, pp. 6–13.

Lucas, J. S., Southgate, P. C., and Tucker, C. S. (2019). Aquaculture: *Farming Aquatic Animals and Plants*. Hoboken: Wiley-Blackwell.

Newton, K., Côté, I. M., Pilling, G. M., Jennings, S., and Dulvy, N. K. (2007). 'Current and future sustainability of island coral reef fisheries', *Current Biology*, 17, pp. 655–658.

Pauly, D. and Zeller, D. (2016). 'Catch reconstructions reveal that global marine fisheries catches are higher than reported and declining', *Nature Communications*, 7, pp. 10244.

Pillay, T. V. R. and Kutty, M. N. (2005). *Aquaculture: principles and practices* Oxford: Wiley-Blackwell.

Ripple, W. J., Wolf, C., Newsome, T. M., Barnard, P., and Moomaw, W. R. (2020). 'World scientists' warning of a climate emergency', *BioScience*, 70, pp. 8–12.

Stickney, R. R. (2017). *Aquaculture: an introductory text*. Boston: CABI.

United Nations (2019). *World population prospects 2019*. New York: UN Department of Economic and Social Affairs, Population Division.

CHAPTER 4

Aquaculture systems as mesocosms

'It is highly unlikely that humans can be totally supported by our own monocultures ... we must quickly learn the techniques of culture of ecosystems.'

Adey and Loveland, *Dynamic aquaria: building living ecosystems* (2007)

Most aquatic ecosystems, by far, are marine. The oceans cover approximately two-thirds of the earth's surface, with the deepest known part being the Challenger deep in the Mariana Trench near Guam (Northern Pacific Ocean) at 10,994 m. Less than 3% of water on earth is freshwater (FW) and only about 1% of this FW (0.03% of earth's water) represents surface FW utilized by aquaculture. The remaining FW is less accessible groundwater and permanently frozen ice. Although global warming and climate change liberate FW by melting of glaciers and polar ice caps, most of this liberated FW quickly mixes with the oceans.

In sharp contrast to the minuscule fraction of surface FW relative to marine environments, most aquaculture is performed in FW. Mariculture is generally limited to coastal areas where it produces relatively low yields or has low ecological sustainability, or both. One exception is mariculture of macroalgae (seaweed), which can produce high yields at high ecological sustainability but requires a large spatial footprint in coastal areas. Like all coastal mariculture, seaweed mariculture competes with many other human uses of coastal environments. Another exception is the use of recirculating aquaculture systems (RAS) for producing marine ornamental species and early developmental stages of marine organisms for restocking.

Intensive and ecologically sustainable mariculture with a small footprint in coastal ecosystems is technologically and logistically significantly more challenging than FW aquaculture. Attempts to relocate mariculture farms offshore are hampered by high costs for maintenance and logistical challenges associated with regular access. In addition, offshore animal mariculture is based on open cages rather than closed RAS and adds strain to pelagic and/or benthic marine ecosystem service capacity by producing organic waste that is difficult to remove in a controlled manner.

4.1 The diminishing boundary between capture fisheries and aquaculture

To achieve the primary aquaculture objectives outlined in Chapter 3, humans control trophic cascades and environmental conditions in aquaculture systems. The extent of such control ranges from minimal (extensive systems) to elaborate (intensive systems) but some level of human control applies to all aquaculture systems.

In contrast to aquaculture, capture fishery is the collection of seafood from wild populations (i.e. from natural, unmanaged resources). Some argue that the boundary between aquaculture and capture fisheries is becoming blurred as wild fish stocks are increasingly impacted and managed by humans. Management is evident by protecting spawning grounds and restricting fisheries to specific periods during the year. These management efforts are compounded by negative impacts of human activities (e.g. pollution, habitat fragmentation and encroachment, and anthropogenic causes of climate change).

Humans increasingly take control over 'the management of earth' to an extent that has irreversibly changed the trajectory of the biosphere on our entire

A Primer of Ecological Aquaculture. Dietmar Kültz, Oxford University Press. © Dietmar Kültz (2022). DOI: 10.1093/oso/9780198850229.003.0004

planet. Such augmented management control is a direct result of technological advancements that have accumulated at an exponential pace over the past two centuries. Human management has pervasively affected macrocosms, mesocosms, and microcosms on earth (Figure 4.1). These terms are relative in scope, that is, the entire earth can be viewed as a microcosm relative to the universe as a macrocosm (Adey and Loveland, 2007). Nevertheless, this book refers to macrocosms in the sense of large ecosystems (e.g. biomes), mesocosms in the sense of intermediate scale ecosystems (e.g. an aquaculture pond or indoor hatchery), and microcosms in the sense of small ecosystems < 1 m^3 in size (e.g. an aquarium).

Ecosystems are natural or human-managed (captive) biological communities consisting of biotic (organisms) and abiotic (inanimate physical environment) parameters that interact within relatively distinct boundaries. Since all ecosystems have been impacted by human activities to variable degree (whether intended or not), concurrent capture fisheries could be interpreted as a minimally managed form of aquaculture being subject to the least (but not lacking) human management. In that sense, capture fisheries represent a most extensive aquaculture system that differs from aquaculture *sensu stricto* in the scale and extent of human intervention.

Capture fisheries entail an entire macrocosm, for example, the North Atlantic Ocean for Atlantic cod (*Gadus morhua*) production. In contrast to capture fisheries, extensive aquaculture is typically associated with a mesocosm, for example, an estuary for oyster grow-out or an atoll for sponge or pearl oyster culture. Whether intended or unintended, human impact on wild stocks of aquatic animals alter their population structure in a way that has been described as human-induced evolution (Allendorf and Hard, 2009) and fisheries-induced evolution (Heino, Díaz Pauli, and Dieckmann, 2015). Thus, human management of global earth's ecosphere, whether intentional or unintentional is already a reality. For this reason, some conservation efforts, although well-intended, may be misaligned with the reality of globalization and ecosystem succession (Section 4.4): 'The concept that species can be saved one by one at best applies only to mammals and a few birds and fish, if at all' (Adey and Loveland, 2007).

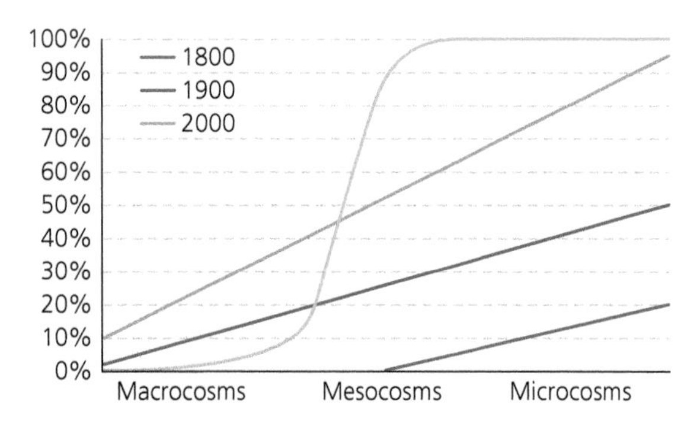

Figure 4.1 A macrocosm contains many mesocosms and a mesocosm consists of many microcosms. Technological advances have increased human ability to manage microcosms and mesocosms during the past two centuries. However, they have also resulted in unintended impacts on macrocosms. The ideal of aquaculture (and other forms of human) management is the ability to selectively control specific mesocosms without adversely affecting the corresponding greater macrocosm (yellow line). The schematic also illustrates generalized trends over time (blue, red, and gray lines) using arbitrary Y axis scale values for the extent of human impact. Actual values for human impact/management depend on how these activities are being assessed.

4.2 Minimizing reliance of aquaculture on ecosystem services

Despite the progress made in developing technologies and generating knowledge for operating closed recirculating aquatic microcosms and mesocosms, current aquaculture approaches are heavily dominated by semi-intensive, open systems that have a large ecological footprint. The biomass of decomposers and primary producers required for generating feeds and recycling waste of commercial animal aquaculture systems is enormous.

Ecosystem services broadly represent benefits provided by ecosystems to humankind for industrial, agricultural/aquacultural, and recreational activities or other purposes. Ecosystem services are critical for the provisioning of clean drinking water, buffering pollution in marine and FW ecosystems, and decomposing organic waste. A steadily increasing world population and accelerated urban development cause a significant decline of natural ecosystems that provide ecosystem services while demand placed upon them is increasing. This mismatch calls for solutions that minimize the reliance on ecosystem services for all purposes, including aquaculture.

One overarching solution is regulatory incentivization that rewards efforts towards decreasing reliance on ecosystem services by offsetting the costs of required technologies. Achieving such incentivization is not trivial as it requires unbiased information for policy makers that is not tainted by funds spent for lobbying efforts or research programmes whose sustenance is dependent on prescribed circumstances. In the increased demand scenario discussed in Section 4.3, intensive, closed systems that require large technological investments are most promising for minimizing reliance on ecosystem services without sacrificing high production yields. Such investments can only be economical if proper value is placed on ecosystem services.

A clear trend of the future is that technologies will be more affordable, more efficient, and easier to implement while ecosystem services will become scarcer and demand for them will become more competitive over time. Therefore, unless the human population declines significantly over the next decades, there is great need for research and development of technologies that enable or facilitate intensive production of human food (not just seafood) in closed mesocosms with highly controlled input and output of matter and energy.

Such technologies will facilitate responsible stewardship of available ecosystem services. More research is also needed to find better solutions for other challenges associated with intensive, human-managed production systems. Examples include ethical aspects of animal care, minimizing disease susceptibility, and human-induced evolution.

Reliance of aquaculture on ecosystem services can be minimized by addressing shortcomings of current recycling and filter technologies for treatment of water polluted with animal waste. For example, it is often challenging to discriminate desirable particulate matter (e.g. photosynthetic planktonic microorganisms or bacterial decomposers) from particulate waste. Ideally, intensive, closed RAS would achieve a perfect (or near-perfect) balance between feed containing all necessary input of matter and energy (other than light) and aquatic animal biomass produced as the sole output of matter and energy. This can only be achieved by having a proper ratio and proper metabolic capacity of primary producers, consumers, and decomposers in the system, irrespective of micro- or mesocosm scales.

On a larger scale one would also have to consider the ecological footprint of producing aquaculture feeds, human waste produced from consumption of the seafood harvested, and unconsumed seafood waste. Thus, efforts for designing highly sustainable aquaculture systems cannot succeed when isolated from efforts for recycling other types of anthropogenic waste.

4.3 Extensive versus intensive aquaculture: which is more sustainable?

Aquaculture systems differ greatly regarding the level of human control and management. On one end of the spectrum are intensive systems with maximal human control of both inputs and outputs and high yields. The opposite end of the spectrum encompasses extensive systems with minimal human control. In general, extensive systems are open, largely dependent on ecosystem services, and have a large spatial footprint per yield. Ecological sustainability of semi-intensive systems increases greatly when such systems are at least partly closed and rendered less dependent on ecosystem service capacity (Figure 4.2).

There are six general types of aquaculture systems:

1. Extensive, open systems.
2. Semi-intensive, semi-open, static or low-flow pond systems.
3. Semi-intensive, semi-closed (water re-use) pond systems with wastewater treatment ponds.
4. Semi-intensive, open, high-flow raceway and tank systems.
5. Semi-intensive, open cage and net-pen systems.
6. Intensive or semi-intensive, closed RAS.

Most current aquaculture systems are semi-intensive static or flow-through pond systems (system 2) that rely to a large extent on ecosystem services but have greater stocking densities and smaller spatial footprints per yield than extensive systems.

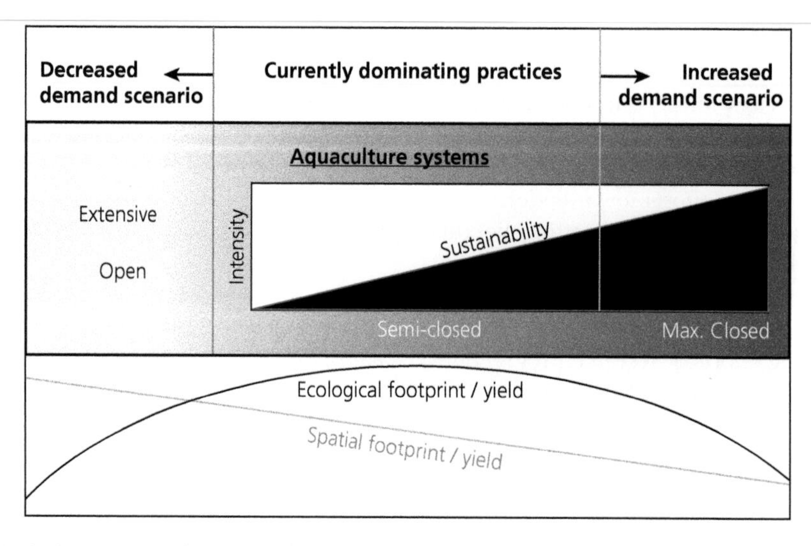

Figure 4.2 Relationship between types of aquaculture systems, their yield, ecological footprint, and spatial footprint. In a maximally closed system (RAS), the technology component is most prevalent, ideally only requiring addition of food and sunlight and harvest of the aquaculture organisms produced. However, since technology is used to maintain abiotic water quality parameters in such systems, additional inputs/outputs need to be considered, including those for management of dissolved gases, pH, and filtration. Filtration outputs will need to be collected in systems that are skewed towards culture of consumers even if biological filtration (decomposer capacity) is in equilibrium to remove particulate waste and some inorganic nutrients. Note that most current aquaculture systems are semi-closed and semi-intensive and that their sustainability is negatively correlated with production yield.

Whether intensive or extensive systems are used for aquaculture and what level of management is exerted both depend on a variety of factors. The most important factors are the capacity for ecosystem services to support aquaculture practices at the chosen site, capacity for up-front investment into technology for intensification and management, level of demand for the aquaculture product, species being cultured, traditional/cultural customs, and consumer preference.

Which of the different aquaculture systems is most promising and sustainable for future development? While the answer depends somewhat on what species is being considered, several generalizations exist based on two different scenarios. In an increased-demand scenario, population growth and demand for seafood continues at its current trajectory or plateaus soon and remains high. In such a scenario, the current level of reliance of aquaculture practices on ecosystem services is unsustainable. This means that intensification of aquaculture systems in an ecologically sustainable manner, that is, by developing maximally closed systems, should be a priority.

Ecologically sustainable intensification can best be achieved with systems that have a minimal ecological and spatial footprint while having maximum production capacity. RAS effectively isolate human-controlled production systems from the external environment. Achieving such isolation while maintaining high production capacity is far from trivial and depends on significant research and development. In addition, large up-front investments are needed for such highly intensive systems to afford reliable technologies that balance trophic levels and effectively eliminate waste from uncontrollably entering natural ecosystems. Moreover, investments must also be made in efficient monitoring and backup solutions to prevent catastrophic losses in the event of technical failure.

The second scenario assumes a decreased demand in seafood, either because of a decrease in world population and/or a shift in consumer preference towards ecologically more sustainable vegetarian food sources. In this case, the available ecological and spatial footprint for extensive aquaculture production systems may be compatible with both demand for seafood and sustainable ecosystem service capacity in carefully selected regions. Production capacity in such extensive, open systems will be relatively low because of lack of supplemental feeding and fertilization, but ecosystem service capacity will not be strained and impact on biodiversity and community structure in affected areas will be chiefly limited to competition for space. This second scenario has proven effective historically for thousands of years until the twentieth century, when industrialization led to a rapid increase in world population and intensification of aquaculture (Chapter 2).

In both scenarios, we must depart from currently dominating semi-intensive open systems that have large ecological and spatial footprints. Depending on these two scenarios, the goal of aquaculture should be to either develop more closed, intensive systems with small ecological and spatial footprints or expand open, extensive systems that rely on a large spatial footprint but without supplemental feeding or fertilization and, thus, without unduly altering the succession of natural ecosystems (Figure 4.2). The demand for seafood will dictate the suitability of these approaches.

One could envision to apply these two scenarios for development of aquaculture variably in different areas of the world. However, there exist evident mismatches between projected high and low demand for seafood, import and export of seafood by different countries, and common practices and technologies accessible in different areas of the world. While projected population size increases most in developing countries, the potential for implementing closed intensive RAS is highest in industrialized countries. Conversely, spatial capacity for open, extensive aquaculture is highest in developing countries.

Effects of globalization are readily evident from an ecological perspective and in many aspects of human society. Moreover, the vast knowledge and resources developed for human management of ecosystems during the previous century must be utilized more effectively to achieve more desirable, predictable global consequences in lieu of unintended, destructive outcomes. Currently, many management practices in one part of the world are inevitably linked to their effects on the rest of the world. Therefore, global planning and implementation of future aquaculture development must be raised to a new level to transcend national and continental boundaries more effectively.

4.4 Microcosms for ornamental aquaculture

As pointed out in Section 4.3, closed intensive RAS and water treatment and re-use pond systems are most ecologically sustainable for meeting a high seafood demand. However, unique technological challenges must be overcome in developing such systems, especially for mariculture. Many technological advances have been made for culturing ornamental aquatic organisms in small, closed circulation systems that represent microcosms, that is, systems that encapsulate *in miniature* the characteristic qualities or features of the natural environments of these organisms. Technological advances in the aquarium industry have enabled intensive, ecologically sustainable, closed RAS on a microcosm scale by controlling and tightly balancing the input and output of matter and energy.

Even though seafood production is the single most important objective of aquaculture, the growth of ornamental aquaculture has outpaced the development of other branches of aquaculture in recent decades. FW aquaria represent models of sustainability for aquaculture. Many FW aquaria microcosms are maximally closed in that input and output of matter are tightly controlled. Many ornamental FW aquaria contain sufficient filter capacity for recycling and responsible discharge of waste.

Potent technologies exist for mechanical, biological, and chemical filtration of animal waste and for maintaining other aspects of water quality (e.g. temperature, pH, hardness). Moreover, many ornamental FW fish are amenable to cultivation in captivity because their native habitats are relatively small bodies of water. Most (90%) ornamental FW species are now produced in captivity by aquaculture (Chapter 12).

Marine aquaria maintained by hobby aquarists also represent microcosms. However, lighting and water quality are much more challenging to manage in marine than FW microcosms. Lighting is particularly important (and energy-expensive) for establishing mini-reef microcosms with invertebrates that rely on symbiotic photosynthetic algae. In addition, salinity and proper ionic and trace element balance in the water represent critical parameters for mariculture in closed RAS microcosms. The oceans represent well-balanced and stable systems regarding salinity and ionic/elemental composition. Such balance is more difficult to achieve in human-managed microcosms that have low buffering capacity.

Evaporation must be compensated for daily variation in solar heating and biochemical cycles of major and minor elements require management by using proper mechanical, biological, and chemical filtration units that are adequate for seawater (SW). For example, replacement of evaporative loss requires ion-free water, prepared by reverse osmosis or double-distillation. Sumps, protein skimmers, chemical dosing systems (e.g. pH dosing pumps, salinity dosing systems), many types of biofiltration approaches, wet-dry filters, multi-tiered trickle filter systems, under-gravel filters, water pumps and flow regulators, and efficient aeration, heating and lighting technologies have been invented and perfected to support SW microcosms. These technologies are now available for scaling-up to mesocosm aquaculture.

In addition, tireless efforts of countless hobby and professional aquarists involved in ornamental aquaculture have generated knowledge about the biology and health management of numerous aquatic organisms and optimization of their feeds. These (often empirical) efforts provide insight into combinations of organisms and corresponding feeds that are well suited for achieving a balanced nutrient cycle in closed aquatic microcosms. For example, many species of detritivores (e.g. sea cucumbers) have been explored in marine microcosms and some are now used for larger mesocosm-scale integrated multitrophic aquaculture (IMTA). Many of these advances come from the need to maintain closed, maximally self-sustained microcosms for marine aquaria in the absence of regular access to large amounts of SW.

4.5 Mesocosms for economically and ecologically sustainable aquaculture

Commercial aquaculture farms can learn much from ornamental microcosms and technologies developed and pioneered for smaller aquaria. However, several key differences between ornamental microcosms and production-level mesocosms need to be considered besides their scale. An important difference is the density of organisms cultured, which is generally higher in intensive production mesocosms. Moreover, many commercial aquaculture farms produce carnivores that occupy high trophic levels and are extremely resource-intensive regarding both feeds as well as wastewater treatment.

In addition, commercial mesocosms are subject to large fluctuations in biomass within the system due to rapid growth and regular removal of large quantities of the organisms produced. An exception are some aquaculture systems that are not geared towards seafood production (e.g. pearl oyster farms). These differences render balancing nutrient cycles in commercial aquaculture mesocosms more challenging than in ornamental microcosms.

However, mesocosms have a larger inherent capacity for buffering fluctuating environmental conditions than microcosms, which affords more time to respond to failure of technology in intensive recirculating systems. Even though the emergent properties of natural systems render scaling-up from microcosms to mesocosms difficult, current efforts to develop larger, mesocosm scale RAS are encouraging.

The Biosphere II project (Figure 4.3) is an example for a sophisticated mesocosm that has been studied in great detail (Marino and Odum, 1999). Some zoological parks and public aquaria also contribute increasingly to efforts for generating mesocosms that represent self-sustained systems, which provide more suitable habitat to captive animals by more closely resembling their natural environment.

Maintaining a good balance of input versus output of matter and energy and nutrient cycles in such experimental mesocosms is not trivial and is difficult to achieve for extended periods of time. Firstly,

Figure 4.3 Biosphere 2 campus of the University of Arizona in Oracle, AZ. Photo reproduced from Wikimedia Commons under CC0-1.0.

nutrients and waste are represented by an enormous number of chemically diverse compounds that are metabolized in many ways. Secondly, cycles need to be carefully balanced not only for major (nitrogenous, phosphorous, and calcareous) compounds but also for many different micronutrients with chemical structures that include trace elements.

Virtually all elements in the periodic table are present in SW and marine organisms have evolved to perform optimally at specific concentrations of specific compounds comprised of these elements. Even though homeostasis of some compounds is more critical than others, an imbalance in any aspect of the overall nutrient cycle has the potential to destabilize a mesocosm or other ecosystem. Therefore, future research needs to: 1) comprehend all essential aspects of nutrient cycles for mesocosms of interest, and 2) develop technologies to maintain those cycles properly balanced.

An important aspect of natural ecosystems is their dynamic equilibrium described as ecosystem succession. Succession refers to the self-organization and continuous evolution of natural ecosystems. Ecosystems are not static but constantly change, that is, they are in succession from one state to another. Ecosystem change that is noticeable to us may happen over decades, centuries, or longer periods. No ecosystem remains the same over time. This property is important when considering the design of maximally closed mesocosms for aquaculture. Ecosystem succession is also critical for well-planned conservation efforts.

Succession in human-managed aquaculture systems is mostly undesirable, rather the goal is to set up a stable system that maximizes yield while minimizing reliance on ecosystem services. However, when establishing aquaculture mesocosms in the first place, they must undergo some succession to equilibrate and attain what is considered an optimal and robust steady state for aquaculture purposes (maximal yield and recycling capacity, minimal ecosystem service dependence).

To maintain this optimal steady state over time, further succession of aquaculture systems must then be actively counteracted by corresponding human management. For aquaculture mesocosms, we have much to learn about how to initially accelerate succession for attaining a desirable steady state and then counteract it to prevent or delay departure from that state.

4.6 Balancing trophic levels in mesocosms

A key property of any ecosystem, whether natural or human made, is its trophic (feeding) structure, which defines the flow of matter and energy through the system. The 'eat and be eaten' paradigm lies at the heart of any ecosystem and literally represents the ultimate goal of seafood production. All ecosystems consist of organisms that represent a fine balance of three main trophic levels: primary producers, consumers, and decomposers. Consumers are commonly classified into several subcategories such as herbivores, detritivores, first-level

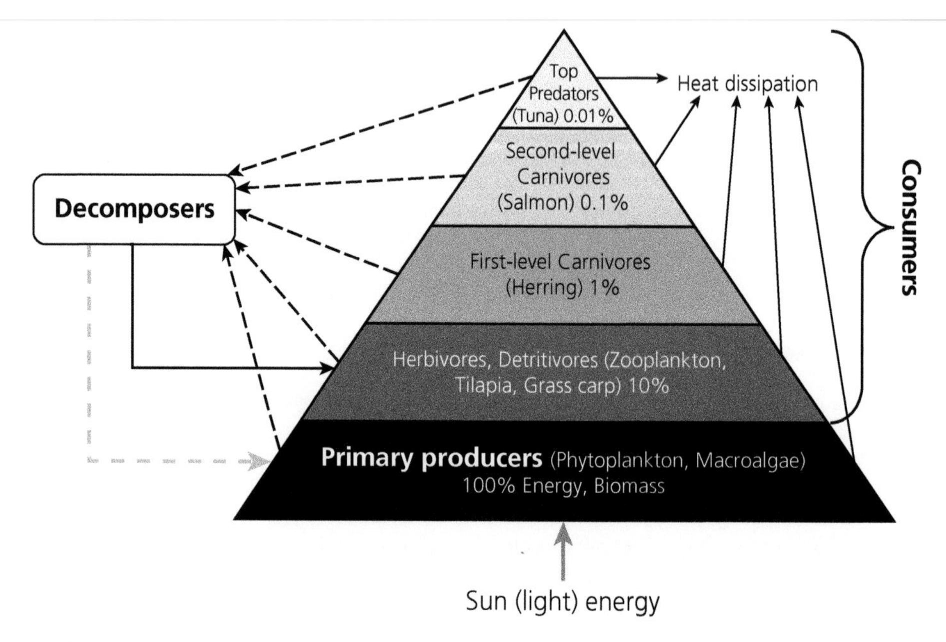

Figure 4.4 Simplified relationship between trophic levels in an ecosystem. A Trophic Level Transfer Efficiency (TLTE) of 10% was assumed for all levels. Dashed black lines indicate dead biomass and a bold grey dashed line indicates the flow of inorganic plant nutrients.

carnivores, second-level carnivores, and top-level predators. The relationship between these trophic levels is often visualized as a food pyramid (Figure 4.4). The concept of food pyramids draws on a pioneering study by Howard T. Odum and conceptualizes the transfer of matter and energy through trophic levels (Odum, 1957).

The transfer efficiency of matter and energy across different trophic levels is calculated as wet weight (biomass) and chemical energy content of the higher trophic level relative to the biomass and chemical energy content of the lower trophic level serving as the food source. Trophic level transfer efficiency (TLTE) is typically 10–20% (Adey and Loveland, 2007). Most of the energy and matter transferred to the next higher trophic level is 'lost' in the form of metabolic waste and kinetic and thermal energy. Energy lost from trophic cycles by heat dissipation is compensated for by photoautotroph primary producers that convert energy in sunlight to chemical energy.

The flow of matter and energy through an ecosystem must be balanced to match inputs and outputs and maintain equilibrium. Such balance can only be achieved by proper proportions of primary producers, consumers, and decomposers. Although trophic pyramids illustrate this principle well, they represent highly simplified models. In reality, relationships between trophic levels are much more complex and better modelled as trophic webs (food webs).

When designing RAS, it is important to keep in mind that, while heterotroph consumers are often the focus of aquaculture efforts, bacterial decomposer communities are the main driving force for aquatic food webs, as they serve as the major sinks for carbon, energy, and nutrients (Alongi, 1994). Generally, ecosystem complexity is correlated with their robustness and stability (Conrad, 1976; O'Neill *et al.*, 1986). However, trophic level structure and biodiversity is less complex in aquaculture systems, especially monocultures, than in wild ecosystems. Large public aquaria may be an exception.

While species diversity is not the sole determinant of ecosystem robustness (Wilmers, 2007), a sufficient number of species representing all essential trophic levels must be present to facilitate ecosystem persistence. Therefore, closed aquaculture systems are best achieved by pursuing suitable polyculture combinations rather than monoculture. For polycultures focusing on animals produced as seafood, high yields can only be achieved by culturing a high density of consumers. To balance trophic levels, corresponding amounts of feeds, primary producers, and decomposers are required. This balance can be achieved by implementation of biological filter technologies that are highly space efficient and substitute for natural ecosystem services.

Space-efficient technologies are more difficult to achieve for light-dependent primary producers than decomposers. Anaerobic denitrifying bacteria that convert nitrate to atmospheric nitrogen are an alternative to primary producers for removing excess nitrogen from the system without lighting. Nevertheless, elements other than nitrogen, including phosphorous, calcium, and sulfur, cannot be readily converted into a gas that escapes into the atmosphere. Therefore, suitable microbes and/or chemical filters are needed for closed aquaculture mesocosms to assimilate waste into harvestable biomass or sequester it in filter cartridges.

Waste compounds assimilated into microbial biomass are effectively recycled into a source of organic nutrients for herbivorous and omnivorous consumers. During waste assimilation, primary producers generate oxygen in the presence of light, which further improves water quality. Microalgae represent the primary producers of choice for balancing trophic levels in aquaculture. Some space-efficient designs such as algal turf scrubber (ATS) technology (Figure 3.9) consist of decomposers (bacteria) and primary producers (green microalgae) that form dense mats. These microbial mats contribute the trophic levels that are needed to close trophic cycles in highly intensive animal aquaculture mesocosms.

Microbial mats are promising solutions for RAS, but they can also be used for restoration of eutrophic public waters and for human and agricultural wastewater treatment. A key consideration is to engineer the microbial mats to be as space efficient as possible and provide them with maximal access to light. The biomass produced by growing microbial mats must be harvested periodically to balance the input of feeds for (and output of waste from) animals in aquaculture mesocosms.

If consumers (animals) and microbial mats consisting of decomposers and primary producers are balanced well in closed aquaculture mesocosms, then the main inputs into the system are food and juvenile stages of aquaculture animals and the main outputs are seafood harvest and microbial mat biomass. The harvested microbial mats can be used for direct supplemental feeding of herbivores, for producing feed pellets, as fertilizer for agriculture, for biofuel production, and for a variety of other purposes. If toxic waste has been accumulated in the microbial mat biomass, then it can also be deposited into long-term geological carbon and energy sinks (e.g. fossilized or sub-fossilized).

Alternatives to microbial assimilation as a means of wastewater treatment include chemical filters and reverse osmosis devices, both commonly used for RAS. They are usually combined with mechanical filters and biological filters containing decomposers. Chemical filter and reverse osmosis units can be periodically backwashed to collect nutrients that are then used as fertilizer in other systems (agriculture, etc.). Saturated and exhausted chemical filter cartridges can also be stored in relatively inert, long-term geological sinks, for example, underground trapping in rock formations or old mines. In essence, chemical filters uncouple aquaculture systems from primary producers.

Such uncoupling could be viewed as an advantage of chemical filters over the use of light-dependent ATS or other types of biofilters that rely on primary producers. However, it comes at the expense of having to dispose of saturated filter cartridges. Moreover, reliance on chemical filters reduces oxygen production by primary producers and chemical filter cartridges may not quantitatively capture all micronutrients present in wastewater. In addition, reverse osmosis systems are unfeasible for intensive mariculture and matching their capacity to the requirements of intensive FW RAS as well as maintaining them represents a significant challenge.

Key conclusions

- The boundary between aquaculture and capture fisheries is becoming blurred as wild populations of aquatic organisms are ever more impacted (unintended) and managed (intended) by humans who have irreversibly and to an unprecedented extent changed the trajectory of the biosphere.
- Most current aquaculture systems are semi-intensive, open systems that depend to a very high degree on natural ecosystem services.
- A mismatch between a significant human-induced decline of natural ecosystems on the one hand and increased human demand for ecosystem services on the other hand calls for solutions that minimize the reliance on ecosystem services for all purposes, including aquaculture.
- The most ecologically sustainable aquaculture systems for a growing human population are intensive, closed RAS.
- Low-yield, extensive aquaculture systems are also sustainable in suitable areas, but they are not economical on a global scale unless the human population declines significantly.
- Since TLTE is only 10–20%, aquaculture systems for top-level carnivores are least energetically efficient and most resource intensive.
- Knowledge generated and technologies developed for culture of ornamental organisms in aquaria at microcosm scale have advanced ecologically sustainable aquaculture at larger mesocosm scales.
- Intensive, closed RAS are human-managed mesocosms that require constant balancing of energy and the chemical composition of all matter in inputs and outputs.
- Ecosystem succession must be managed when establishing aquaculture mesocosms to maintain optimal conditions for growth, health, and development of aquaculture organisms.
- Balancing trophic levels in closed aquaculture mesocosms requires proportional matching of primary producers, consumers, and decomposers to minimize reliance on ecosystem services.

References

Adey, W. H. and Loveland, K. (2007). *Dynamic aquaria: building living ecosystems*, 3rd edn. Boston: Academic Press.

Allendorf, F. W. and Hard, J. J. (2009). Human-induced evolution caused by unnatural selection through harvest of wild animals. *Proceedings of the National Academy of Sciences*, 106, pp. 9987–9994.

Alongi, D. M. (1994). The role of bacteria in nutrient recycling in tropical mangrove and other coastal benthic ecosystems. *Hydrobiologia*, 285, pp. 19–32.

Conrad, M. (1976). 'Patterns of biological control in ecosystems', in Patten, B. C. (ed.) *Systems analysis and simulation in ecology*, vol 4. New York: Academic Press, pp. 431–456.

Heino, M., Díaz Pauli, B., and Dieckmann, U. (2015). Fisheries-induced evolution. *Annual Review of Ecology, Evolution, and Systematics*, 46, pp. 461–480.

Marino B., and Odum, H. T. (1999). Biosphere 2: introduction and research progress. *Ecological Engineering*, 13, pp. 3–14.

O'Neill, R. V., Deangelis, D. L., and Waide, J. B. (1986). A hierarchical concept of ecosystems. *Monographs in Population Biology 23*, Princeton: Princeton University Press.

Odum, H. T. (1957). Trophic structure and productivity of Silver Springs, Florida. *Ecological Monographs*, 27, pp. 55–112.

Wilmers, C. C. (2007). Understanding ecosystem robustness. *Trends in Ecology and Evolution*, 22, pp. 504–506.

CHAPTER 5

Integrated multitrophic polycultures

'Allied with technocentric sentiments where human ingenuity is expected to provide solutions to environmental problems and social development needs there is a danger that efforts will focus on piecemeal management interventions and technology deployment as opposed to fundamental reform of damaging policies and practices.'

Bunting, *Principles of sustainable aquaculture* (2013)

Sustainable aquaculture will only be possible globally if regulatory structures that govern how human society operates shift significantly to place proper value on ecosystem services. The trends in this regard are clearly noticeable and encouraging. However, progress is slow and current incentives for maintaining ecosystems as a common property resource are no match for disproportionally large profits obtained by comparably few entrepreneurs from exploiting ecosystem services.

Responsible environmental stewardship and rehabilitation of ecosystems by wetland restoration, replanting of mangroves, re-establishing salt marshes, re-planting of embankments, or other activities do not yield harvestable products or profits from a business perspective. Although these activities have obvious value for the common good it is difficult to measure their value against profits made from exploiting corresponding ecosystem services or benefits gained from utilizing them (e.g. recreational).

Indicators of aquaculture sustainability include a species' suitability for aquaculture domestication, potential for invasiveness and disease transfer, trophic level, and food/energy conversion ratio as well as the ecological, energy, and carbon footprints of the facility (Pullin *et al.*, 2007).

5.1 Polycultures are more ecologically sustainable than monocultures

Except for domestication and invasiveness potentials of a species, these sustainability indicators can be managed more favourably in polycultures than monocultures. However, monocultures are generally more profitable and, therefore, dominate aquaculture.

Examples for integrated multitrophic mariculture are shrimp–oyster, salmon–sea cucumber, and grey mullet–seabream polycultures. Integrated multitrophic freshwater (FW) aquaculture examples include carp–prawn, carp–mullet, and tilapia–prawn polycultures. In these forms of aquaculture multiple trophic levels are integrated into a single culture system. Often not the primary trophic levels (primary producers, consumers, decomposers) but secondary trophic levels (e.g. herbivores, omnivores, detritivores, carnivores) are integrated into polycultures.

Polycultures represent more ecologically sustainable aquaculture practices than the more commonly used monocultures, especially if they are integrated at multiple trophic levels and with regular fallowing periods for affected areas. The main reason is that polycultures generally require fewer overall inputs of feeds and fertilizer and produce less waste

A Primer of Ecological Aquaculture. Dietmar Kültz, Oxford University Press. © Dietmar Kültz (2022). DOI: 10.1093/oso/9780198850229.003.0005

when measured against the biomass of harvested products.

Integrated polycultures represent a combination of trophic levels making them less dependent on ecosystem services than monocultures. Nevertheless, monocultures dominate aquaculture because they are most profitable in current regulatory circumstances. The benefit of polycultures consisting of different fish species is generally proportional to the difference in trophic levels between the species. Information on trophic levels of aquaculture fish species is provided in FishBase (www.fishbase.org).

The ecological value of the polyculture is also proportional to the diversity of feeding niches occupied by the cultured species. For example, the ecological value of polyculture (relative to monoculture) is lower for carp–tilapia polyculture than for tilapia–crayfish or carp–mussel polycultures since the feeding niches of carp and tilapia are more similar than those of these two invertebrates (crayfish, mussels).

Furthermore, the ecological value of polycultures depends on the ratio at which multiple species are co-cultured. For tilapia–crayfish polyculture, the value depends on how much direct or indirect food waste from tilapia culture can be utilized by crayfish, which is directly proportional to the tilapia–crayfish ratio and the hydrological conditions in the culture area (water current, etc.). Polycultures can reduce reliance on fish meal and fish oil by utilizing the biosynthetic capacity of multiple trophic levels for generating the corresponding nutrients (Box 5.1).

Box 5.1 Fish meal and fish oil in aquaculture

Feeds for aquaculture animals often rely on essential nutrients such as polyunsaturated fatty acids (PUFAs) that are present in high amounts in fish meal and fish oil (Box 17.3). These feed ingredients are produced by capture fisheries of wild pelagic herring, sardines, anchovies, and other small fish species that are also used for human consumption.

Lowering the use of fish meal and oil is important since the annual catch of wild fish processed into fish meal and fish oil is approximately 5 million tonnes (Boyd et al., 2019). Producing fish from fish, especially if those fish could also be processed into seafood for human consumption, is ecologically wasteful as the energy gain in food is minimal or non-existent while the ecological footprint is dramatically raised.

Fish meal and fish oil are also derived from capture fisheries by-catch, which represents everything caught other than the primary target species. Unfortunately, such indiscriminatory bycatch now represents a valuable source of secondary income from capture fisheries since it generates profit for producing fish meal and fish oil, which reinforces the praxis of indiscriminatory bycatch.

Aquaculture is not the only market for fish meal and fish oil. Most of the fish meal is processed into feeds for poultry and swine. However, the use of fish meal by aquaculture has increased from 10% of all fish meal produced in 1988 to 35% in the early 2000s (Costa-Pierce, 2014). In recent years, this trend has been reversed due to large efforts in developing more sustainable formulated feeds for aquaculture. Alleviating dependence of aquaculture on fish meal is critical for improving its sustainability and for discouraging indiscriminatory collection of capture fisheries bycatch.

Nevertheless, whether in poly- or monocultures, heterotrophic fish and aquatic invertebrates produce and excrete nitrogenous waste (mostly in the form of inorganic ammonia) and chemically diverse organic waste. The overall amount of this waste is a function of the overall animal biomass and metabolism in the system, irrespective of whether a single or multiple species are cultured.

In the most common aquaculture systems (i.e. open systems), the recycling of ammonia and organic waste relies on ecosystem services provided by decomposers (bacteria, fungi) and primary producers (green plants). Although open integrated multitrophic aquaculture (IMTA) represents a marked improvement over monocultures, the ecological sustainability of open water IMTA is still low if animals are cultured at high density.

For example, animal polycultures of consumer species such as salmonids or other carnivorous fish grown in open cages in coastal waters combined with sessile invertebrates (molluscs or echinoderms) below the cages improves ecological sustainability compared to monocultures. However, even though these invertebrates contribute to improvement of water quality either by counteracting turbidity or by utilizing detritus, these animals generate ammonia and organic metabolic waste that requires decomposers (bacteria, fungi) and primary producers for trophic recycling.

The amount of decomposer and primary producer biomass required for proper trophic recycling of animal waste is generally much greater than that of the animal biomass and must be contributed by ecosystem services in open IMTA systems. Thus, quantitative integration of culturing seaweed or other sessile plants and artificial reefs for propagating bacterial decomposers into open IMTA systems is very challenging. Such efforts are complicated by limited photosynthetically suitable space underneath animal net cages and by water currents that rapidly disperse waste beyond the area used for open systems.

For small-scale IMTA systems such effects are diluted out by spreading the soluble animal waste throughout the vast expanse of the ocean. However, numerous examples (plastics, etc.) should have taught us by now that it is not wise to use the oceans as a waste bucket for undesirable by-products of human activities. This practice only exacerbates the problems and defers dealing with detrimental consequences (if they are not irreversible) to future generations.

In addition, open IMTA has a relatively large spatial footprint. Therefore, open-water (cage-based) IMTA mariculture does not represent one of the most ecologically sustainable solutions for high-yield aquaculture of the future. Nevertheless, small-scale IMTA may prove useful for local markets if the world population and global demand for seafood decline significantly (scenario 2 in Section 4.2).

5.2 Multitrophic polycultures of animals with plants

Animal–plant polycultures are ecologically sounder than polycultures limited to multiple animal species because two of the three primary trophic levels are represented (consumers and primary producers). The third level (decomposers) may be partially bypassed in animal–plant polycultures since algae and other plants can take up ammonia directly and metabolize it to increase their biomass.

However, plants also produce ammonia from nitrate rather than relying on its uptake. Moreover, ammonia is not the only animal waste released into the water and decomposers are required for recycling a large variety of other waste compounds. Plants utilize inorganic compounds as nutrients and much of the organic waste produced by animals is decomposed into inorganic nutrients such as nitrate and phosphate by bacteria and fungi.

Duckweeds (family *Lemnaceae)* and watercress (*Nasturtium spec.*) are commonly integrated with aquaculture of fish. These FW plants cover the water surface to provide shelter for fish and reduce eutrophication by assimilating nutrients and limiting light supply to microalgae in the water. Duckweeds can serve directly as a food source for herbivores and omnivores such as tilapia. Watercress can be harvested and processed for human consumption.

Mangrove–shrimp culture is practised in Indonesia, the Philippines, Thailand, and Vietnam. It serves a dual purpose for producing seafood (shrimp) while also restoring mangrove forests to rebuild ecosystem service capacity in the long term. Mangrove polycultures mitigate widespread mangrove losses that have resulted from unsustainable aquaculture development in the past. Mangrove replanting programmes have had mixed success and are improved by including a larger variety of mangrove species to better reproduce natural ecosystems and restore coastal mangrove wetlands.

For mariculture, seaweed is often used to integrate animal and plant aquaculture. Polyculture of Atlantic salmon with the red macroalgae *Porphyra purpurea* in Canada's Bay of Fundy has contributed to assimilating the waste produced by the salmon, which was estimated to include 1225 tonnes of nitrogen and 245 tonnes of phosphorus in 2001 (Chopin *et al.*, 2007).

Other examples for animal–seaweed polycultures include various filter-feeding bivalves and seaweed species, which have been explored in tropical and temperate regions. Integrated polycultures have been developed for Blue mussels (*Mytilus edulis*) with the red macroalga Irish moss (*Chondrus crispus*) in Canada (Chopin *et al.*, 2007) and pearl oysters (*Pinctada martensi*) with the red macroalga *Kappaphycus alvarezii* in Japan (Qian *et al.*, 1996).

Seaweed use in animal–plant polycultures minimizes eutrophication caused by algal blooms by competing with microalgae and cyanobacteria for the inorganic nutrients that are produced by the animals and decomposers that break down animal waste.

Seaweed assimilates these inorganic nutrients and converts them to biomass. In addition to this role for bioremediation, seaweed represents a harvestable product to diversify aquaculture revenue. It is used for seafood (e.g. sushi wrapping with nori/red laver) and for producing agar and alginates, processed into feed for shrimp and fish, and utilized for diverse other purposes (Chapter 8).

Integrating seaweed culture with fish or shrimp aquaculture is complicated by the inhibitory effect of turbid wastewater on seaweed growth. This pitfall can be remedied by installation of settlement or mechanical filters in decoupled systems and/or polyculture with filter feeders (e.g. oysters, other molluscs, or sponges).

Filter-feeding molluscs and other filter feeders have the added benefit of serving as bioindicators of water quality since they accumulate contaminants such as heavy metals or other toxins and pathogens in their biomass. Regular tests for such bioaccumulation of contaminants have a dual role for ensuring food safety as well as diagnosing water quality problems.

Water quality problems can also be diagnosed by regular monitoring of taxonomic community composition in polycultures that employ bioremediation organisms (decomposers) in biofilters. Shifts in species representation in microbial decomposer and primary producer communities are indicative of changing water quality.

A very common form of multitrophic polyculture is the combination of aquaculture with rice farming. This form of polyculture has been practised for centuries in parts of Asia and is still a dominant form of aquatic animal–plant polyculture. Carp, tilapia, shrimps, and prawns are often used in combination with rice farming.

Integrated rice–shrimp polyculture is common in parts of Bangladesh and India, but such polyculture is not limited to Asia. In the US, crayfish (*Procambarus clarkii*) is co-cultured with rice, utilizing Louisiana wetlands for double cropping. Animal aquaculture combined with rice farming represents an intersection for integrating aquaculture with agriculture (IAA, Section 5.6).

However, unlike other IAA, the integration of aquaculture with rice farming is performed in a coupled system while for many other agricultural crops the integration is decoupled by collecting wastewater from aquaculture farms and pumping it to adjacent areas for irrigating agricultural crops and orchards.

5.3 Pros and cons of polycultures

Polycultures have many advantages besides being generally more ecologically sustainable. Financial risks can be dispersed by diversification of products and markets, economies of integration can be harnessed, and earlier returns on investments can be achieved when culturing organisms requiring long pay-off periods (e.g. sturgeon aquaculture for producing caviar).

Sturgeon–plant polycultures can reduce financial risks associated with the long maturation time of sturgeon females needed for caviar production. Income from crops irrigated with sturgeon aquaculture wastewater can be used to offset investments into caviar production during sturgeon maturation, which can take up to a decade.

In addition, products of polycultures and integrated multitrophic production systems are often preferred by the public who are willing to pay a premium for these products, which helps offset higher costs (Bunting, 2013). The reason for the favourable public perception of polycultures over monocultures is their generally greater ecological sustainability and reduced utilization of common property ecosystem services.

Semi-intensive monocultures of carp and tilapia species in fertilized earthen ponds continue to be the most productive and profitable sector of aquaculture. However, monocultures are less ecologically sustainable than polycultures because of their larger ecological footprint and higher demand on common property ecosystem service resources.

Assessment of absolute ecological footprints required for wastewater recycling from aquaculture farms is very difficult in practice because the productivity and capacity of ecosystems to assimilate waste is not static but highly dynamic. It depends on many dynamic environmental parameters (e.g. seasonal climate) and local variation within ecosystems.

Nevertheless, the relative ecological footprint of monocultures is generally substantially larger than that of polycultures. Smart integration of multiple trophic levels into polycultures decreases the ecological footprint of aquaculture and makes more efficient use of available space to reduce negative consequences of aquaculture.

Multitrophic polycultures of omnivores that tolerate low water quality, have low trophic levels, and grow well on plant-based feeds represents a meritorious strategy for lowering the ecological footprint of aquaculture relative to carnivore monocultures. Such polycultures can be maintained in coupled and decoupled systems (Box 5.2).

Box 5.2 Coupled versus decoupled aquaculture and aquaponics systems

Polyculture integration can be achieved in coupled systems and decoupled systems. A typical coupled system would be a pond in which herbivorous fish are cocultured with omnivorous or carnivorous species to make better use of resources by exploiting different feeding niches and reducing reliance on ecosystem services.

Decoupled systems consisting of units that are positioned in proximity but can be managed somewhat independently regarding inputs and outputs. They are most promising for future development of aquaculture. Aquaponics systems in which the flow of wastewater from fish tanks to hydroponics tanks and vice versa is discontinuous and can be regulated by a valve or pumps represent an example for a decoupled polyculture system.

Aquaponics represents a special case of an integrated multitrophic recirculating aquaculture system (RAS). Aquaponics systems combine the hydroponic culture of terrestrial plants such as lettuce and basil with aquaculture of fish (commonly tilapia). The wastewater from fish tanks is used as a nutrient source for the plants.

Coupled aquaponics has the same advantages and is subject to the same challenges as other coupled polyculture systems. Notably, problems of matching nutrient capacity in fish waste with the requirements of plants is challenging in a coupled aquaponics system. Decoupling permits adjustment of the amount of water that is being exchanged between the fish tanks and hydroponics system.

Decoupling also affords the possibility to pre-treat wastewater by mechanical filtration to remove insoluble material, and in biological filter units containing decomposers to optimize inorganic nutrient availability for plants. Elaborate management tasks have thus far prevented a wide-spread adoption of aquaponics on a commercial scale.

Technically, RAS that employ biological filters are also multitrophic integrated polycultures. They combine the culture of animals (even if it is only a single species of consumer) with that of decomposers (bacteria) and sometimes primary producers (microalgae). RAS setups are coupled or decoupled.

In a decoupled RAS, animal wastewater is collected to periodically feed biofilters, wastewater treatment lagoons, or ATS. The algal/bacterial biomass can be harvested from ATS or bioflocs (Section 5.5) periodically and processed into feeds, fertilizer, biofuel, and other products while the water cleaned by the bacteria and algae can be periodically recycled to replace wastewater in the fish tanks.

Decoupling of polyculture systems requires careful planning and increases management associated with proper balancing the inputs and outputs of each culture unit. Moreover, separate culture units must be maintained for each species or community of species integrated within the system, which adds costs and is labour-intensive. Depending on the distance between decoupled culture units, transporting inputs and outputs (e.g. water, feeds) between the different units adds logistical challenges and increases costs and reliance on proper functioning of associated technologies (e.g. water pumping systems).

These drawbacks of decoupled polyculture systems are offset by their many advantages. Decoupling inputs and outputs for each species or group of species in the system enables buffering and separate adjustments of changes in biomass for each of the units. For example, changes in biomass of fish and ATS or bioflocs due to growth and cropping can be compensated for by regulating how much water is being exchanged between the fish tanks and biological wastewater treatment units.

Waste-nutrient mismatches between different units can be balanced by proper supplementation and/or chemical filtration in decoupled systems. Mechanical filters can be operated between units to periodically remove solid waste and collect sludge.

Another argument for decoupled systems is that decoupling can disrupt pathogen life cycles and prevent their spread if appropriate water sterilization procedures (UV, chlorine, or ozone) are implemented. Thus, decoupled systems lower the risk of disease transmission because the spread of pathogens can be more readily contained, and biosecurity measures are easier to implement.

Many hatcheries already employ decoupled polyculture of microalgae and/ or artemia to produce live food for fish and shrimp. When combined with biofilters containing appropriate decomposer communities, such systems can be closed or semi-closed to minimize usage of clean water, reduce feeds and fertilizer, and lower the discharge of wastewater. Decoupling polyculture units also affords opportunities for alleviating potential constraints due to different temperature optima for the different species cultured.

Despite its advantages, polycultures are still much less common than monocultures for several reasons. Managing polycultures is more labour intensive, requires greater knowledge and training, has higher operational costs, and uses more infrastructure resources than monocultures. In addition, most chemical pesticides used for production of crops are harmful to aquatic animals and their use is greatly limited in animal–plant polycultures, which reduces the yield of produced crops.

5.4 Decomposers: a critical link in multitrophic polycultures

Bacteria (prokaryotic decomposers) have a much greater metabolic diversity and adaptability than animals and plants. Bacteria and eukaryotic decomposer (fungi) communities can metabolize the full spectrum of organic waste compounds excreted by animals. Therefore, decomposer community species composition and metabolic profiles are sensitive bioindicators of water quality. Knowledge about the physiology and biochemistry of decomposer species and their relative representation enables deduction of specific forms of organic wastes and their quantity in the water.

As mentioned, algae and other plants can take up ammonia, which is the major form of nitrogenous waste produced by aquatic animals. However, the recycling of organic animal waste into the full spectrum of inorganic nutrients taken up by plants relies on decomposers.

Bacterial decomposer species diversity, numbers, and metabolic capacity is much greater than that of all eukaryotes (animals, plants, fungi, etc.) combined. However, eukaryotic fungal species also have important decomposer functions in aquatic ecosystems, including the decomposition of plant litter (Kubicek and Druzhinina, 2007). Decomposer communities shift rapidly in their species composition and metabolic activity to accommodate the spectrum of nutrients available in waste.

Decomposer community structure depends on the types and amount of organic waste compounds present in the water and oxygen availability. If oxygen is limited (hypoxia) or absent (anoxia), then microbial decomposer communities shift from species that prefer aerobic to those preferring anaerobic metabolism.

Anaerobic bacterial metabolism in the presence of large amounts of organic waste can lead to the formation of 'dead zones' in oxygen-deprived sediment layers. Such dead zones are characterized by toxic inorganic compounds derived from anaerobic bacterial metabolism that breaks down organic waste (e.g. H_2S from cysteine).

In smaller semi-intensive aquaculture ponds or areas under marine net-pen cages anoxic zones can form in the sediment. In highly eutrophic bodies of water such as the Baltic Sea and the Black Sea or in very eutrophic lakes, anoxic dead zones extend above the sediment to the deep layers of the open water column.

To avoid the formation of dead zones in aquaculture or in areas inundated with aquaculture wastewater it is imperative to provide sufficient substrate for aerobic bacterial decomposers and maintain oxygen levels near saturation. Substrates for attachment of non-planktonic aerobic decomposing bacteria include sediment grains, plant materials, rocks, dead invertebrate exoskeletons, and insoluble particles that are suspended in the water column.

Depending on the type of sediment, its grain size, the extent of spaces between grains, and the extent of water circulation through those spaces, decomposer communities in sediment are either aerobic or anaerobic. Thus, the nature of sediment in aquaculture ponds and underneath coastal mariculture cages should be considered when assessing ecosystem service capacity.

Organisms growing on the non-sediment substrates mentioned above are collectively referred to as *Aufwuchs* (overgrowth in German) and is often used synonymously with periphyton. *Periphyton* comprises a complex community of microorganisms including microalgae, bacteria, fungi, protozoans, and others. It grows on dead (Figure 5.1) and live (Figure 5.2) animals and plants.

Microbial periphyton communities have a large metabolic capacity for fulfilling decomposer and primary producer roles in trophic cycles. Their biomass and capacity are directly proportional to the surface of substrate they colonize. This principle is utilized in biological filters, which consist of porous substrates with very large surface areas.

The same principle also applies to artificial reefs, which provide an enlarged surface area for periphyton. Periphyton is critical for recycling animal waste into inorganic nutrients and biomass that serves as a food source at the base of trophic networks. Given its importance in this process, research that characterizes periphyton, its species composition, metabolic processes of pertinent species, and dynamic changes in species composition and metabolism in response to animal wastewater should be promoted.

Periphyton can be used as a bioindicator of water quality (Kireta *et al.*, 2012). Moreover, identification of microbial communities, individual species, and metabolic processes that foster recycling of animal waste will improve biological filters and enable better assessment of habitat-specific ecosystem service capacity. In addition, aquaculture of periphyton focused on wastewater treatment using algal turf scrubbers (ATF) and bioflocs can be further improved by informed management of microbial species composition and metabolic capacity in periphyton films.

In contrast to periphyton, *bioflocs* refer to a self-organizing decomposer community that does not require a substrate (such as a filter resin or artificial reef or particles with a large surface area).

Artificial reefs have been proposed to integrate and increase decomposer with primary producer capacity in open aquaculture systems. In that sense, artificial reefs fulfil the same function that biological filters have in closed recirculating systems. Artificial reefs help sequester and recycle animal waste to a more localized area by providing sufficient substrate for attached decomposers to match the animal waste load from fish and invertebrates with the biomass and metabolic capacity of local decomposer communities.

Figure 5.1 Periphyton on dead staghorn coral branches has an important nutrient recycling function in coral reefs that are exposed to inflow of nutrient-rich waters. Photo by author.

Figure 5.2 Periphyton on live seagrass consists of decomposers (bacteria, fungi), microalgae, and other microorganisms that represent a critical component of trophic networks. Photo by author.

To minimize dispersal of animal waste, artificial reefs are most effective when positioned near the source of the waste. However, depending on hydrological conditions at the aquaculture site, the ratio of waste dissipation from the site versus assimilation by local decomposer communities may differ greatly. Moreover, the extent of artificial reef required to account for the waste produced by intensive cage mariculture and corresponding alteration of existing ecosystems must be considered when planning such highly integrated aquaculture projects.

Periphyton and decomposer communities growing on abiotic substrates must be removed periodically if animal waste levels are high and bacterial/algal biomass grows rapidly. Periodic removal has been optimized for ATS and bioflocs, which are human-managed microbial communities used for wastewater treatment that operate on the same principle as periphyton. Bioflocs, ATS, and other bioremediation solutions for wastewater treatment fill the niche of decomposers (bacteria) and primary producers (microalgae) in aquaculture systems (Box 17.4).

Two main challenges must be addressed when implementing such biological wastewater treatment approaches. First, the capacity of decomposers and primary producers must be matched to the amount of animal waste produced, which is difficult to achieve in a space-efficient manner. Second, microbial communities must be matched to the type of animal waste produced.

Even though common types of nitrogenous, phosphorous, and sulfurous waste are usually dealt with effectively by many microbial communities, mismatches in micronutrients need to be balanced by either supplementing those that cannot be fully recycled or chemically removing organic waste that cannot be metabolized by available decomposers.

For example, sodium silicate is essential for diatoms present in periphyton and decomposing planktonic diatoms. Any mismatch of silicate provided in animal waste versus what is required in biofilters or bioremediation tanks inoculated with diatoms needs to be balanced.

Addressing these challenges through corresponding research has strong merit since closed, recirculating systems reduce the ecological footprint and increase ecological sustainability of aquaculture considerably. For example, intensive eel aquaculture in a recirculating system in Taiwan reduced the use of space thirtyfold and of water hundredfold (Chao and Liao, 2007).

High-intensity RAS demand high levels of technology and energy and require a high skill level for management (Martins *et al.*, 2010). They utilize education, knowledge, and technological advances available in the twenty-first century to the fullest extent to offset the negative environmental impacts of human development and population growth that parallel the intellectual and technological advances.

Research and development of RAS should be better incentivized by policymakers to offset high costs. The use of renewable, clean energies rather than fossil fuels (solar, wind, geothermal, biofuel) for powering RAS should also be incentivized. Clearly, much research, development, and regulatory legislation is yet to be implemented to render RAS commercially feasible and attractive on the scale needed for accommodating a high demand of seafood (scenario 1 in Section 4.2).

5.5 Bioflocs technology

Bioflocs are floating aggregated communities of decomposing bacteria and other decomposers (fungi); these also contain primary producers (microalgae) and low trophic level heterotroph consumers (protozoa, zooplankton, small invertebrates). In addition to these organisms, bioflocs contain food remains and faecal matter that are aggregated into flocs.

Bioflocs technology aims towards creating a bacteria-dominated ecosystem that minimizes waste by quantitative recycling of organic waste produced by aquaculture animals. This technology has first been used in sewage treatment plants and is increasingly adopted for wastewater treatment in aquaculture.

Bioflocs technology requires efficient aeration devices to maintain a high dissolved oxygen (DO) concentration in the water and keep the aggregated communities of decomposers (bioflocs) suspended in the water column. It can be utilized in coupled systems but is most efficient and stable when used in de-coupled systems (Box 5.2).

Bioflocs are dominated by heterotrophic bacteria that rapidly oxidize and convert organic animal waste to bacterial biomass. They reduce the need for autotrophic phytoplankton in recycling animal waste (chiefly ammonia) and, therefore, minimize algal blooms and eutrophication. Bioflocs biomass represents a food source for detritivorous aquaculture animals. Therefore, this technology contributes to balancing the trophic cycle in aquaculture systems and minimizing reliance on external ecosystem decomposer services.

Major advantages of bioflocs technology include its suitability for maximizing the productivity of RAS and minimizing the disposal of organic waste from semi-intensive ponds.

The main challenges of bioflocs technology are related to the dynamic nature of ecosystems, which are in a state of constant succession. This succession is accelerated in aquaculture where animals grow very quickly and the amount and composition of feeds and organic waste changes rapidly. Maintaining a dynamic equilibrium that matches the decomposer capacity of bioflocs communities with the organic waste quantity and composition of aquaculture wastewater is technically challenging and represents an important area of research.

5.6 Integration of aquaculture with agriculture

Integration of aquaculture into polycultures is not limited to farming multiple aquatic species. Aquaculture has been integrated with other uses of available water to make better and more efficient use of this precious natural resource, as well as other resources. Multi-purpose utilization of water in arid areas with limited water availability includes first use of water for animal aquaculture followed by secondary use of the wastewater discharge for irrigation of agricultural crops and orchards.

Similarly, aquaculture has been combined with agroforestry and silviculture of mangroves. Other integrated uses of water for aquaculture include conservation efforts (e.g. to provide habitat for migrating and water birds) and recreation (e.g. for anglers, sailors, wind surfers).

Aquaculture has also been integrated with terrestrial agriculture utilizing animal manure to fertilize fish ponds, for multi-purpose use of irrigation channels and ditches as well as water storage reservoirs, and for utilizing saline groundwater or wastewater from industrial and desalinization processes (Bunting, 2013). Such integration affords more efficient use of resources but also requires more demanding management.

Integration of aquaculture with agriculture (IAA) by utilizing manure from farming chickens and other bird species, cattle, pigs, goats, and sheep serves to fertilize ponds for growing algae and other aquatic plants as food for herbivorous fish. However, this form of polyculture is on the decline because of public health and food safety concerns (De Silva and Wang, 2019).

Moreover, organic waste from animal agriculture adds to the waste produced by aquaculture fish and requires aerobic bacteria for decomposition into inorganic compounds that are accessible as nutrients for plants.

Aerobic bacteria consume oxygen, as do animals and plants. Plants only produce oxygen by photosynthesis during the day when light is available, although they continue to consume oxygen by respiration at night. Therefore, fertilization of aquaculture ponds has limits of intensification that need to be carefully managed.

Oxygen consumption by aerobic bacteria and plants in the pond is proportional to their biomass and capacity to process the organic animal waste. An increase in that biomass causes eutrophication and lowers DO levels at night when plant photosynthesis does not occur, but respiration continues. Consequences of such hypoxia or even anoxia are algal crashes and fish mortality. The fertilization level at which such negative consequences occur can be raised when using aeration devices. However, aeration does not prevent eutrophication and requires investment into technology and energy.

Key conclusions

- Integrated multitrophic polycultures represent an ecologically more sustainable form of aquaculture than monocultures; their economic feasibility relies on proper incentivization by policy makers.
- Polyculture, combined with a shift to culturing omnivores and herbivores instead of carnivores, reduces the unsustainable use of fish meal and fish oil by commercial aquaculture.
- The ecological value of any polyculture is proportional to the diversity of feeding niches occupied by the cultured species. The most ecologically sustainable form of polyculture combines all primary trophic levels (primary producers, consumers, decomposers) at appropriate ratios in a fully closed system.
- Decomposers represent critical links for polyculture systems. They fulfil an essential role in nutrient cycles and are cultured in biofilters and artificial reefs as ATS, periphyton, and bioflocs. Research on biochemical capacities of bacterial decomposer communities affords new opportunities for wastewater treatment.
- Polyculture systems can be coupled or decoupled. Although coupled systems have some advantages, decoupled systems appear more promising for future development of sustainable aquaculture.
- Aquaculture integration is not limited to polycultures of aquatic organisms but should also be pursued to link aquaculture and alternative uses of water, including drinking, sanitarian, recreational, conservation, industrial, and agricultural uses.
- Integration of Aquaculture and Agriculture (IAA) is practised in two forms: 1) wastewater produced by aquaculture is utilized for irrigation of crops and orchards; and 2) animal manure produced by agriculture is used to fertilize semi-intensive aquaculture ponds.

References

Boyd, C. E., McNevin, A. A., and Tucker, C. S. (2019). 'Resource use and the environment', in Lucas, J. S., Southgate, P. C., and Tucker, C. S. (eds.) *Aquaculture: farming aquatic animals and plants*. Oxford: John Wiley & Sons Ltd, pp. 93–112.

Bunting, S.W. (2013). *Principles of sustainable aquaculture: promoting social, economic and environmental resilience*. Routledge.

Chao, N.-H. and Liao, I. C. (2007). 'Sustainable approaches for aquaculture development: looking ahead through lessons in the past', in Bert, T. M. (ed.) *Ecological and genetic implications of aquaculture activities, methods and technologies in fish biology and fisheries*. Dordrecht: Springer, pp. 73–82. https://doi.org/10.1007/978-1-4020-6148-6_4

Chopin, T., Yarish, C., and Sharp, G. (2007). 'Beyond the monospecific approach to animal aquaculture—the light of integrated multi-trophic aquaculture', in Bert, T. M. (ed.) *Ecological and genetic implications of aquaculture activities, methods and technologies in fish biology and fisheries*. Dordrecht: Springer, pp. 447–458. https://doi.org/10.1007/978-1-4020-6148-6_25

Costa-Pierce, B. A. (2014). 'The Ahuapua'a aquaculture ecosystem in Hawaii', in in Costa-Pierce, B. (ed.) *Ecological aquaculture:* the evolution of the blue revolution. Oxford: Wiley-Blackwell, pp. 30–43.

De Silva, S., Wang, Q. (2019). 'Carps', in Lucas, J. S., Southgate, P. C., and Tucker, C. S. (eds.) *Aquaculture*. Oxford: John Wiley & Sons, Ltd, pp. 339–362. https://doi.org/10.1002/9781118687932.ch23

Kireta, A., Reavie, E., Sgro, G., Angradi, T., Bolgrien, D., Hill, B., and Jicha, T. (2012). Planktonic and periphytic diatoms as indicators of stress on great rivers of the United States: testing water quality and disturbance models. *Ecological Indicators*, 13, pp. 222–231. https://doi.org/10.1016/j.ecolind.2011.06.006

Kubicek, C.P., Druzhinina, I.S. (eds.). (2007). 'Fungal decomposers of plant litter in aquatic ecosystems', in Druzhinina, I. S. and Kubicek, C. P. (eds.) *Environmental and microbial relationships: the mycota*. Berlin: Springer, pp. 301–324. https://doi.org/10.1007/978-3-540-71840-6_17

Martins, C. I. M., Eding, E. H., Verdegem, M. C. J., Heinsbroek, L. T. N., Schneider, O., Blancheton, J. P., d'Orbcastel, E. R., Verreth, J. A. J. (2010). 'New developments in recirculating aquaculture systems in Europe: a perspective on environmental sustainability', *Aquacultural Engineering*, 43, pp. 83–93. https://doi.org/10.1016/j.aquaeng.2010.09.002

Pullin, R. S. V., Froese, R., and Pauly, D. (2007). 'Indicators for the sustainability of aquaculture', in Bert, T. M. (ed.) *Ecological and genetic implications of aquaculture activities, methods and technologies in fish biology and fisheries*. Dordrecht: Springer Netherlands, pp. 53–72. https://doi.org/10.1007/978-1-4020-6148-6_3

Qian, P.-Y., Wu, C. Y., Wu, M., and Xie, Y. K. (1996). 'Integrated cultivation of the red alga *Kappaphycus alvarezii* and the pearl oyster *Pinctada martensi*', *Aquaculture*, 147, pp. 21–35. https://doi.org/10.1016/S0044-8486(96)01393-2

CHAPTER 6

Domestication of aquaculture species

'Most farmed fish are far less domesticated than terrestrial livestock and there is enormous scope for their domestication, with genetic enhancement and improved husbandry.'

Pullin, Froese, and Pauly, *Indicators for the sustainability of aquaculture* (2007)

6.1 Trait selection during domestication

For millenia, humans have been breeding plants and animals, resulting in the domestication of specific strains (or breeds) of these species by artificial (human-controlled) selection. Domestication of organisms favours the expression of traits that improve the performance of these organisms to support human society. Such traits include tameness, rapid growth, and fitness (reproductive success) in captivity.

Additional traits that benefit human society while increasing the ecological sustainability of agriculture and aquaculture are increased feed efficiency, reduced waste excretion, increased disease resistance, and lower stress during confinement and handling.

The genetic makeup of domesticated populations of a species reproduced over multiple generations in captivity changes from that of wild populations, even if no intentional efforts are made. The reason for this change is the very different set of selection pressures experienced in captivity compared to those found in natural habitats. Significant genetic (and epigenetic) changes associated with domestication can occur rapidly and manifest themselves within just a few generations in fish (Dunham, 2011). Epigenetics refers to changes in hereditary information that are not based on DNA sequence, for example, heritable patterns of DNA methylation and histone posttranslational modifications.

Domestication programmes differ regarding prioritization of the traits (and corresponding selection conditions) mentioned, but commercial programmes emphasize traits that increase production (growth, reproduction, disease resistance, etc.). Some beneficial traits are sometimes inversely correlated (e.g. growth and stress tolerance) and the best possible trade-offs in performance of one trait versus the other must be selected when both are essential.

For example, climate change may necessitate selection for high growth in stressful environments (e.g. those with elevated salinity and temperature). However, because of energy partitioning into the biological processes underlying growth and stress tolerance, these traits may be inversely correlated.

Likewise, growth and disease resistance to Taura Syndrome Virus are inversely correlated in white-leg shrimp (*Litopenaeus vannamei*) and selection for one of these traits will negatively affect the other (Argue *et al.*, 2002). Compromises required for animal welfare should also be factored into breeding programmes. Therefore, modern breeding programmes are complex and consider not just a single trait for selection but multiple traits.

One of the main criteria for selecting a species for domestication is the ease of achieving reproduction in captivity. Thus, for many domesticated species artificial selection is now accompanied by methods that control their reproduction.

A Primer of Ecological Aquaculture. Dietmar Kültz, Oxford University Press. © Dietmar Kültz (2022). DOI: 10.1093/oso/9780198850229.003.0006

In sexually reproducing species, such methods are based on intramuscular or intraperitoneal injection of gonadotropic (reproductive) hormones that induce gonadal maturation and ovulation. An alternative to hormonal control of reproduction is to select for genetic strains whose reproduction can be readily controlled by less-invasive cues such as photoperiod and/or temperature.

Although aquaculture species have been domesticated for a relatively short time, many aquatic species are currently under consideration for genetic selection of desirable traits, either by traditional breeding and artificial selection or by genetic engineering.

These efforts often have the primary goal to increase productivity and the growth performance of domesticated aquatic species. However, they are equally promising for improving the ecological sustainability of production by increasing feed efficiency and reducing trophic level (alleviation of dietary constraints) and waste excretion of domesticated animals. In addition, animal welfare concerns and ethical considerations can be addressed by genetic selection that improves health, well-being, and disease resistance under captive and crowded conditions.

6.2 Domestication by breeding-based artificial selection of traits

Traditional artificial selection of desirable traits is achieved through successive rounds of breeding. This approach can be lengthy since it stretches over multiple generations. If animals with long generation times are domesticated, then this process is slow. However, it has proven very effective, as evidenced by the artificial selection of many breeds of dogs over the past 10,000 years.

Shorter-lived species can be domesticated more rapidly. Therefore, domestication of many aquatic species with relatively short lifespans proceeds more rapidly than that of terrestrial mammals with long generation times.

Genetic selection is the enrichment of specific genotypes and corresponding phenotypes (traits) in distinct strains or lines of organisms. Traditional selection of aquaculture strains has focused on production traits such as growth and meat quality. More recently, selective breeding has also been applied to lower the trophic level of aquaculture organisms. In this case, food preference is the trait of interest and changing it from a carnivorous, fish meal-based diet to herbivorous plant-based diets is the goal.

Current efforts as regards diet on sea bass, trout, and other carnivorous species have had mixed success and may benefit from longer periods of selection. Selection for feeding at lower trophic levels can ameliorate the worrisome trend of preferentially consuming high-level predator species (e.g. salmonids), which propelled aquaculture development until the turn of the millennium (Figure 6.1).

Alternatively, a shift in the choice of current aquaculture enterprises and end-user preference from carnivorous to herbivorous species that can be produced with plant-based feeds would improve ecological sustainability by lowering the trophic level of aquaculture species without the need for extensive genetic modification of feeding habits and metabolism.

6.2.1 Phenotype selection and crossbreeding

Different breeding-based strategies for artificial selection have been developed. Most commercial efforts on aquaculture fish, crustaceans and molluscs have focused on phenotypic mass selection (i.e. selection of those individuals from a

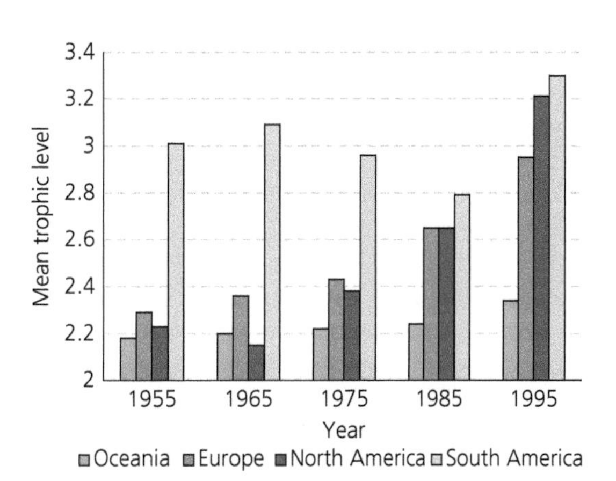

Figure 6.1 Approximate mean trophic levels of aquaculture species produced in Oceania, Europe, North America, and South America. Data source: (Pullin *et al.*, 2007).

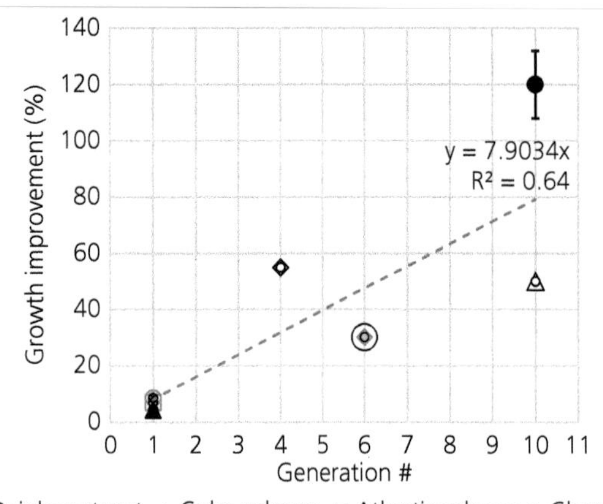

Figure 6.2 Improvements of growth rate in some aquaculture species that have been achieved by selective breeding. The average growth improvement per generation is 7.9% although it varies substantially by species and the relationship between trait improvement and generation number is not truly linear. Data source: (Dunham, 2019).

large population that perform best regarding the trait(s) of interest). This strategy must be pursued over multiple successive generations to enable statistical estimation of the response to selection. Within the first ten generations, it is possible to obtain an average 7.9% increase in growth per generation for various species of fish, molluscs, and crustaceans (Figure 6.2).

To counteract inbreeding depression after multiple rounds of phenotypic selection, multiple genetic lines can be selected for the same or different traits and then cross-bred after several generations. Moreover, outbreeding of inbred strains with an ancestral population can be performed to increase genetic diversity with continued selection for the trait(s) of interest.

Intraspecific crossbreeding of different strains or genetic lines of the same species counteracts inbreeding depression very effectively and often resets the inbreeding coefficient to zero in the first (F1) generation of offspring (Dunham, 2019). However, crossbreeding does not always improve the performance of aquaculture strains. The outcome depends on the genetic composition of the parental strains, which must be chosen carefully.

Crossbreeding can be performed among different strains of the same species as well as between different species. Crossing different species is known as hybridization. The increased genetic diversity resulting from hybridization can result in heterosis and hybrid vigour, although such cases are more of an exception than the rule.

Heterosis refers to a difference in performance of offspring relative to either parent regarding specific traits (growth, size, etc.), and this difference can be positive or negative. In commercial aquaculture, positive heterosis of production traits is desirable. The heterosis concept is broader than (but inclusive of) hybrid vigour, which is a term to describe effects of hybridization on the fitness of offspring.

In some cases, such as crossing different tilapia species, hybridization generates fertile offspring. However, often inter-species hybridization causes infertility of the offspring. Infertility of aquaculture stocks can be desirable to prevent unwanted reproduction, stunting, and colonization of wild habitat by escapees.

Often, the heterosis effect is strongest in the F1 generation, which is why some commercial crosses consist only of the F1 generation derived from crossbreeding non-hybrid parents. Interspecific hybridization (crossing different species) does not always yield offspring that are suitable for aquaculture because heterosis may benefit traits other than typical production traits such as growth and meat quality.

Interspecific hybridizations have been attempted for many combinations of species. However, yielding F1 offspring that are more suitable for aquaculture than either parent is more the exception than the rule (Pillay and Kutty, 2005). Nevertheless, notable examples where this is indeed the case include crosses between different tilapia species, Channel catfish (*Ictalurus punctatus*) and Blue catfish

(*I. furcatus*), Striped bass (*Morone saxatilis*) and White bass (*M. chrysops*), African catfish (*Clarias gariepinus*) and Thai catfish (*C. macrocephalus*), and Silver carp (*Hypophthalmichthys molitrix*) and Bighead carp (*H. nobilis*).

Molluscs can also be hybridized to show positive heterosis in production traits. For example, hybrids of the freshwater pearl mussels *Hyriopsis cumingii* and *H. schlegelii* exhibit an increase in pearl size (23%), higher pearl quantity (32%), and increased frequency of large pearls (3.7-fold) (Dunham, 2019).

Hybridization sometimes results in monosex offspring or progeny with a highly skewed male to female ratio. Such outcome is beneficial if sexual dimorphism exists in production traits or other traits of interest (e.g. ecological sustainability traits). For example, caviar production utilizes sturgeon females and for some species of fish there are significant growth differences between the sexes, which renders one sex preferable over the other for aquaculture. Furthermore, unwanted reproduction can be controlled by aquaculture of monosex populations.

Hybridizing Striped bass (*Morone saxatilis*) and Yellow bass (*M. mississippiensis*) produces all female offspring. However, completely monosex hybrid populations are rare and often the sex ratio is highly skewed, rather than exclusively biased, towards female or male. Many tilapia hybrids show a sex ratio that is skewed towards predominantly male (e.g. Nile tilapia (*O. niloticus*) × Blue tilapia (*O. aureus*), Nile tilapia (*O. niloticus*) × Wami tilapia (*O. urolepsis*), and Mozambique tilapia (*O. mossambicus*) × Wami tilapia (*O. urolepsis*)). For these interspecific tilapia crosses, the presence of even a few females among hybrid offspring significantly impedes efforts to control unwanted reproduction.

6.2.2 Genotype selection using genetic markers

An alternative to phenotypic selection is genotype selection. This method does not select for the phenotypic traits of interest per se, but for genetic markers (specific DNA sequences) that are correlated with those traits.

Genetic markers have been historically developed using linkage maps, which have been gradually replaced by physical maps. The most comprehensive and informative physical genetic map of an organism is its complete genome sequence. Generating the whole genome sequence of an individual is now feasible for under $US 1000 but costs rapidly add up if many individuals must be analysed.

Moreover, mapping genetic differences between individuals is not trivial as true single nucleotide polymorphisms (SNPs) and other mutations must be reliably distinguished from noise associated with precision limits of sequencing technologies. Therefore, genome-wide association studies (GWAS) are often performed to identify specific SNPs in a small number of representative individuals that can later be used for targeted low-cost and high-throughput genotyping of a very large number of individuals.

The main advantage of genotype selection is that it can be performed rapidly on all offspring at a very early life stage before the phenotype (trait) of interest has been expressed. For example, fin clips can be collected from all fry of tilapia at a very young age. The DNA can then be extracted from fin tissue and genotyped for the DNA marker that is associated with the trait of interest. Once the genotyping result is known, then only those fry that harbour the desired genetic marker(s) need to be raised to adulthood to build the genetic stock.

This approach saves significant resources as animals selected against do not need to be raised. Moreover, genotype selection can also be used for a lethal selection screen that is applied to a representative part of a clutch (e.g. for testing environmental stress tolerance). Siblings from the remainder of the clutch that share genetic markers with high-performing individuals sacrificed in the selection screen can then be identified by genotyping fin clips and propagated.

One drawback of genotype selection is that a sufficiently robust correlation between the desired phenotype and corresponding unambiguous genetic markers need to be established in the first place. For establishing robust correlations, several generations and well-documented pedigrees are often needed to develop reliable genetic markers for the trait of interest. Consequently, development of reliable genetic markers requires significant resources.

6.2.3 Induction and selection of altered karyotypes

Chromosomal manipulation represents somewhat of an intermediate between traditional breeding-based artificial selection and genetic engineering. In this approach, human intervention alters the karyotype (the complement of chromosomes containing the genome) by exposing gametes (milt or eggs) to extreme environmental stress.

Since the karyotype (genome) is altered by stress-induced evolution, the resulting organisms are genetically modified. The GMO label is not applied for aquaculture organisms produced by chromosomal manipulation because genetic engineering tools are not used. Likewise, traditional breeding-based selection also results in genetic modification of organisms by shifting allele frequencies in populations but also in the absence of genetic engineering (molecular) tools.

Instead of using molecular tools for the manipulation (as in genetic engineering), chromosomal manipulation utilizes environmental stresses as tools. According to the post-modern synthesis genome theory of evolution, this exposure results in rearrangements of the karyotype to facilitate stress-induced evolution (Heng, 2019; Mojica and Kültz, 2022).

Some of these rearrangements can be described as genome chaos and chromothripsis, which represent genome reshuffling phenomena during which many chromosome breaks occur and the pieces are reassembled into novel combinations. Under certain conditions copy numbers of all chromosomes change. For example, transient exposure of eggs from some fish species to severe hydrostatic pressure or temperature stress shortly after fertilization inhibits the expulsion of one set of maternal chromosomes and induces triploidy in normally diploid species.

Triploid animals are often sterile and can be used to control unwanted reproduction more effectively than steroid hormones, although often some diploid embryos that are able to reproduce remain after treatment. As an added benefit, some triploid fish (e.g. Channel catfish (*I. punctatus*)), are larger, have greater feed efficiency, and yield greater carcass mass than diploid conspecifics at comparable early age. However, this is an exception and in most cases triploid fish do not have improved growth rates during the period of their life cycle that is relevant for production (i.e. before sexual maturity (Dunham, 2019)).

Triploidy can also be induced in commercially important molluscs and crustaceans. Meat quality in Pacific oysters (*Crassostrea gigas*) is significantly improved in triploid animals compared to normal diploids because more energy is partitioned into muscle tissue and less into gonadal development. Some triploid oysters can reverse to a diploid karyotype in some of their germ cells (gametes). Triploidy induction in the Whiteleg shrimp skews populations towards female sex, which is desirable for aquaculture production since female shrimp grow faster than their male counterparts.

In addition to eggs, fish milt (spermatocytes) can also be exposed to severe environmental stress. Transient exposure of milt to high doses of ionizing radiation destroys sperm chromosomes without rapidly killing the sperm cells. If such irradiated milt is used to fertilize eggs, then only one set of chromosomes (the maternal set) is present in the fertilized egg, which then has a haploid karyotype.

Haploid embryos usually do not survive unless spontaneous duplication of the chromosome set of the egg occurs, which is rare. But when it happens, spontaneous diploidization yields viable offspring with two identical copies of the maternal genome. Although not used on a commercial scale, diploid animals that are genotypically all females may represent a viable alternative to using steroid hormones for controlling sexual differentiation.

Other technologies for producing monosex populations of aquaculture fish (gynogenesis, androgenesis) have been explored and some have the potential to be commercialized. Examples include genetically all female XX chromosome rainbow trout produced by sex reversal and breeding approaches and genetically all male Nile tilapia produced by YY male chromosome technology, which improves production yield and reduces unwanted reproduction.

6.3 Domestication by genetic engineering

Genetic engineering approaches are now being increasingly considered as an alternative to traditional breeding-based genetic modification of organisms, which is very slow and resource-intensive. Like breeding-based artificial selection, genetic engineering has the goal to improve stocks by increasing growth, feed efficiency, animal health/welfare, and disease resistance while decreasing trophic level, waste excretion, ecological footprint, and disease susceptibility.

In addition, aquatic organisms are genetically modified for producing pharmaceutical compounds that improve human health. Importantly, genetic modification of aquaculture strains may reduce the need for chemical management of diseases (e.g. antibiotics) and reproduction/growth (e.g. hormones).

Controversial aspects of aquaculture with the pros and cons currently being widely debated include the use of genetic engineering tools for domestication and stock improvement. There are many ways of genetic engineering to yield GMOs, which are organisms whose genome (hereditary DNA complement) has been modified by any genetic engineering approach that uses molecular tools (enzymes and artificial nucleic acid constructs). In contrast, organisms that have been genetically altered by non-molecular approaches, for example, selective breeding that is based on naturally occurring genetic variability and stress-induced evolution, are not considered GMOs.

The next two sub-sections briefly summarize the main types of genetic engineering approaches to facilitate a more informed and granulated assessment of pros and cons of GMOs. DNA has the same chemical structure in all organisms—it is a polymer consisting of four nucleotides (A,T,C,G). The order in which these nucleotides are arranged in the polymer differs between species.

In addition to this sequential order of nucleotides, higher-order structures of DNA packaged into eukaryotic chromosomes and chromatin also differ between species. Thus, even though the primary chemical building blocks of DNA are identical in all organisms, species differences in sequence and higher-order DNA structure determine phenotypic differences between species.

Genomic DNA consists of genes and regulatory sequences outside of the protein-coding regions of genes. Genes encode proteins that determine the structure and function, that is, the overall phenotype (sum of all traits) of the organism. Regulatory sequences control how much of a gene is transcribed into mRNA, which is translated into the corresponding protein.

GMOs can be genetically engineered to:

1. knock out (functionally inactivate) an intrinsic gene or regulatory region;
2. mutate (change the DNA sequence) of an intrinsic gene or regulatory region;
3. add an extrinsic piece of DNA from the same species to the genome (e.g. to increase gene copy number or alter gene regulation); and
4. add an extrinsic piece of DNA (artificial or from a different species) to the genome.

Protein-coding genes can be targeted and edited (modified) using genetic engineering. But genes are not the only sequences in the genome that control phenotypes. There are other possible targets for genetic engineering. Non-coding DNA such as regulatory promoters, enhancers, silencers, repressors, insulators, and other genetic elements that coordinate the extent of expression of specific genes in specific tissues under certain environmental conditions can also be targeted and edited to change their function.

Genetic engineering of organisms is commonly performed on one- or two-cell stage embryos. At that early time of development, the genome is manipulated by delivering genetic engineering tools into the cell(s). Mosaicism (only *some* cells in the organism harbour a modified genome) can result, requiring crossbreeding after the initial manipulation. Moreover, for diploid organisms it is possible that one allele has been altered while the other remains wild-type (unaltered).

For the sake of brevity, I will not elaborate on those possibilities. The take-home message is that a single genetic engineering intervention performed on early embryos changes the genome permanently such that the organism developing from this embryo (and its offspring) represents a GMO strain.

Genetic engineering commonly uses recombinant DNA—a DNA polymer that contains at least two different DNA pieces in an artificial combination that is not found in nature. Often, these different pieces are derived from different species, for example, plasmids used as vectors in genetic engineering contain bacterial and eukaryotic (plant, animal, etc.) DNA.

For example, targeted knock out can be achieved by using recombinant DNA plasmids that consist of a mixture of bacterial DNA and various pieces of DNA from other organisms. Plasmids used for this purpose include resistance genes for bacterial propagation, bacterial regulatory sequences (origin of replication), regulatory sequences of eukaryotes (e.g. promoters), coding sequences for a bacterial nuclease, and sequences encoding small RNAs.

Some genetic engineering approaches do not utilize recombinant DNA as a tool. For targeted knock out of genes, it is possible to avoid introduction of recombinant DNA during genetic engineering by delivering small guide RNAs (sgRNAs) in combination with engineered bacterial proteins (nucleases) into embryos.

6.3.1 Genetic engineering by non-specific DNA insertion

One main approach of genetic engineering is non-specific insertion of foreign DNA into the genome. Non-specific DNA insertion can be spontaneous or enhanced by transposase enzymes. Transposases are enzymes that facilitate horizontal (lateral) gene transfer between different genomic loci and species.

Transposons (transposable elements, jumping genes) represent pieces of DNA that were first discovered in maize (McClintock, 1950; Ravindran, 2012). They are flanked by inverted terminal repeats (ITRs), which are short, repetitive DNA sequences that are common in genomes of many organisms and recognized by transposase enzymes.

For genetic engineering, a transposon (or plasmid encoding it) is delivered into cells or embryos along with a plasmid that contains a DNA sequence intended to be added to the genome (the genetic cargo), which is flanked by ITRs. The genetic cargo can contain additional copies of a wild-type gene, a gene from a different species, a dominant negative mutant of a gene, or a sequence encoding short antisense or interfering RNAs. This sequence is swapped into the genome by the transposase utilizing the corresponding ITRs (Figure 6.3).

Spontaneous integration of DNA lacking ITRs into the genome of a host cell is a rare but regular event happening in the order of one cell out of a million cells that have taken up that DNA (depending on cell type and DNA structure). Therefore, spontaneous integration is only feasible when millions of cells are amenable to efficient gene delivery and a selectable marker is included and retained in the stably integrated DNA fragment.

A selectable marker is a foreign gene encoded in the recombinant DNA plasmid and expressed in the host cell to generate a protein, which enables the cell to survive in the presence of a selection agent. A selection agent is a chemical inactivated or degraded by the selectable marker protein, which protects cells that have stably integrated the recombinant DNA containing the selectable marker gene.

All other cells are killed if cultured in the presence of the selection agent for several weeks since plasmid DNA is lost within a short period unless stably integrated into the genome. This method is limited to cell lines and organisms for which totipotent embryonic stem cells can be grown in vitro. For example, totipotent stem cells from mice can be cultured in large numbers in vitro and transplanted into embryos after selection of cells that have spontaneously integrated the DNA of interest and adjoined selectable marker.

Transposon-mediated integration of DNA containing ITRs is several orders of magnitude more efficient than spontaneous integration, which is why this approach is more suitable in combination with microinjection as a gene delivery method for early embryos. Other advantages of transposon-mediated integration of recombinant DNA include the ability to control the exact length and sequence of DNA that is integrated, and the ability to limit integration to genomic loci that contain the transposase-specific acceptor sequences (the ITRs).

However, because many ITRs are spread throughout the genomes of most organisms, a drawback

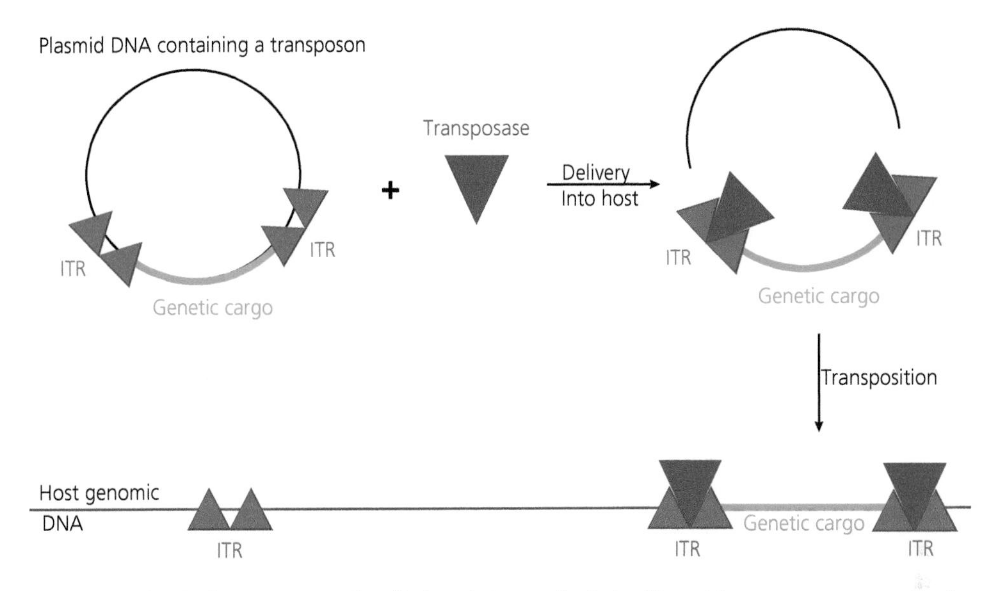

Figure 6.3 Genetic engineering by transposase-mediated horizontal gene transfer. A plasmid containing a transposon representing the genetic cargo (external DNA to be inserted into the genome) flanked by short, inverted terminal repeat (ITR) sequences is co-delivered with a transposase enzyme (or plasmid encoding it) into the host. ITRs are also present in many places of the host's genomic DNA and any of these places can serve as an acceptor site for the genetic cargo. Genetic cargo is transpositioned (swapped) between plasmid and genomic DNA by the transposase enzyme. Several transposase systems are commonly used in research for genetic engineering, including piggyBac, SleepingBeauty, and Tol2. Each of these systems consists of a transposase and transposase-specific ITRs.

in common with spontaneous integration is that copies of genetic cargo may be integrated concurrently into multiple genomic loci. Thus, the ratio of transpose versus genetic cargo plasmid molecules delivered into the host must be carefully optimized and transgene copy number should be determined in each GMO strain (e.g. by Southern blot).

In addition, transposase-mediated DNA insertion into host genomes has the potential side-effects of excising intrinsic transposable elements from the host genome at higher-than-normal rates. Moreover, spontaneous integration of plasmid vector outside the transposon flanked by ITRs is possible, but unlikely since the incidence of spontaneous integration is orders of magnitude lower than the frequency of transposition events.

6.3.2 Genetic engineering by gene targeting/editing

The second main type of genetic engineering used for creating GMOs is locus-specific gene targeting or gene editing. Nucleases are enzymes used for gene targeting/editing, and cut DNA to form a double strand break (DSB). There are three artificial DNA cleavage systems devised that consist of a nuclease and an element that targets the nuclease to a specific site in the genome.

These cleavage systems are zinc finger protein nucleases (ZFNs), transcription activator-like effector nucleases (TALENs), and clustered regularly interspaced short palindromic repeat (CRISPR)-associated (Cas) nucleases (Figure 6.4). ZFNs were developed before TALENs and CRISPR/Cas systems, and the latter is now the most-used gene editing system.

Gene editing occurs by directing nucleases to a specific genomic locus of interest where they cut DNA. ZFNs and TALENs are proteins that consist of a DNA binding domain that has been engineered to bind a specific DNA target sequence and a nuclease domain that cuts DNA.

Cas nucleases non-covalently interact with an sgRNA, which can be engineered to direct the nuclease to the desired target site in the genome. The DNA DSB resulting from the cut made by either of these DNA cleavage systems at the target site is repaired by the endogenous DNA repair machinery present in host cells.

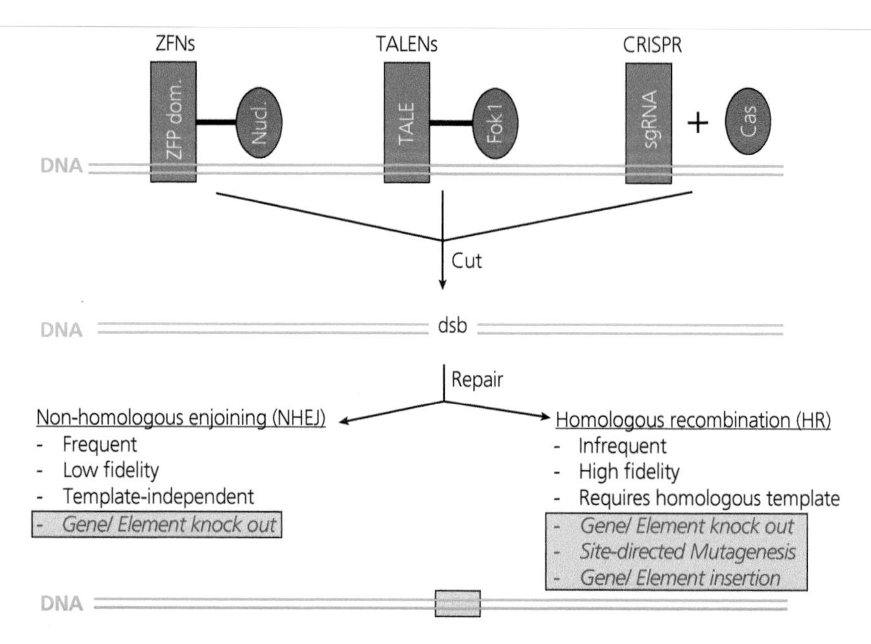

Figure 6.4 Genetic engineering by gene targeting with artificial nuclease-dependent DNA cleavage systems. Three DNA cleavage systems are shown in the order of their discovery (from left to right) with DNA-binding domain indicated in blue, and nuclease indicated in red. Zinc finger nucleases (ZFNs) are comprised of a zinc finger protein (ZFP) DNA-binding domain and a nuclease domain that are covalently fused into a single protein. TALE nucleases (TALENs) are also covalent fusion proteins harbouring a DNA-binding domain and a nuclease domain. For TALENs, the DNA-binding domain consists of repeats of transcription activator-like effectors (TALE) of bacteria that infect plants, and the nuclease domain is derived from the bacterial nuclease Fok1. In contrast to ZFNs and TALENs, clustered regularly interspaced short palindromic repeat (CRISPR)-associated (Cas) nucleases are not recombinant fusion proteins but bacterial proteins that interact with a separate small guide RNA (sgRNA) non-covalently. In the CRISPR system, the gRNA substitutes for the DNA-binding domain of nuclease fusion proteins. DNA double strand breaks (dsb) introduced at a specific genomic locus by either of these cleavage systems are repaired using host-specific DNA repair mechanisms (non-homologous end joining, NHEJ, and homologous recombination, HR). Plasmids used for gene targeting can be designed to achieve gene and genetic element mutagenesis or insertion if homology arms for HR flank the genetic cargo in the recombinant DNA plasmid and a negative selection system is used to reduce spontaneous integration at random loci.

Two main DNA DSB repair mechanisms can be used for repairing the cut: non-homologous end joining (NHEJ) is highly prevalent in most cell types of many species, does not require a template, and has low fidelity. The low fidelity of NHEJ results in insertions or deletions of one or more nucleotides (indels) at the cut site, which inactivates (knocks out) a gene if the reading frame is disrupted.

The reading frame of a gene refers to the arrangement of DNA nucleotides in triplets, each of which encodes a single amino acid in protein-coding genes. If a nucleotide triplet in DNA is mutated to a singlet or doublet resulting from indel formation, then the reading frame for producing the protein from that gene is disrupted.

CRISPR/Cas9 utilizing NHEJ to knock out a gene has been used in many species of fish, for example, the gene encoding the slc45a2 peptide transporter in Nile tilapia (Figure 6.5). Another example is the knock out of the myostatin gene, which encodes a protein that suppresses growth, in several species of fish.

Off-target effects can result from unspecific cleavage by the artificial nuclease system. For CRISPR/Cas systems, off-target effects can be minimized by optimizing the sgRNA sequence and the sgRNA-to-Cas nuclease molar ratio. If off-target effects due to indel formation at non-target loci disrupt a functional gene or regulatory element, then they may cause undesirable changes in phenotype.

The second DNA DSB repair mechanism utilized by host cells for repairing the DNA DSB made by artificial DNA cleavage systems is homologous

Figure 6.5 Red colour phenotype of a Nile tilapia (*Oreochromis niloticus*) for which the CRISPR/Cas9 system was used to target and knock-out the slc45a2 peptide transporter gene. The reddish slc45a2 knock out fish is shown below a typical grey wildtype control fish. This phenotype indicates pleiotropic effects on traits such as body colour other than the primary target trait (nutrient/peptide transport). This genetically engineered tilapia strain inherited the germline mutation to the offspring. Photo kindly provided by Dr Avner Cnaani (ARO, Beth Dagan, Israel).

recombination (HR). This mechanism is much less prevalent than NEHJ in many organisms, has high fidelity, and requires a template for repair. Diploid organisms (including humans and many other animals) have two copies of each chromosome and the template for HR repair is normally provided by the undamaged copy.

For genetic engineering, the HR repair template is provided in the form of homology arms, which are two pieces of DNA that are complementary to the regions flanking the DNA DSB targeted. A plasmid that contains genetic cargo flanked by the homology arms surrounding the cut site is delivered into the host along with the artificial DNA cleavage system and serves as the HR repair template. Any sequence enclosed between the two homology arms in the plasmid can be designed to either mutate the target sequence, insert additional copies of that sequence, or insert single or multiple copies of foreign sequences.

Because HR is generally much less prevalent than NHEJ, spontaneous integration of plasmid DNA into random (off-target) genetic loci is possible. Therefore, selection or reporter systems are employed when utilizing HR for repairing cuts induced by artificial DNA cleavage systems.

For cell lines (e.g. cultures of totipotent stem cells) a negative selectable marker (e.g. thymidine kinase) that is positioned on the plasmid but outside the region flanked by the homology arms is often used to kill cells that have undergone spontaneous, non-specific integration of the plasmid, rather than HR-mediated specific integration of the region flanked by the homology arms.

The negative selectable marker encodes a protein that interferes with an essential aspect of cell metabolism to induce cell death. For embryos, the negative selectable marker can be substituted with a reporter gene encoding a fluorescent protein (GFP) to rapidly screen for unwanted spontaneous integration into off-target loci.

6.4 GMOs in aquaculture

The genetic engineering approaches described here have been successfully (some extensively) used in research with many different species of fish. These approaches are essential for establishing cause–effect relationships between genotypes, specific environmental and developmental contexts, and phenotypes.

They enable researchers to dissect gene regulatory, protein, and metabolite networks that govern how organisms function in diverse environments. The use of genetic engineering for research is subject to strict regulation and confinement of GMOs and requires detailed protocol approval by commissions providing academic oversight.

Confinement of GMOs must be guaranteed, and the numbers of GMO individuals generated must not be higher than necessary for answering the specific research question. GMO broodstock is maintained in low numbers in some designated research facilities to be available for propagation and distribution to researchers who require these model organisms to study specific biological processes or diseases.

For example, GMO zebrafish lines are maintained at the European Zebrafish Resource Center (EZRC) to reduce the cost of research by avoiding duplication of effort for generating GMO lines with identical or very similar modifications of genomes.

Despite their common use for mechanistic research that establishes functions of genes in different environmental contexts and elucidates the scientific laws that govern how genotypes interact with environments to produce phenotypes, the application of genetic engineering for commercial aquaculture is very limited.

Unlike research institutions, most commercial aquaculture produces seafood for human consumption and thus food safety needs to be considered. In addition, confinement of GMO aquaculture strains is much more difficult to accomplish on a commercial scale than on a limited research scale.

Initial attempts of genetic engineering of aquatic animals date back as far as the late 1980s and have used spontaneous integration of mammalian growth hormone genes into salmonid and carp genomes. Later, fish growth hormone genes, sometimes in combination with strong heterologous (foreign) promoters, have been inserted into the genomes of salmonids, carps, catfishes, and tilapia.

There are advantages of GMOs that render them attractive for commercial aquaculture. A major advantage over traditional selection-based breeding programmes is that domestication of aquaculture species by genetic engineering is much faster as it does not rely on pedigrees and multiple generations. This advantage is particularly obvious for species with long generation times that require many years for reaching sexual maturity (e.g. salmon).

Genetic engineering has proven to increase production yields for several species while reducing costs for maintenance relative to harvest yield. Most commercial attempts to utilize GMOs for aquaculture have focused on this aspect to increase seafood production and profitability.

However, GMOs also have potential for reducing the environmental impact of aquaculture by increasing feed efficiency and altering metabolism to reduce waste. Genetic engineering of GMOs also has potential for increasing stress tolerance, immunity, and disease resistance, which would reduce the use of antibiotics and chemical baths and the cost of immunizations.

One aspect of ecological aquaculture where GMOs could contribute, and which is often overlooked, is the genetic engineering of microorganisms (decomposing bacteria and microalgae). Microorganisms have great potential for genetic engineering of metabolic pathways that increase the efficiency of bioremediation (wastewater treatment).

However, confinement of genetically modified microorganisms represents an even greater challenge than that of large plants and animals. Nevertheless, traditional selection of microbial genotypes already takes place in every biological filter and is driven by wastewater chemistry and biofilter substrate properties as major selection agents.

The only GMO aquaculture species that had been approved for human consumption in the US (2015) and Canada (2019) is a strain of Atlantic salmon (*Salmo salar*) produced by Aquabounty Inc. in 1989. This transgenic salmon contains a heterologous (foreign) growth hormone gene driven by a heterologous promoter.

These transgenic salmon grow much larger and faster than normal salmon. They are triploid and reproductively sterile, although a small fraction of transgenic fish remains diploid and can reproduce to propagate the transgene. There are also reports of hybridization between transgenic Atlantic salmon and wild Brown trout (*Salmo trutta*). Even though it took over a quarter century for approval, legal controversy over its marketing as seafood still prevails (Egelko, 2020).

A GMO tilapia strain (FLT01) also produced by Aquabounty has been approved in Argentina in 2018 without classifying it as a GMO (The Fishsite, 2018). The decision to not GMO label a genetically engineered strain of fish with a permanently altered genome clearly illustrates that public perception of GMO is unfavourable. However, limiting consumer choice by omission of information does not properly address the concerns associated with GMOs.

Arguing for non-GMO classification based on the premise that no DNA sequences were used as engineering tools disregards that other molecular tools (enzyme, RNA) were used for genetic engineering and that the genome of these organisms has been modified by insertions and/or deletions of nucleotides that alter certain genomic loci.

Such genetic modification disrupts the function of specific genes and can alter how gene regulatory networks operate just as effectively as transgenes. Side effects may include a change in invasiveness (stress tolerance), change in disease susceptibility and pathogen transmission, and changes in metabolism altering the nutritional value. Such indirect, context-specific changes in GMOs reach beyond the immediate genetic locus that is altered. They may be advantageous or detrimental regarding their value for humans.

It is impossible to predict all potential consequences of a specific genetic engineering modification for all environmental and developmental contexts that the organism can experience. Thus, the risks and benefits need to be assessed and weighed carefully when developing GMOs. Corresponding evaluation and assessment processes include a significant statistical element that considers the most likely development × environment × genome interactions, not unlike clinical trials of new pharmaceutical drugs or vaccines.

Development of better assessment procedures for evaluating GMO risks and benefits that are transparent, unbiased, and instil consumer confidence represents a key requirement for commercialization of GMO technology. Moreover, risks and benefits of GMOs should always be assessed relative to risks and benefits associated with the corresponding parental non-GMO strain(s).

While GMOs have significant potential for increasing the efficiency of seafood production for human consumption, the use of this technology for commercialization of ornamental GMOs is debatable. GloFish® contain transgenes encoding fluorescent proteins and are produced for sale to hobby aquarists. Even though there are no food safety concerns associated with ornamental GMOs, producing them meets a luxury demand, rather than being a necessity.

Given the rapid research developments in the field of genetic engineering it is likely that better approaches will continue to become available. Replacement of outdated GMO strains with better strains should always be an option, which requires effective confinement regulation and enforcement. Escapes of aquaculture GMOs represent a concern as competitors and predators of native species that could alter the balance of wild ecosystems.

Commonly referred to ecological risks of GMOs include their superior abilities for inter—and intraspecific competition and interbreeding with conspecifics or congeners resulting in undesirable consequences of escapes from captivity (Chao and Liao, 2007). For instance, GMO salmon containing an extra copy of growth hormone outgrew and outcompeted wild-type salmon when kept in the same enclosure by showing strong agonistic feeding and cannibalistic behaviours (Devlin *et al.*, 2004).

Even if complete confinement of GMOs were possible, other concerns need to be addressed. Concerns about food safety, animal welfare, and property rights add to the anxieties about unwanted interactions of farmed and wild species. All of these concerns are reflected in low consumer acceptance of GMO products.

Concerns about side effects of genetic engineering prevail even in the absence of the genetic off-target effects discussed earlier. They arise from pleiotropic effects of genetic loci and the regulation of transcriptional networks as integrated systems rather than on a per-gene basis.

Pleiotropic effects, that is, effects of manipulation of a gene on traits other than the intended trait, are common as genes of complex organisms are multifunctional and (with few exceptions) phenotypes (traits) are complex and multigenic. The degree to which pleiotropic effects are manifested depends on developmental and environmental contexts and the specific interactions of the manipulated gene with other genes within gene regulatory networks.

For example, transfer of a growth hormone gene to create transgenic GMO fish does not only affect growth, but also body shape and composition, head morphology, feed efficiency, metabolic rate, disease resistance, reproduction, stress tolerance, carcass yield, osmoregulation, sexual maturity swimming ability, predator avoidance, and other traits (Pillay and Kutty, 2005). Pleiotropic effects on nutritional quality include decreased body protein, lipid, and energy content and increased water content per unit mass (Molfese, 2015).

Pleiotropic effects have the potential to impair animal welfare. For example, in transgenic salmon and other fish expressing an extra copy of a growth hormone gene, reports exist of acromegaly and other skeletal and muscle deformities (Frewer *et al.*, 2013). There are also reports of compromised swimming ability, feeding behaviour, and hypoxia (low oxygen) tolerance in transgenic fish (Kaiser, 2002).

Food safety concerns about GMOs are mainly centred around the unpredictability and context-specificity of pleiotropic effects of genetic engineering on metabolism, which could theoretically promote the accumulation of allergenic or toxic metabolites. Even though pleiotropic effects can also be beneficial, predicting and comprehensively analysing them is a daunting task. They may manifest themselves only in specific developmental and/or environmental contexts and may not be immediately evident as obvious organismal phenotypes.

A comprehensive diagnosis of pleiotropic effects requires comparative, systems-level analyses of transcriptome (mRNA complement of cells), proteome (protein complement of cells), and metabolome (metabolite complement of cells) networks in appropriate environmental and developmental contexts. These molecular phenotype networks are suitable for comprehensively mapping pleiotropic effects because they consist of thousands of molecular network nodes that govern whole-organism phenotypes including physiology, morphology, and behaviour. Such analyses should be performed in GMOs and their corresponding parental strains as part of each assessment strategy.

Overall, genetic engineering technologies are powerful tools with high potential for improving aquaculture and other production systems. However, comprehensively understanding the impact of these technologies on domestication of aquaculture animals requires significantly more research, in particular research that elucidates how manipulation of specific genetic loci affects the behaviour of gene regulatory networks, molecular and organismal phenotypes, and genome evolution on a system-wide scale.

Large knowledge gaps remain that mandate a precautionary approach to commercialization of GMO technologies (Molfese, 2015). Whether GMOs represent a better solution for ecologically sustainable aquaculture with a lower risk/reward ratio than alternative solutions is difficult to evaluate and requires elaborate assessment on a case-by-case basis. Such assessment is especially important to avoid the creation of additional problems, like ecological invasiveness, which may be impossible to remedy and could overshadow any benefits in the long term.

Utilization of GMO technologies for commercial enterprises has the potential to irreversibly alter wild ecosystems, and, like the enthusiastic and uncontrolled exports of gametes and embryos for many aquaculture species that started more than a century ago and continued for several decades, should be avoided. Gamete and embryo exports have resulted in large shifts of species composition in many wild habitats and invasive species issues that are irreversible, (e.g. for tilapia, salmonids, carp, mussels, crayfish, and many others).

6.5 Domestication effects during stock enhancement and sea ranching

Stock enhancement and sea ranching are two forms of aquaculture that differ from other forms of aquaculture. Both aquaculture practices have in common that they are limited to raising early developmental and juvenile stages of aquatic species in hatcheries, which are then released into wild habitat.

Stock enhancement has the objective to supplement wild stocks of aquatic organisms that have been depleted, are threatened, or endangered. The goal of stock enhancement is usually either to augment capture fisheries or conservation of the depleted species or population. Marine fishes are commonly used for stock enhancement, but corals and other marine invertebrates are also cultured for this purpose (Section 9.2.3).

Stock enhancement programmes are usually non-profit enterprises that are funded by state or federal governments. Examples include salmon stock enhancement programmes in the US and Canada designed to counteract the loss of spawning habitat from construction of hydroelectric dams.

Sea ranching is normally performed with migratory species that display strong homing behaviour. After breeding and raising offspring to the juvenile stage in a hatchery, they are released at a convenient location, ideally close to the hatchery to minimize handling and transport stress. The homing behaviour of migratory species then causes these released animals to return to or near the site of release, where they are harvested.

Unlike stock enhancement programmes, sea ranching aquaculture is usually practised by producing seafood that can be harvested after a period of grow-out in natural habitat. Sea ranching is most common for anadromous fishes that spawn and develop in freshwater and grow-out in the ocean, including several salmonid species (Chapter 14).

One of the drawbacks of sea ranching compared to more conventional forms of aquaculture is that ownership cannot be enforced during the grow-out period and animals may be subject to capture fisheries by parties other than the owners of the sea ranching hatchery.

Both stock enhancement and sea ranching have the opposite goal regarding confinement than other forms of aquaculture; the objective is to release all animals that are being produced. Since confinement is not an option for stock enhancement and sea ranching, the influence of domestication of aquaculture stocks used for these forms of aquaculture on wild stocks must be minimized by other means.

Concerns about genetic dilution and alteration of wild populations are widespread and need to be addressed when releasing large numbers of juveniles for a particular species into natural ecosystems. These concerns are warranted because domestication effects on aquatic species have been observed in only a few generations.

Moreover, phenotypic plasticity (imprinting of behaviour, morphology, and/or physiology during development) can affect fitness in only a single generation. Therefore, releasing large numbers of juveniles raised in a human-controlled environment introduces domesticated phenotypes into wild stocks and alters allele frequencies within the gene pool of natural populations.

An allele represents a specific variant of DNA sequence at a given genomic locus (specific site in the genome), which affects either the regulation of that locus, the function of the corresponding protein if the locus is within the coding region of a gene, or the interaction of that locus with other loci. For diploid organisms who harbour two copies of the genome, any individual can either have two identical alleles (homozygous) or two different alleles (heterozygous) at any given locus.

To illustrate this, individual 1 may harbour alleles A and B (heterozygous), while individual 2 harbours alleles C and C (homozygous), and individual 3 harbours alleles C and D (heterozygous), and so on, all at the same genomic locus. Thus, within an entire population consisting of many individuals there may be many different alleles (genetic variants) for a given genomic locus.

Allele frequency refers to the representation of specific alleles (A, B, C, D in the example) in the entire population. This frequency will change greatly if a nearly extinct wild population consisting of few individuals is swamped with many hatchery-reared offspring that all originate from the same parents that are both homozygous for the same allele. For example, if both parents harbour alleles D and D, then all hatchery-reared offspring will harbour alleles D and D and allele frequency will shift greatly in favour of D.

If the frequency of specific alleles is favoured in hatchery-raised offspring, then those alleles will be over-represented and all other DNA sequence variants (alleles) at that locus will be underrepresented in admixed populations. Thus, the gene pool in the wild population would be diluted by swamping out genetic variation in favour of over-represented alleles.

Depending on what genes are affected, differences in allele frequencies may ultimately alter organismal phenotypes (physiological, morphological, behavioural traits) including fitness, competitiveness, disease resistance, and other traits that could impact the dynamics of the entire ecosystem.

Concerns about genetic dilution/alteration of natural populations by stock enhancement or sea ranching are highest when the corresponding natural population is small, relative to the number of hatchery-raised offspring released. This effect represents a dilemma for conservation efforts that could backfire for highly depleted or nearly extinct populations when stock enhancement is used.

Habitat restoration that promotes natural reproduction always represents the better option for conservation of such populations. Even if stock enhancement is used, habitat restoration and properly mitigating the primary causes for population decline represents a prerequisite for long-term success of any stock enhancement programme.

To minimize negative consequences of stock enhancement and sea ranching aquaculture on wild populations several management precautions can be implemented. These include an assessment of the carrying capacity of the ecosystem targeted for release and a thorough understanding of the reasons for the decline of the population to be supplemented with hatchery-reared offspring (Drawbridge, 2002).

The carrying capacity of the ecosystem refers to the availability of habitat and resources of sufficient quality and quantity to accommodate the mass release of juveniles produced in hatcheries. The ecosystem carrying capacity is dynamic, difficult to assess, and must be analysed on a case-by-case basis as comprehensively as possible.

In addition to ecosystem carrying capacity, a thorough understanding of the population structure, genetic variability, and effective population size (individuals contributing to reproduction) of the wild population to be supplemented should be developed prior to restocking efforts. This guides the design of breeding and husbandry practices for restocking programmes.

In general, as many different broodstock animals as possible (at least hundreds) should be used to produce hatchery offspring and they should be mated as single pairs. Furthermore, collection of broodstock animals is preferred from different locations and at different times in habitats where the population targeted for release occurs. This praxis maximizes the representation of the gene pool (genetic variety) of the target population in the broodstock used for hatchery production of offspring.

If the target population is extinct, then broodstock can be collected from a population that occupies similar habitat. In addition, broodstock for restocking programmes should always be collected from the wild and never propagated beyond a single generation of offspring. Genotyping broodstock used for restocking enables assessment of the contribution from each parent to the gene pool and permits monitoring of restocking effects on genetic diversity of wild populations by using genetic markers.

While genetic markers are used to identify hatchery-raised fish, various physical markers, including physical tags, fin clips, and internal transponders, also play a part. However, recent genetic markers are most reliable for identification, and they do not require any physical alteration of the organisms that could potentially impact their fitness.

The most complete and informative genetic marker is the whole genome sequence, which can be attained from just a tiny clip of a fin or other tissue. Every individual has a unique genome sequence and allelic variation at all (or almost all) loci can be used to identify individuals.

GWAS compare complete genomic sequences in multiple individuals that differ in specific phenotypes and correlate variation at specific genetic loci with those phenotypes. Although it is becoming less expensive to generate whole genome sequences, this complete genetic marker is still not suitable for routine high throughput genotyping of thousands of individuals. Therefore, other genetic markers are currently more commonly used, including allozymes, restriction fragment length polymorphisms, minisatellites, and microsatellites.

Microsatellite markers are now the most popular genetic markers as they represent the best compromise between cost and specificity. Microsatellites are repetitive genomic loci that are not part of protein-coding genes. They represent short (two to five base pair) repeats in genomic DNA.

DNA is a polymer consisting of the four different bases adenine (A), guanine (G), cytosine (C), and thymine (T). Each base is paired with its complementary base (A with T, and G with C) in a double-stranded DNA polymer (hence base pairs).

The order of arrangement of these base pairs determines the DNA sequence. A microsatellite example would be the sequence ATCATCATCATC (paired with the complementary sequence TAGTAGTAGTAG into a DNA double-strand). This example represents a 4× repeat of the sequence ATC.

Microsatellite loci are highly polymorphic meaning that the number of repeats varies between different alleles. For example, in genome A, a microsatellite locus could consist of four repeats (ATCATCATCATC) while in genome B the same locus comprises five repeats (ATCATCATCATCATC) and in genome C that locus comprises only two repeats (ATCATC).

The difference in sequence length at a microsatellite locus can be identified and used for diagnostic purposes (e.g. twelve base pairs for genome A, fifteen base pairs for genome B, and six base pairs for genome C in the example). The preceding example assumes haploid genomes containing only a single copy for each locus. For diploid genomes two variants are possible in each individual for each microsatellite locus, for example, 12 and 15 bp, or 6 and 12 bp, or 12 and 12 bp, etc., for the example locus.

Microsatellite loci are highly polymorphic (variable) in the number of short repeats they contain because DNA polymerases, the enzymes that copy DNA during cell division (mitosis and meiosis) are prone to 'slipping' when encountering redundant sequence stretches consisting of short repetitive DNA. The incidence of such DNA polymerase 'slipping' is high enough to introduce significant polymorphism into a large population (on a short evolutionary time scale) but low enough to conserve stable repeat numbers at a given locus over several generations.

Typically, ten or more microsatellite loci are combined into a genetic marker, which results in very high statistical probability of identifying a specific individual in a population consisting of many millions. Monitoring programmes based on genetic markers are designed to be very powerful and effective in assessing dynamic changes of population diversity in wild populations and tracking the genetic contributions of hatchery-reared offspring.

However, monitoring changes in genetic diversity and managing (correcting) them represent two very different sides of the coin. Reversing adverse effects of a restocking programme on the gene pool may be virtually impossible. More research on the potential and rate for generating evolutionary novelty in stressful, suboptimal environments is urgently needed for species that are endangered and/or considered for restocking programmes.

Bert and colleagues (2007) suggested that predicting the effects of restocking programmes on the genetic diversity and fitness of natural populations is impossible. Therefore, restocking should only be considered a viable option when it is essential to maintain the population under consideration and all other options have been exhausted (Hedgecock and Coykendall, 2007).

Key conclusions

- Domestication is the change in genotype and phenotype of a population that results from the different selection pressures experienced in captivity compared to those found in natural habitats.
- Domestication causes significant genetic (and epigenetic) changes in aquatic animals that manifest themselves within just a few generations.
- Domestication is often biased towards artificial selection of production traits such as growth, but also has potential for reducing the ecological footprint of production and improving animal welfare.
- Traditional breeding-based artificial selection utilizes phenotypes, crossbreeding, and/or genetic markers (genotypes) that are highly correlated with desirable phenotypes (traits).
- Desirable traits can be rapidly induced during domestication by imposing environmental stress on gametes to alter and select karyotypes by manipulation of chromosomes and ploidy (chromosome copy number).
- Genetic engineering represents a rapid way of manipulating the genome during domestication by using molecular tools that permanently alter the DNA of the engineered embryo and its offspring.
- Two types of genetic engineering are common: 1) insertion of foreign DNA into the genome, which can be random or transposase-mediated, and 2) gene targeting/editing, which requires targeted nucleases.
- Genetic engineering creates genetically modified organisms (GMOs), which are invaluable models for

basic biological and biomedical research to establish causality between genotype and phenotype.

- The commercial production of GMOs has certain benefits, but these need to be weighed on a case-by-case basis against many pitfalls; pleiotropy of genes renders assessment of GMO impacts on wild species, ecosystems, and food safety difficult, which is reflected in low consumer acceptance.
- Stock enhancement and sea ranching differ from other forms of aquaculture by releasing, rather than confining, hatchery-raised offspring; they require a very elaborate genetic management plan.
- Post-release monitoring of genetic diversity of wild populations subject to restocking is facilitated by physical and genetic markers but predicting impacts of restocking on wild populations is not possible.

References

Argue, B. J., Arce, S. M., Lotz, J. M., and Moss, S. M. (2002). 'Selective breeding of Pacific white shrimp (*Litopenaeus vannamei*) for growth and resistance to Taura Syndrome Virus', *Aquaculture, Genetics in Aquaculture VII*, 204, pp. 447–460. https://doi.org/10.1016/S0044-8486(01)00830-4

Bert, T. M., Crawford, C. R., Tringali, M. D., Seyoum, S., Galvin, J. L., Higham, M., and Lund, C. (2007). 'Genetic management of hatchery-based stock enhancement', in Bert, T. M. (ed.) *Ecological and genetic implications of aquaculture activities, methods and technologies in fish biology and fisheries*. Dordrecht: Springer, 123–174. https://doi.org/10.1007/978-1-4020-6148-6_8

Chao, N.-H. and Liao, I. C. (2007). 'Sustainable approaches for aquaculture development: looking ahead through lessons in the past', in Bert, T. M. (ed.) *Ecological and genetic implications of aquaculture activities, methods and technologies in fish biology and fisheries*. Dordrecht: Springer, 73–82. https://doi.org/10.1007/978-1-4020-6148-6_4

Devlin, R. H., D'Andrade, M., Uh, M., and Biagi, C. A. (2004). 'Population effects of growth hormone transgenic coho salmon depend on food availability and genotype by environment interactions', *Proceedings of the National Academy of Sciences of the United States of America*, 101, pp. 9303–9308. https://doi.org/10.1073/pnas.0400023101

Drawbridge, M. A. (2002). 'The role of aquaculture in the restoration of coastal fisheries', in Costa-Pierce, B. A. (ed.) *Ecological aquaculture*. Oxford: Blackwell Science, 314–336. https://doi.org/10.1002/9780470995051.ch11

Dunham, R. (2019). 'Genetics', in Lucas, J. S., Southgate, P. C., and Tucker, C. S. (eds.) *Aquaculture: farming aquatic animals and plants*. Oxford: Wiley-Blackwell, 127–155.

Dunham, R. A. (2011). *Aquaculture and fisheries biotechnology: genetic approaches*, 2nd edn. Oxford: CABI.

Egelko, B. (2020). 'Genetically engineered salmon production illegally approved by FDA, judge in S.F. rules', SFChronicle.com [online].

Frewer, L. J., Kleter, G. A., Brennan, M., Coles, D., Fischer, A. R. H., Houdebine, L. M., Mora, C., Millar, K., and Salter, B. (2013). 'Genetically modified animals from life-science, socio-economic and ethical perspectives: examining issues in an EU policy context', *New Biotechnology*, 30, pp. 447–460. https://doi.org/10.1016/j.nbt.2013.03.010

Hedgecock, D. and Coykendall, K. (2007). 'Genetic risks of marine hatchery enhancement: the good, the bad, and the unknown', in Bert, T. M. (ed.) *Ecological and genetic implications of aquaculture activities, methods and technologies in fish biology and fisheries*. Dordrecht: Springer, 85–101. https://doi.org/10.1007/978-1-4020-6148-6_5

Heng, H. H. (2019). *Genome chaos: rethinking genetics, evolution, and molecular medicine*. London: Academic Press.

Kaiser, F. (2002). 'Anticipation in migraine with affective psychosis', *American Journal of Medical Genetics*, 110, pp. 62–64. https://doi.org/10.1002/ajmg.10382

McClintock, B. (1950). 'The origin and behavior of mutable loci in maize', *Proceedings of the National Academy of Sciences of the United States of America*, 36, pp. 344–355.

Mojica, E. A., and Kültz, D. (2022). '*Physiological mechanisms of stress-induced evolution*'. *Journal of Experimental Biology*, 225, jeb243264. 1-13. https://doi:10.1242/jeb.243264.

Molfese, C. (2015). 'A review of genetic engineering in current aquaculture practices', *Journal of Aquaculture & Marine Biology*, 2(1), pp. 00013. https://doi.org/10.15406/jamb.2015.02.00013

Pillay, T. V. R. and Kutty, M. N. (2005). *Aquaculture: principles and practices*, 2nd edn. Oxford: Wiley-Blackwell.

Pullin, R. S. V., Froese, R., and Pauly, D. (2007). 'Indicators for the sustainability of aquaculture', in Bert, T. M. (ed.) *Ecological and genetic implications of aquaculture activities, methods and technologies in fish biology and fisheries*. Dordrecht: Springer, 53–72. https://doi.org/10.1007/978-1-4020-6148-6_3

Ravindran, S. (2012). Barbara McClintock and the discovery of jumping genes. *Proceedings of the National Academy of Sciences*, 109, pp. 20198–20199.

The Fishsite (2018). 'Gene edited tilapia secure GMO exemption'. https://thefishsite.com/articles/gene-edited-tilapia-secures-gmo-exemption (accessed 15 January 2021).

PART 2

Biology and Culture of Aquatic Species

CHAPTER 7

Overview of aquaculture species diversity

'Genomes are more dynamic than had been assumed, lineages less coherent through time, evolution less tree-like.'

Ragan, McInerney, and Lake, in *Phil. Trans. Royal Soc. London B, vol. 364* (2009)

7.1 Introduction

Part II of this book illustrates how the general principles, trends, challenges, and opportunities outlined in Part I are practically applied to the culture of select economically important aquatic species. The biology and ecological context of these species in their natural habitat is compared to the conditions under which they are maintained during domestication in aquaculture systems. This comparison illustrates the varying demands of different species and how they are being met in aquaculture by utilizing a range of aquaculture systems and approaches (extensive versus intensive, open flow-through versus closed recirculation, etc.).

Opportunities for increasing the ecological sustainability of aquaculture approaches for key species are highlighted. Although the beginning of Part II emphasizes the importance of plant aquaculture, its focus is on animals to reflect their current dominance in aquaculture. It begins with a general overview of aquaculture species' placement within the network of life, followed by a discussion of select aquaculture species starting at the base of this network. The selection of aquaculture species considered is not exhaustive but is built on their economic relevance and suitability for exemplifying and illustrating the general concepts introduced in Part I.

7.2 Basic concepts of taxonomy and phylogenetics

We humans tend to classify all things around us to keep track of our experiences and structure knowledge in a way that facilitates learning and is most accessible and useful for us. This is how the human brain works, and taxonomy (the science of classifying life forms into groups, or taxa) is no exception. Like any other classification system, taxonomy simplifies knowledge by introducing rules that streamline observations. The trade-off for gaining the ability to comprehend large amounts of information is that exceptions to and deviations from the postulated rules may be neglected or misinterpreted. Think of classification systems as compressing information, not unlike the ways in which digital data are compressed into zip files or raw images are compressed into jpeg files by applying a set of rules that simplify information.

Carl von Linné devised the first biological taxonomic system and also introduced the unambiguous binomial nomenclature that refers to a species by its genus name followed by the species suffix (Linné, 2003). The dual species name is in Latin and italicized by convention, for example, *Salmo salar* is Latin for saltwater salmon, which denominates Atlantic salmon. Taxonomy utilizes knowledge generated by phylogenetics, but these

A Primer of Ecological Aquaculture. Dietmar Kültz, Oxford University Press. © Dietmar Kültz (2022). DOI: 10.1093/oso/9780198850229.003.0007

two scientific disciplines are not identical. While taxonomy is mainly concerned with the generation of the most intuitive classification system for all life forms, phylogenetics (systematics, cladistics) is primarily concerned with the study of evolutionary (ancestral) relationships between life forms.

Ideally, classification systems should closely reflect phylogenetic (evolutionary) relationships between different life forms and (for the most part) this is the case, but there are exceptions. For example, many prokaryotes are phylogenetically much more distant from each other than all eukaryotes differ from each other, yet prokaryotes are commonly classified into only two kingdoms (archaebacteria/archaea and eubacteria/bacteria) while eukaryotes are often classified into four kingdoms (protista, fungi, plants, animals) (Figure 7.1).

Nevertheless, research on the phylogenetic relationships between life forms (phylogenetics) greatly informs taxonomy and contributes to steady improvements of the classification system of life. Species are the smallest major category used in taxonomy although sub-species and populations are sometimes distinguished. Figure 7.1 shows the six major taxonomic categories above the species level.

The boundaries between taxonomic categories are not clearly discernible. Rather they are fluid and somewhat arbitrary, that is, they are based on definitions that represent widely agreed upon scientific conventions. Even the species concept is not unambiguous and boundaries (e.g. hybrids and transitional states) between populations and species on the one hand, and species and genera on the other hand are somewhat fluid rather than rigid. However, a discussion of these complex concepts, which include speciation by karyotype evolution, exceeds the scope of this book (Wilkins, 2018, Heng, 2019).

Phylogenetics is increasingly based on DNA sequence comparisons, either whole genome sequence comparison or comparison of closely related (homologous) regions of genomes. Phylogenetics studies that utilize DNA sequence (genotype) comparisons for deriving ancestral (phylogenetic) relationships between different species are now much more common than those that utilize phenotype comparisons. In contrast to genotypes, phenotypes are any traits that constitute the structure and function of organisms other than DNA sequence, including morphological appearance, behavioural patterns, and physiological functions. Phenotypes differ in scale ranging from molecular (e.g. the abundance of specific proteins) to organismal (e.g. body colour). Until the mid-twentieth century, morphological phenotypes such as the colour and shape of specific body parts

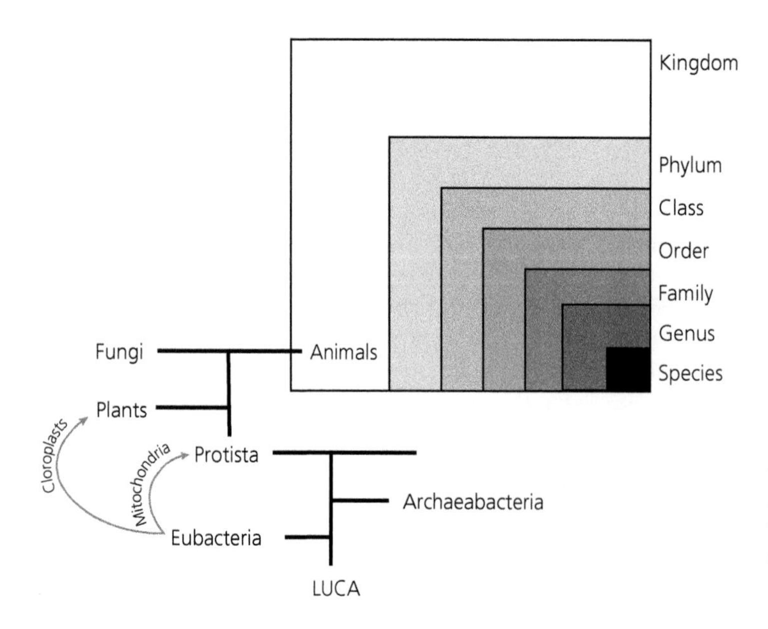

Figure 7.1 Simplified phylogenetic relationship of the six kingdoms of life (Eubacteria, Archaebacteria, Protista, Plants, Fungi, and Animals). LUCA = Last Common Universal Ancestor. Major taxonomic classification categories are illustrated for the animal kingdom, i.e. a kingdom consists of phyla, a phylum consists of classes, and so on. The boxes representing the major taxonomic categories are not drawn to scale. For instance, the species box is much larger than what it should be for a single species as the animal kingdom comprises several million species.

or exoskeletons were used as the main criteria for both phylogenetics and taxonomy.

The *'tree of life'* is a metaphor that is commonly used to describe phylogenetic relationships of organisms to justify their taxonomic classification. This classical metaphor reflects the historical trajectory of the evolution of different life forms properly for the most part but has been subject to gradual refinement ever since Barbara McClintock's ground-breaking discovery of transposable elements, or transposons (McClintock, 1950). The branches of the tree of life are not just splitting into ever more lineages over time like the branches of a growing tree. Instead, some branches are also linked by connectors that render the appearance of the phylogeny more like a network than a tree. What are those connectors giving rise to a *'network of life'*?

There are two mechanisms that comprise branch connectors: introgression by hybridization of two distinct species, and transposon- or plasmid-mediated horizontal transfer of hereditary information (DNA or RNA) between discrete species. For example, transposons (DNA transfer) and retrotransposons (RNA transfer and reverse transcription to DNA) that were acquired by horizontal gene transfer and then multiplied make up half of our (human) genome. Thus, transfer of genomes from parent to offspring is not the only mode of transferring hereditary information and phylogenies resemble networks more than trees. What does the network of life look like? An answer to this question requires more research on horizontal links (branch connectors) between taxa. Figure 7.2 provides a very crude and extremely simplified schematic of the timeline of evolution of the major clades (groups) of organisms.

This figure also depicts an example for yet another mechanism giving rise to lateral branch connectors in phylogenetic trees: endosymbiosis, which likely gave rise to the first primordial ancestors of eukaryotes (Martin *et al.*, 2015). Because of common basic features that are shared between all extant life forms, many biologists support the hypothesis that they were all derived from a single ancestor, the Last Universal Common Ancestor (LUCA) (Futuyma and Kirkpatrick, 2017).

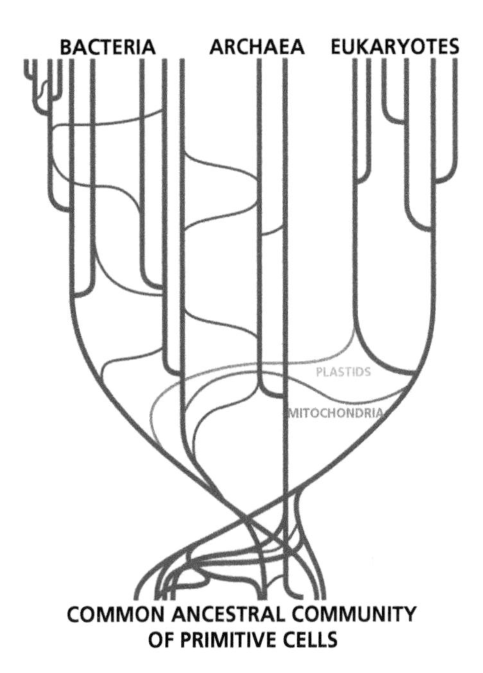

Figure 7.2 A simple network depicting the phylogenetic (ancestral) relationships between the three major domains of life (bacteria = eubacteria, archaea = archaebacteria, eukaryotes = eukarya). Domains are a classification category above the level of kingdoms. Organisms are related by 1) vertical transcend from their parental ancestors and 2) lateral transfer of hereditary material (DNA, RNA) such as introgression through hybridization, transposon-mediated horizontal gene transfer, plasmid-mediated horizontal gene transfer, or endosymbiosis. Endosymbiosis gave rise to eukaryotes (bacterial endosymbiont evolved into mitochondria) and plants (bacterial endosymbiont evolved into chloroplasts/plastids). Reproduced from Wikimedia under CC BY-SA 4.0 (Smets and Barkay, 2005).

7.3 Aquaculture species within the network of life

The number of animal species used in aquaculture is far greater than those used for any other commercial farming or as non-aquatic pets. The number of commercial aquaculture species exceeded 350 in 2017, an almost tenfold increase since 1950 (Figure 7.3). In praxis this number is even higher as many species are not counted but grouped in the category not elsewhere included (nei), which contains hybrids and species that have only been classified at the genus or higher taxonomic levels. The Food and Agriculture Organization of the United Nations (FAO) statistics recorded 622 'species items' used in commercial aquaculture in 2018, which

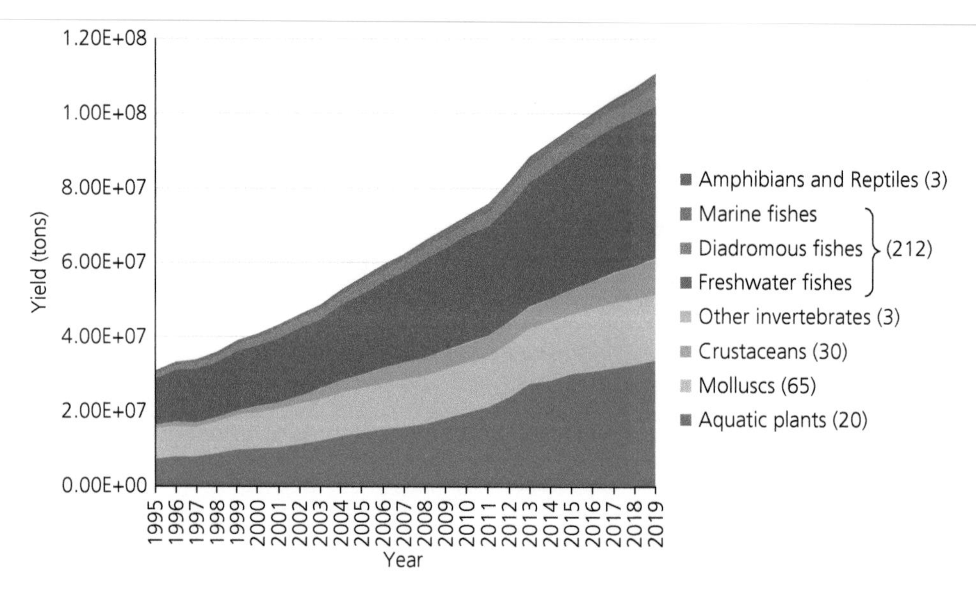

Figure 7.3 Yield of aquaculture products from major taxa. The number of aquaculture species reported for each taxon is indicated in parentheses for the year 2017. Only species used for seafood production were included. Species classified under the category "not elsewhere included" (nei) were not considered even if they are produced as seafood because their species classification is unreliable. *Data sources: OECD.stat and Metian et al. (2020).*

include 466 individual species, seven interspecific hybrids of fish, and ninety-two, thirty-two, and twenty-five species groups classified at the higher taxonomic levels of genus, family, and order, respectively. These numbers represent conservative approximations that underestimate the actual number of species used in commercial aquaculture by several hundreds, even when not considering many species that are either not reported, categorized as 'not elsewhere included' (nei), represented as one of thousands of ornamental species, used for aquaculture research experiments, kept in home aquaria, or cultivated as live food (FAO, 2020).

The large number of aquatic species considered for commercial aquaculture is partly due to the great diversity of aquatic environments and climatic conditions and the rich aquatic species diversity in different parts of the world where aquaculture has been practised for many centuries. In general, diversification of species and other aspects of aquaculture is considered beneficial for its sustainability as it increases the resilience of ecological system performance, that is, the capacity to persist in a robust state when conditions change. However, not all species are well suited for aquaculture and numerous attempts of culturing a new species have failed such that aquaculture efforts had to be abandoned. It has been estimated that aquaculture efforts for approximately 25% of all species tested have failed and have been abandoned within a period of five years (Teletchea and Fontaine, 2014). The reasons for such failure often represent a combination of lack of knowledge about the biology of the species, suboptimal conditions in captivity, and unfavourable economic factors.

Although there are many more commercial aquaculture species than terrestrial agriculture species (in particular for animals), they only represent a tiny fraction of the diversity of life on earth. But compared to agriculture, aquaculture animals are much more diverse not only in species number, but also in phylogenetic diversity of the species used. In other words, aquaculture organisms occupy a greater territory within the network of life than do terrestrial agriculture species.

Most aquaculture organisms are metazoans (multicellular animals). The remainder are aquatic plants such as macroalgae (Chapter 8), not considering the many species of decomposing bacteria and microalgae that are cultured in biological filters, turf scrubbers, bioflocs, and other tools used for wastewater treatment. Aquaculture

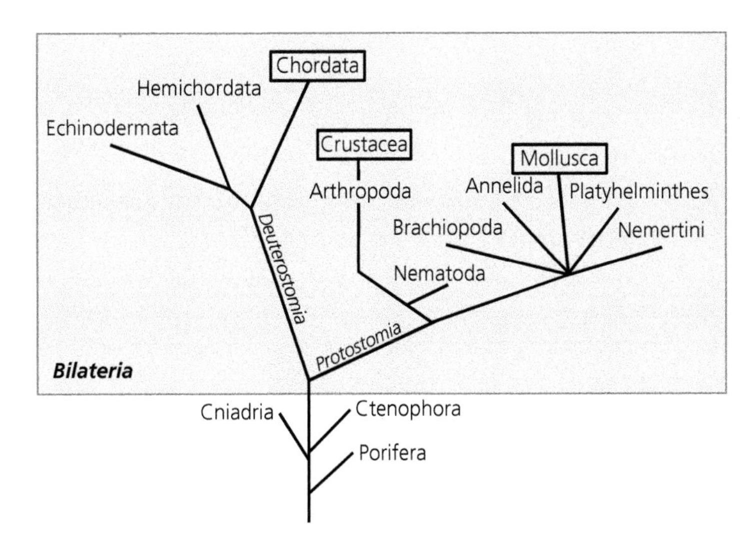

Figure 7.4 Phylogenetic tree of the animal kingdom based on 18S ribosomal RNA (rRNA) sequence divergence between animal phyla (Adoutte *et al.*, 1999). Note that lateral branch connectors to other forms of life are not considered in this common way to simplify the depiction of phylogenetic relationships. The three major animal taxa used for aquaculture are enclosed in boxes. Fish are a class of the phylum chordata.

animals represent many of the extant animal phyla, including porifera (sponges), cnidarians (corals, anemones, jellyfish), polychaete and other worms, molluscs (oysters, mussels, clams, snails), crustaceans (shrimps, prawns, crabs, crayfish), echinoderms (sea urchins, sea cucumbers), and chordates (fishes, frogs, crocodilians) (Figure 7.4).

Much of the aquaculture of molluscs, crustaceans (a subphylum of arthropods), echinoderms, and fish accounts for producing seafood. However, not all aquaculture animals are produced as seafood for human consumption. Rather, their use for aquaculture is based on alternative objectives (Chapter 3). Sponges are primarily produced for their skeletons and for harvesting pharmaceutically active compounds. Cnidarians (corals, anemones, jellyfish, etc.) are commonly produced for conservation restocking and as ornamental species. Polychaetes and other worms are produced as bait for sportfishing, and alligators are produced primarily to harvest the hides.

7.4 Predominant aquaculture species

The largest number of different aquaculture species is found in Asia, with China being the country having the most diverse aquaculture species portfolio (Metian *et al.*, 2020). On other continents aquaculture is dominated by fewer species, which have been tailored towards more streamlined consumer demand and acceptance and for which production systems have been fine-tuned to a greater extent. Despite the relatively high overall diversity of species used for aquaculture, most commercial production systems prefer the same few species. Only twenty species of fish accounted for over 83.6% of total fish aquaculture production in 2018 (FAO, 2020). In addition to fish, which represent the greatest diversity of aquaculture species, two other animal taxa, molluscs and crustaceans, contribute significantly to overall aquaculture production yields (Figure 7.5).

Fishes represent a class of the subphylum vertebrates (phylum chordates), crustaceans are a subphylum of the phylum arthropods, and molluscs represent a phylum on their own. Because of these different levels of taxonomic classification of fishes, molluscs, and crustaceans I collectively refer to them as taxa, that is, groups of organisms that occupy different taxonomic levels.

Figure 7.5 provides a condensed overview of the commonly produced aquaculture taxa. Besides the three animal taxa, seaweed contributes significantly to overall aquaculture. Other animal phyla (sponges, cnidarians, echinoderms, etc.) are represented by many aquaculture species but their production volume is minimal compared to fish, molluscs, crustaceans, and aquatic plants (seaweed). In general, freshwater aquaculture yields more seafood if seaweed, which is also processed into products other than seafood and has much lower monetary value than animal aquaculture products, is not considered.

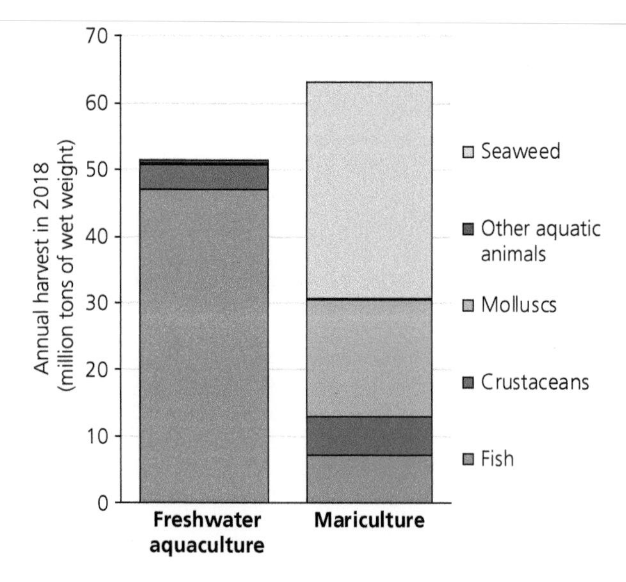

Figure 7.5 Major taxa used for commercial seafood aquaculture as indicated by global (worldwide) harvests in 2018 (million tons of wet weight). Note that seaweed harvest includes uses other than seafood. Data source: (FAO, 2020).

Figure 7.5 clearly illustrates that fish dominate in aquaculture and that, by far, most fish are produced by inland aquaculture in freshwater systems. However, the dominance of fish in freshwater aquaculture has been gradually reduced from 97.2% at the turn of the millennium to 91.5% in 2018, due to an increase in coastal mariculture. The top seven aquaculture fishes are all freshwater species and all of them, except Nile tilapia (*Oreochromis niloticus*), are carps.

More crustaceans are produced by mariculture than in freshwater but the difference between inland and coastal production is not as big as that for fish. Five million tons of a single crustacean species, the Whiteleg shrimp (*Litopenaeus vannamei*), are harvested annually, which rivals the aquaculture yields for any single of the top fish species. Mariculture of Whiteleg shrimp accounts for more than half of the global crustacean production while most of the remaining top crustacean species are freshwater crayfish and prawns.

In contrast to fish, almost all aquaculture production of molluscs is contributed by coastal mariculture. The leading species are cupped oysters, with the Pacific (Japanese) cupped oyster (*Crassostrea gigas*) contributing the most for any single mollusc species. Freshwater mollusc aquaculture is negligible compared to mariculture and primarily performed for purposes other than seafood production. For example, aquaculture of the freshwater triangle shell mussel (*Hyriopsis cumingii*) produces pearls, and aquaculture of the European pearl mussel (*Margaritifera margaritifera*) aids in the conservation of this species.

As mentioned, mariculture of seaweed (macroalgae) is difficult to compare to that of animal aquaculture in the context of seafood production. While some seaweed species are primarily produced as seafood, including *Undaria pinnatifida*, *Porphyra spec.*, and *Caulerpa spec.*, others are also harvested for alternative purposes. This multipurpose use is characteristic for many seaweed species including Japanese kelp (*Laminaria japonica*), the most produced species. Agar and alginates extracted from kelp and other seaweed are used for industrial purposes and as food additives. The rapid increase in seaweed production, which has more than tripled since the turn of the millennium, has been attributed in large part to carrageenan extraction for the production of food thickeners (FAO, 2020). The second-most produced species group of seaweed (*Eucheuma spec.*) is preferentially used for this purpose. The next chapter provides more detail on plant aquaculture and seaweed farming. Chapters 9 to 15 focus on animals that occupy increasingly higher levels in the network of life.

Key conclusions

- Aquaculture species diversity is much greater than that of farmed terrestrial species and these species occupy more diverse positions within the network of life.
- The species that form the network of life are classified by the discipline of taxonomy founded by Carl von Linné, who introduced the unambiguous binomial nomenclature of species.
- Taxonomic classification of species within higher-level categories in the network of life is based on their phylogenetic (ancestral) relationships that can be revealed by comparing their genomes.
- Many aquatic animals are used for aquaculture, but three taxa dominate: fishes (class of subphylum vertebrates), molluscs (a phylum), and crustaceans (subphylum of phylum arthropods).
- Although many aquatic species are being cultured, only a handful of fish, mollusc, and crustacean species are produced at high quantity and account for most of global seafood production.
- Freshwater fishes, in particular carps, contribute most to aquaculture yields.
- A single marine species of crustaceans (Whiteleg shrimp) accounts for more than half of crustacean aquaculture, while the remainder is contributed mostly by freshwater crayfish and prawns.
- Molluscs are farmed almost exclusively by mariculture, with cupped oysters such as the Pacific (Japanese) cupped oyster dominating mollusc aquaculture for production of seafood.
- Aquaculture production of marine seaweed has tripled since the turn of the millennium but its contribution to seafood production is difficult to estimate because of alternative uses.

References

Adoutte, A., Balavoine, G., Lartillot, N., and de Rosa, R. (1999). Animal evolution: the end of the intermediate taxa? *Trends in Genetics*, 15, pp. 104–108. https://doi.org/10.1016/S0168-9525(98)01671-0

Food and Agriculture Organization of the United Nations (FAO) (2020). *The state of world fisheries and aquaculture 2020*. Rome: FAO.

Futuyma, D. and Kirkpatrick, M. (2017). *Evolution*, 4th rev edn. New York: Oxford University Press.

Heng, H. H. (2019). *Genome chaos: rethinking genetics, evolution, and molecular medicine*, London: Academic Press.

Martin, W. F., Garg, S., and Zimorski, V. (2015). 'Endosymbiotic theories for eukaryote origin', *Philosophical Transactions of the Royal Society B: Biological Sciences*, 370, pp. 20140330. https://doi.org/10.1098/rstb.2014.0330

McClintock, B. (1950). 'The origin and behavior of mutable loci in maize', *Proceedings of the National Academy of Sciences of the United States of America*, 36, pp. 344–355.

Metian, M., Troell, M., Christensen, V., Steenbeek, J., and Pouil, S. (2020). Mapping diversity of species in global aquaculture. *Reviews in Aquaculture*, 12, 1090–1100. https://doi.org/10.1111/raq.12374

Ragan, M. A., McInerney, J. O., and Lake, J. A. (2009). 'The network of life: genome beginnings and evolution', *Philosophical Transactions of the Royal Society of London B: Biological Sciences*, 364, pp. 2169–2175. https://doi.org/10.1098/rstb.2009.0046

Smets, B. and Barkay, T. (2005). Horizontal gene transfer: perspectives at a crossroads of scientific disciplines. *Nature Reviews. Microbiology*, 3, 675–678. https://doi.org/10.1038/nrmicro1253

Teletchea, F. and Fontaine, P. (2014). 'Levels of domestication in fish: implications for the sustainable future of aquaculture', *Fish and Fisheries*, 15, pp. 181–195. https://doi.org/10.1111/faf.12006

von Linné, C. (2003/1735). Systema Naturae 1735: *Facsimile of the* first edition with an introduction and an English translation of the 'observationes'. Translated by M. S. J. Engel-Ledeboer and H. Engel. Leiden: Brill.

Wilkins, J. S. (2018). *Species: the evolution of the idea*, 2nd edn. Boca Raton: CRC Press.

CHAPTER 8

Environmentally sustainable plant aquaculture

'As a result of experimental work undertaken to clarify the situation and to try to bridge the gap in the life-history of Porphyra umbilicalis (L.) Kütz, *it has been fully established that, while at times under conditions of culture a few spores remain ungerminated and enlarged, due possibly to the abnormal conditions, the vast majority of the spores germinate a few days after liberation into very distinctive branched filamentous growths.'*

Drew, in: *Nature, vol. 164* (1949)

8.1 Marine macroalgae (seaweed)

The chapter epigraph from Kathleen Drew's classic 1949 *Nature* paper represents the discovery of the microscopic phase (the conchocelis) in the biphasic life cycle of purple laver seaweed (nori) and the dawn of commercial seaweed aquaculture. Today, aquaculture accounts for the vast majority (96.5%) of the total harvest of wild-collected and cultured aquatic plants (31.2 million tons in 2016) (FAO, 2020). Despite this great amount of aquatic plants produced in 2018, which accounts for more than a quarter of all aquaculture production yield in wet weight, the global value of plant aquaculture was less than ten percent ($13 billion) of the value of animal seafood produced for human consumption ($150 billion). Nevertheless, plant aquaculture is economically lucrative as evidenced by a twenty-fold increase of aquatic plant harvest in fewer than fifty years.

Seaweed aquaculture has a rich, century-long history in Asia. Compared to other forms of aquaculture it is more ecologically sustainable (Chapter 4 and Section 8.4) and gaining increased global consideration. In 2016, five major species/ genera contributed over 26 of the 31.2 million tons of seaweed produced globally. The remaining 5 million tons were contributed mostly by *Kappaphycus spec.*, *Gracilaria spec.*, *Chondrus spec.* (Rhodophyceae), *Sargassum spec.* (Phaeophyceae), *Ulva spec.*, and *Caulerpa spec.* (Chlorophyceae).

Although yields for several forms of plant aquaculture (e.g. ornamentals or plants used for wastewater treatment) are not routinely included in overall production estimates, the culture of marine macroalgae, commonly named seaweed, accounts for most plant aquaculture. Relatively few species/genera of seaweed are exploited on a commercial scale with red macroalgae (Rhodophyceae) and brown macroalgae (Phaeophyceae) being more commonly used for aquaculture than green macroalgae (Chlorophyceae). Examples of red macroalgae are *Mastocarpus jardinii* (Figure 8.1a) and *Kappaphycus spec.* (Figure 8.1b), while brown macroalgae include *Laminaria japonica* (Figure 8.1c), and green macroalgae include green sea tang (*Ulva spec.*) (Figure 8.1d).

The main purpose of seaweed aquaculture is production of seafood. Unlike other seafood, seaweed is commonly (but not exclusively) consumed in a dried form (e.g. as sushi wrapping) or applied as a food additive or nutraceuticals (nutritional supplements and health foods). Seaweed is rich in certain nutrients, including minerals (iodine, iron, calcium, potassium, and selenium) and vitamins (A, C, and $B_{1,2,6,12}$). It represents a valuable non-fish

A Primer of Ecological Aquaculture. Dietmar Kültz, Oxford University Press. © Dietmar Kültz (2022). DOI: 10.1093/oso/9780198850229.003.0008

Figure 8.1 Economically important seaweeds classified as red (A, B), brown (C), and green (D) macroalgae. **A)** *Mastocarpus jardinii*, photo by author. **B)** *Kappaphycus spec.*, photo reproduced with kind permission from Miguel Sepulveda. **C)** *Laminaria japonica*, photo reproduced with kind permission from Prof Delin Duan. **D)** *Ulva spec.*, photo by author.

source of natural omega-3 poly-unsaturated fatty acids (PUFAs). The concentration of some important trace minerals, for example, calcium and iron, is higher in seaweed than in fruits and vegetables. In addition, the protein content of dried seaweed is high for some species, e.g. 35% for dried purple laver/nori (*Porphyra spec.*) (Pillay and Kutty, 2005). Traditional species such as *Laminaria japonica* (kelp, kombu), *Undaria pinnatifida* (kelp, wakame), and *Porphyra spec.* (purple laver, nori) are primarily harvested for human consumption. Several species of green macroalgae of the genus *Caulerpa* are also produced as seafood and commonly referred to as sea grapes and green caviar because of their grape- and fish egg-like appearance (Figure 8.1).

The application of seaweed as a food additive and many of its industrial applications are based on its high contents of specific carbohydrates with emulsifying (gelling, thickening) properties that can be extracted in large quantities. Such carbohydrates include agar (a mixture of agarose and agaropectins), carrageenan (various sulfated polysaccharides), alginates (salts consisting of metal cations and alginic acid), and mannitol (a sugar alcohol used as a food sweetener and diuretic drug).

Since the turn of the millennium the production of other species for extraction of carrageenan, agar, and alginates has greatly increased, in particular in Indonesia (Figure 8.2). Seaweed-producing countries include China, Indonesia, South Korea, the Philippines, Vietnam, Malaysia, Chile, and Japan (Kim, 2020). The major seaweed import countries are China, Japan, and the US. Besides its main use as seafood, seaweed is important for biomedical applications such as treating nutrient (e.g. iodine) deficiencies and parasitism (e.g. as a vermifuge). Additional uses of seaweed include the production of compounds for the cosmetics industry, emulsifiers, pharmaceuticals, animal feeds, and biofuel.

8.1.1 Seaweed farming methods

Seedlings for culturing seaweed are obtained in two principal ways, first, by collection from natural habitat and, second, by controlled production in nurseries. The traditional method is harvest from natural habitat. To collect seedlings from natural habitat, nets are placed into intertidal or subtidal areas at the time when spores are released. The

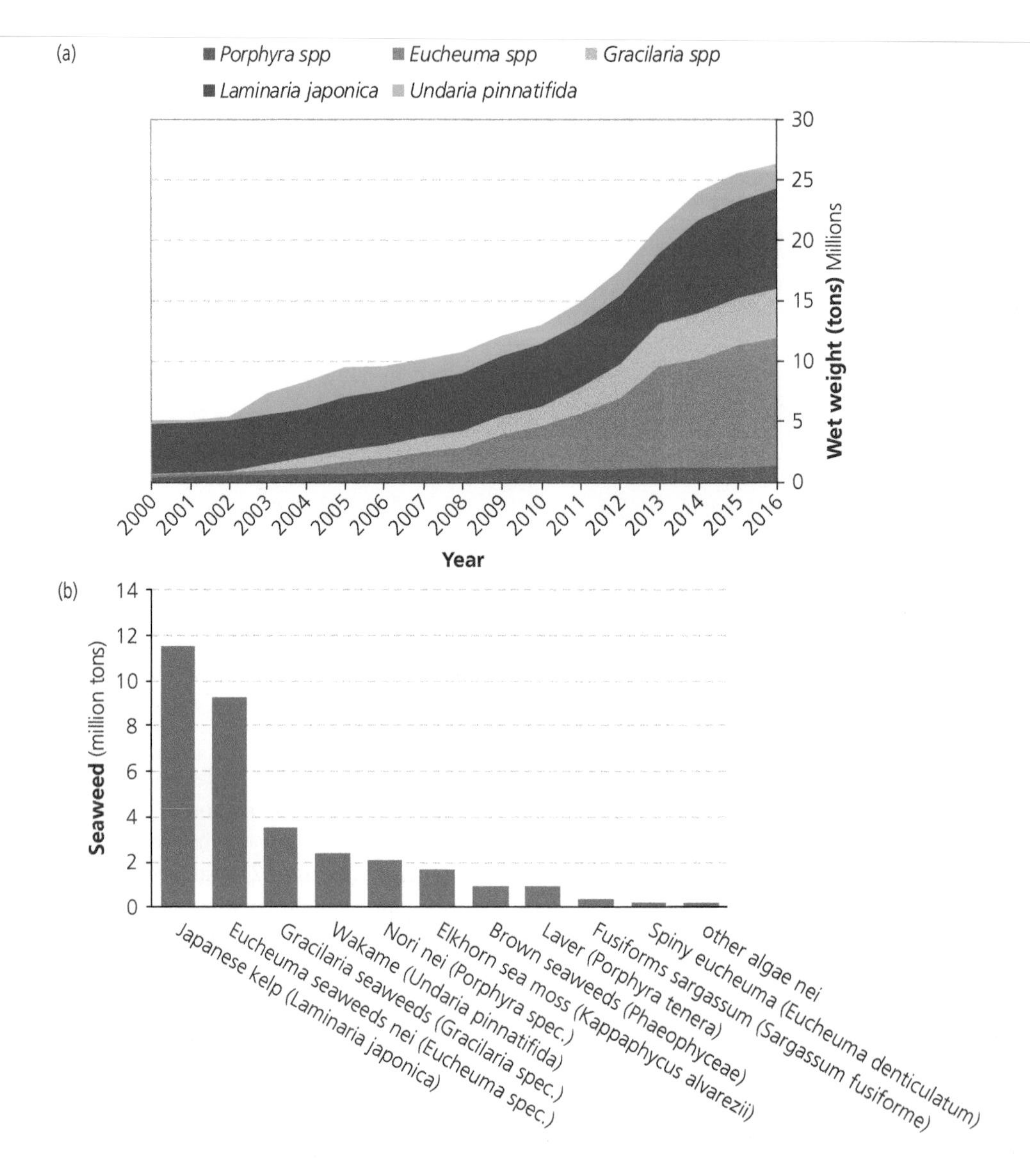

Figure 8.2 Aquaculture production of economically important seaweeds since the turn of the millennium. **A)** Three genera of red macroalgae (Rhodophyceae) and two species of brown macroalgae (Phaeophyceae) accounted for most of the global seaweed aquaculture harvest. **B)** Five additional species accounted for almost all the remaining seafood production in 2018. Data source: (FAO (UN), 2020).

timing is critical for this method to succeed and must accurately match the window of spore release. Seedlings will attach to the nets, which can then be translocated to grow-out areas.

Although this traditional method is still being used, the practice of producing seedlings under controlled conditions in indoor hatcheries is now widespread and some aquaculture farms specialize in this aspect of seaweed farming. Thus, seaweed aquaculture is commonly performed in two distinct stages: an indoor hatchery stage, which is not mandatory but greatly increases productivity, followed by an extensive or semi-intensive outdoor grow-out stage in intertidal and sub-tidal coastal habitat. Alternative to sexual reproduction, vegetative (asexual) propagation of many commercial seaweed species is also common. In such cases, occasional outbreeding by sexual reproduction helps to maintain genetic diversity of the cultured strains to support environmental and disease resilience (Box 8.1).

Three main outdoor grow-out systems are used for seaweed aquaculture: line culture, net culture, and pond culture. Suspension culture systems such as line and net cultures are more environmentally friendly than bottom culture. *Line culture* is the preferred method for most (95%) of seaweed aquaculture. With this method, an approximately 50-m-long rope is anchored to the seafloor on each end and kept on the water surface by buoys for at least half of its length. From this central, floating part of the long line, multiple vertical ropes, each approximately one metre long, are dropped by attaching a weight (e.g. a rock) to their ends. These vertical ropes are spaced about one metre apart along the entire part of the long line that is kept afloat near the surface. Each vertical rope (dropper) has up to thirty plants attached for extensive or semi-intensive grow-out.

Net culture utilizes anchored poles or rafts to hold long nets that are 1–2 m wide at a certain depth. When using poles the depth of seaweed submersion varies depending on the tides, whereas submersion depth is constant when using rafts. *Pond culture* utilizes coastal ponds in which the salinity, temperature, and water quality can fluctuate significantly. Only species that have wide tolerance towards such fluctuations are amenable to pond culture, e.g. *Gracilaria spec.* Pond culture of seaweed is practised in Taiwan and Southeast Asia and often utilizes ponds repurposed from the production of milkfish or marine shrimp. This allows more control over managing things like fertilization, monitoring for disease, etc. In addition, wastewater can be managed more readily since ponds are more contained than open coastal waters.

Most of the seaweed grown by aquaculture is harvested for processing but some thalli (foliose tissues) are also collected to produce new seedlings. Thalli from the macroscopic phase are collected, transferred to small containers filled with seawater, and squeezed to stimulate release of spores or gametes. Alternatively, release of spores or gametes can be stimulated by air-drying the thalli, which leads to desiccation, and then placing them in seawater-filled containers to rupture them by osmosis.

For many brown macroalgae such as kelp (e.g. *Laminaria japonica*, *Undaria pinnatifida*), the macroscopic, foliose phase is the diploid sporophyte. It produces haploid spores by meiosis in sori, which are thalli structures consisting of groups of spore-containing sporangia. The spores develop in indoor hatcheries into the haploid gametophyte phase, which lasts only for a few weeks. Cultures of microscopic male and female gametophytes are kept under controlled indoor hatchery conditions (temperature, lighting, etc.) to optimize growth and survival.

The microscopic gametophyte phase or, alternatively, juvenile sporophytes that develop from fused male and female gametes, which are released by mature gametophytes, settle on an attachment substrate after about two weeks of indoor culture. The substrate can be frames of cords, nets, or other light structures with a large surface area. After a brief period of hatchery culture, these substrates are transported to grow-out areas in bays where the diploid sporophyte develops to maturity.

For many red macroalgae such as purple laver/nori (*Porphyra spec.*), the macroscopic, foliose phase is the haploid gametophyte. Individual gametophytes are either male or female. Thalli with abundant gametes are collected from this phase and transferred to indoor containers. Haploid male and female gametes are released from these thalli and fuse to form the carposporophyte within a structure called the cystocarp on the female gametophyte. The carposporophyte produces many diploid carpospores, which are released and attach to mollusc shells or other suitable substrates. They develop into the filamentous, diploid sporophyte known as the microscopic conchocelis phase.

Conchocelis development takes place during the summer months in indoor tanks or raceways. During this indoor phase, temperature, lighting, and water quality can be controlled to provide optimal conditions for conchocelis development. In autumn, the conchocelis phase has matured and releases haploid conchospores, which are generated by meiosis. These conchospores settle on attachment frames of cords, which are placed into the tanks to provide a large surface area for attachment.

Box 8.1 The life cycle of seaweed

The life cycle of macroalgae, including all farmed seaweeds, is rather complex. Therefore, basic research on the reproductive biology of these aquatic plants, which was boosted by Kathleen Drew's pioneering work (1949) was essential for establishing large-scale commercial seaweed aquaculture. Most aquaculture seaweed species have a biphasic life cycle that consists of diploid (two copy numbers of the genome in each cell, maternal and paternal) and haploid (only one copy number of the genome in each cell) phases. Alternating between diploid and haploid phases is referred to as a diplohaplontic life cycle during which a haploid gametophyte phase of the plant alternates with a diploid sporophyte phase. The alternation of these two phases depends on the genome of the species and the environment, which interact to determine the life cycle (Figure 8.3).

Reproduction of seaweed can be sexual by fusion of male and female gametes that are produced by the haploid gametophyte to form a diploid sporophyte, or asexual either by fragmentation or formation of haploid spores by the sporophyte that will develop into the gametophyte. Fragmentation is a common method of seaweed propagation in aquaculture although periodic sexual reproduction is necessary to maintain genetic diversity and environmental resilience. Depending on the species, both phases (haploid and diploid) may be isomorphic (morphologically identical) or heteromorphic (morphologically different). One of the phases of heteromorphic seaweed species is usually microscopically small while the other is the large macroscopic, foliose phase that is harvested in aquaculture. The microscopic phase can either be the haploid gametophyte, for example, for *Laminaria japonica* (Liu *et al.*, 2017), or the diploid sporophyte phase, for example, the conchocelis of *Pyropia yezoensis* (Mikami *et al.*, 2019). For temperate species, the macroscopic phase prefers lower temperatures (10–20 °C) and grows rapidly during the colder season while the microscopic phase is prevalent at higher summer temperatures.

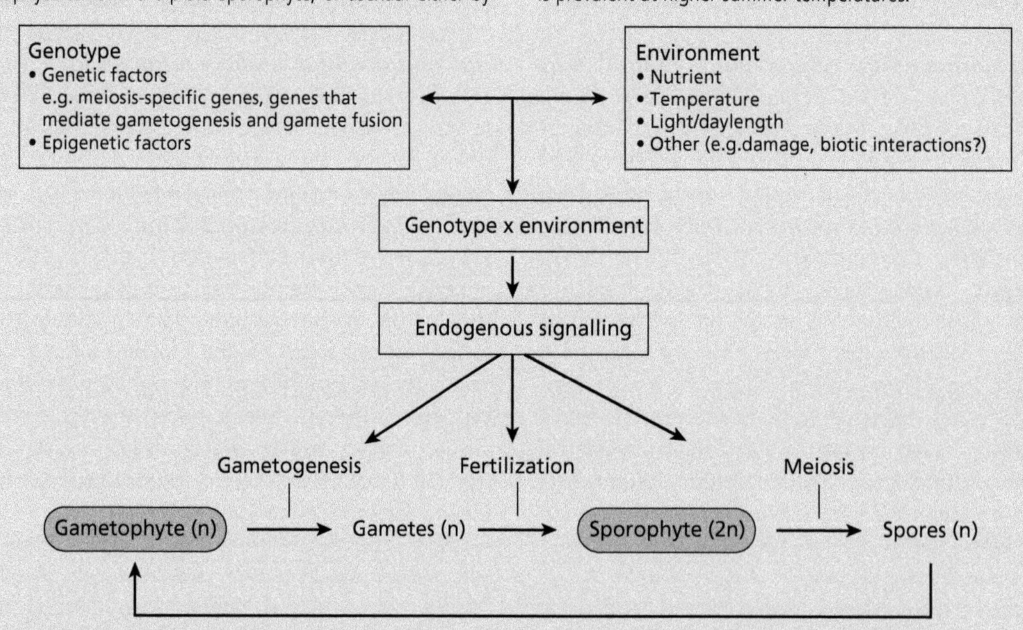

Figure 8.3 Effect of genotype (heritable traits, DNA sequence) × environment interactions on a typical dual stage life cycle of macroalgae (seaweed). The gametophyte stage is haploid (every cell contains a single copy of the genome) and produces male and female gametes that fuse into a diploid sporophyte during sexual reproduction (fertilization). Figure reproduced with kind permission from Prof Olivier DeClerck (Liu et al., 2017).

Attachment frames are convenient for handling and transfer of haploid gametophytes to nursery seawater tanks, where they are grown to a size of 1–2 cm before transferring them to natural grow-out areas such as bays and other intertidal areas. After transfer, the grow-out phase of nori aquaculture consists of the cultivation of the foliose thalli in the field. This phase occurs during the colder winter months and is followed by harvest in spring.

In addition to extensive and semi-intensive culture, more intensive tank or raceway culture of seaweed is also employed for aquaculture in indoor greenhouses. This method is not limited to culturing the microscopic phase of the biphasic life cycle but has also been used for grow-out of the foliose macroscopic phase, although much less commonly than grow-out in natural habitat. Culturing both phases of the life cycle in tanks or raceways is referred to as complete (as opposed to partial) tank cultivation.

Large raceways and greenhouse tanks flushed either continuously or periodically with seawater are used for intensive grow-out of the macroscopic phase of several species, including *Chondrus spec.*, *Porphyra spec.*, and *Caulerpa spec.* These and other species can be cultured to produce high-value nutraceuticals and cosmetics that offset the higher production costs of tank-based systems. Moreover, very high yields can be achieved with such systems by exerting greater control over the environmental conditions (temperature, lighting, competitor species, nutrients, etc.).

8.1.2 Seaweed diseases

Seaweed species used in aquaculture are susceptible to many diseases and must be monitored regularly for pathogens and signs of disease just as aquaculture animals are. There are many reported diseases and currently, cures or treatments are lacking for most of them. Thus, seaweed diseases represent an area of research with tangible broader impacts on the seafood industry.

Ice-ice disease is caused by a bacterial pathogen that affects multiple species of commercial seaweed. If commercially cultured seaweed is stressed by suboptimal culture conditions that deviate from the conditions in its native habitat, then the spread of this disease is accelerated (Largo *et al.*, 2004). Suboptimal conditions can arise in large-scale, dense seaweed aquaculture farms by impairing wave action and currents, limiting access to light, altering sedimentation rate, increasing the activity of fouling organisms, and decreasing local oxygen concentration at night when seaweed photosynthesis ceases but respiration continues. In many cases, relieving stress caused by crowding, poor water quality, and nutrient deficiencies helps mitigate the impact of seaweed diseases.

8.2 Aquatic and semi-aquatic freshwater macrophytes

Freshwater macrophytes are large plants that are either fully submersed or semi-aquatic, that is, only partly submersed in shallow water areas or floating on the water surface. Aquaculture of freshwater macrophytes and their production for human consumption is much less common than seaweed farming. Traditional freshwater aquaculture species include the water chestnut (*Trapa spec.*), watercress (*Nasturtium spec.*), and water spinach (*Ipomoea spec.*).

Two species of watercress (*Nasturtium officinale* and *N. microphyllum*) and their hybrid (*N.* × *sterile*) are cultivated for human consumption in salads or as a vegetable. Despite its name, the hybrid can produce seeds when its female parent is *N. microphyllum*, although reproductive success of hybrids is greatly diminished when compared to the parent species. *Nasturtium officinale* is the most common species used for commercial aquaculture. It is a perennial floating plant that can reproduce sexually (by producing gametes) and asexually (by budding). Commercial aquaculture of watercress sprouts involves a short growth period of only a few days after seed germination, but mature plants are also produced. Watercress is a good source of some vitamins and provitamins (A, B, C, E) and minerals (e.g. iron) and has been used for traditional medicine and gained a reputation as a health food (nutraceutical) in its native range in Europe and Asia.

Watercress grown commercially in aquaculture is invasive and has established feral populations on all continents except Antarctica. Watercress has even colonized the Hawai'ian islands—the most remote land mass in the world. In its non-native distribution range, watercress is considered a nuisance weed because of its fast growth, high environmental resilience, and persistence in wild ecosystems. Nevertheless, watercress grown in aquaculture is susceptible to a variety of diseases (Box 8.2).

8.2.1 Non-food uses of aquatic freshwater plants

Besides production for human consumption, semi-aquatic macrophytes are produced for non-food uses, including wastewater treatment/bioremediation, soil conditioning (production of mulches and organic fertilizers), habitat restoration, and as ornamental plants for the aquarium industry and outdoor ponds. Duckweeds (*Lemna spec.*) such as the common duckweed (*Lemna minor*) and water hyacinth (*Eichhornia crassipes*, Figure 8.4a, c) are floating freshwater plants that are used to treat animal and human wastewater by assimilating ammonia and inorganic nutrients that have been liberated by bacteria from organic waste.

Assimilation of these nutrients into plant biomass can serve multiple purposes. Besides bioremediation, plant biomass can be harvested and processed into mulches, which serve to condition soil

Box 8.2 Watercress diseases

Despite its high environmental resilience, wild watercress is prone to vector-borne diseases transmitted by insects or bivalves as intermediate hosts. For example, watercress infection with the liver fluke (*Fasciola hepatica*) is common in livestock grazing areas and causes fascioliasis liver disease in humans who consume it uncooked. Non-zoonotic diseases of watercress that cannot be transmitted to humans but affect its growth and nutritional value are caused primarily by bacterial and fungal pathogens. Fungal pathogens of watercress include *Cercospora nasturtii* and *Septoria sisymbrii*, which cause leaf spot disease. An example for a bacterial disease is black rot syndrome, which is caused by *Xanthomonas campestris*. Since watercress produced in aquaculture is regularly monitored for diseases it is safer to consume than wild watercress. Watercress is a member of the cabbage family (*Brassicaceae*) and lends itself well for use in aquaponics systems.

for agricultural crops and orchards. Care must be taken that fertilization of soils for growing human food with mulches produced from plants used for wastewater treatment is free of toxins and biologically active compounds that bioaccumulate in crops. If such bioaccumulation is detected, then the plant biomass harvested from wastewater treatment plants can be fermented and processed into biofuel.

Aquaculture is also performed on commercial scale for hundreds of species of freshwater *ornamental plants*, including aquatic plants for the aquarium trade and semi-aquatic plants as outdoor ornamentals. Popular ornamental plants produced by aquaculture are arrowheads (*Sagittaria spec.*), waterweeds (*Anacharis spec.*), anubias (*Anubias spec.*), cape-pondweeds (*Aponogeton spec.*), water crypts (*Cryptocoryne spec.*), burheads (*Echinodorus spec.*), Brazilian waterweed (*Egeria densa*), fanworts (*Camboma spec.*), eelgrasses (*Vallisneria spec.*), water lilies (*Nymphaea spec.*, Figure 8.4a, b), canna lilies (*Canna spec.*), irises (*Iris spec.*) and many others.

Plant aquaculture is also practised for *habitat restoration*, restoration of damaged wetlands, creation of stormwater runoff habitat, and revitalization of coastal habitats, for example, using marine aquatic (e.g. eelgrasses, *Zostera spec.*) or semiaquatic (e.g. various genera of mangroves) and freshwater semiaquatic plants. Common freshwater plants used for aquatic habitat restoration include bulrushes (*Scirpoides spec.*), cattails (*Typha spec.*), water lilies (*Nymphaea spec.*), and pickerelweed (*Pontederia cordata*).

8.2.2 Utilization of semiaquatic floating plants for aquaponics

Aquaponics is an integrated aquaculture approach for combining the culture of freshwater fish or crustaceans with hydroponics of agricultural crops, typically plants from the cabbage (*Brassicaceae*) family. These plants can be grown hydroponically, that is, without soil by immersing the roots in nutrient-rich wastewater produced by the aquatic animals. Aquaponics is based on a principle similar to marine integrated multitrophic aquaculture (IMTA), but it is applied to freshwater systems. Although most plants produced by aquaponics are agricultural

Figure 8.4 Water lilies (left, *Nymphaea spec.*) and water hyacinths (right, *Eichhornia crassipes*) covering a pond near Punalu'u Black Sand Beach on the Big Island of Hawai'i. (**A**). Lower panels show magnified flowers of a **B)** water lily and **C)** water hyacinth. These invasive freshwater plants grow rapidly and are environmentally resilient, traits that are beneficial for aquaculture but also promote invasiveness. Photos by author.

crops (terrestrial species) grown hydroponically, such systems are also utilized for producing aquatic and semiaquatic freshwater plants along with fish or crustaceans on a smaller ecological footprint than monocultures.

Terrestrial crop plants can be substituted by semi-aquatic floating plants such as watercress, water hyacinth, and duckweed in aquaponics systems. These floating plants also have their roots immersed freely in the water. Aquaponics requires significant management skills as it is challenging to balance the amount of waste from fish with the nutrient requirements of the plants. A decomposer component (biological filter) is often needed in addition to mechanical filters for sedimentation of solid waste.

The animal production component of aquaponics systems is limited by the nutrient assimilation capacity of the plant component. Advantages of aquaponics include its higher ecological sustainability compared to monocultures, production of multiple crops targeting different markets, the absence of soil-borne diseases and pathogens, and a small spatial footprint, especially if production systems are vertically integrated. Apart from the high management and operational costs, disadvantages of aquaponics include difficulties in matching animal waste production with biological filtration and plant assimilation capacity and the requirement of a large water surface area for plant growth (Box 8.3).

Box 8.3 Coupled versus decoupled aquaponics systems

Coupled aquaponics refers to an open connection between the animal and plant components of the system. In a coupled system these two components are tightly linked by a single integrated loop. In this single loop, water from the fish tanks is pumped to filtration tanks or sumps and then to the plant hydroponics unit from which it is returned directly to the to fish tanks. Water flow through this loop is determined by the rate of the pump and the direct coupling of all units in the system requires a fine balance of all its components, which can be challenging.

Decoupled aquaponics systems simplify the task of balancing nutrient and energy flow through its individual components. Decoupling the animal and plant components of the system is achieved by separating the recirculating aquaculture system (RAS) for fish from the hydroponic system for plants. Separation is commonly achieved by implementing valves that permit regulation of the extent of water exchange between the two components as needed. Decoupling allows for more flexibility and wastewater treatment in a separate RAS loop, in addition to the hydroponics loop. Such additional wastewater treatment supports an increased density of aquaculture animals. For decoupled systems, the spatial arrangement is also more flexible as the two components can be housed in separate buildings or greenhouses and can be managed at different temperature and light regimes.

8.3 Aquaculture of microalgae

Microalgae are microscopically small plants that are unicellular (single-celled) or comprised of loose aggregations of relatively few cells. They represent a taxonomically very diverse group of organisms representing many phyla comprised of an estimated 40,000 species that inhabit freshwater, marine, and hypersaline (salinity is greater than that of seawater) environments (Paul and Borowitzka, 2019).

Aquaculture of microalgae is common and widespread to aid the bioremediation of wastewater and to produce live food for juvenile aquaculture animals in hatcheries and feeds for agricultural animals. Microalgae also provide nutraceuticals, pigments used for cosmetics, bioactive secondary metabolites, and pharmaceutical compounds for human nutrition and medicine. Moreover, microalgae aquaculture is used for biofuel production and soil conditioning. Despite the prevalence of microalgae aquaculture for these diverse applications, a mere 89,000 tons of farmed microalgae have been reported by only eleven countries in 2016 with the great majority (88,600 tons) reported from China (FAO, 2018). Therefore, it is difficult to weigh the relative economic importance of global microalgae aquaculture and its historical trajectory against that of other aquaculture sectors.

Proper culture of microalgae depends on a thorough understanding of the biology of the species being cultured. Not only do different species vary greatly in their requirements for optimal growth, but also the biochemical composition and, thus, nutritional value of any given species, is highly dependent on the culture conditions. Light, salinity, temperature, and pH are critical abiotic factors that influence the growth and biochemical composition of microalgae. Microalgae and other microorganisms typically grow in four stages:

1. Slow-growth lag phase.
2. Rapid, exponential growth log phase.
3. No-growth stationary phase when cultures become saturated.
4. Negative-growth death phase when cultures are deteriorating.

These phases must be managed carefully by maintaining optimal culture density to avoid large fluctuations in nutritional/economic value, which is a function of the contents of protein (30–60% of dry weight), carbohydrates (5–35% of dry weight), lipid (5–25% of dry weight), and secondary metabolites present in the microalgae.

Biotic factors, such as the presence of bacteria, also affect the yield of desired products from microalgae cultures. Bacteria-free (axenic) cultures are generally preferable and more productive, which is why microalgae starter cultures are preferably grown aseptically in a sterile laboratory environment. Under these conditions, monospecific cultures rather than mixtures of different microalgae species can be achieved. The production of axenic, monospecific microalgae represents a niche aquaculture market geared at supplying hatcheries for aquaculture animals and large-scale microalgae farms with high-quality starter cultures.

Some hatcheries and microalgae farms produce their own starter cultures but the significant added expenses for smaller economies of scale must then be offset by resale income from the product generated. The least expensive alternative to axenic, monospecific microalgae starter cultures are blooms of microalgae that can be induced from mixtures of species present in local phytoplankton. However, phytoplankton blooms consisting of non-axenic mixtures of local microalgae species are not as reliable as axenic monocultures and their nutritional value may vary greatly from batch to batch. To induce such blooms, phytoplankton collected from natural habitat is inoculated into ponds or tanks, to which inorganic fertilizer is added to encourage microalgae growth. This practice is more ecologically sustainable if tanks rather than ponds are used, and tank water is filtered before disposal to avert eutrophication of natural habitat. Two-stage filters are employed to first remove larger zooplankton before harvesting the smaller phytoplankton (microalgae) from tanks with pronounced algal blooms.

Microalgae represent high-value diets for aquaculture animals produced in hatcheries because they contain omega-3 PUFAs and have a high protein content. PUFAs and secondary metabolites such as vitamins (e.g. vitamin E) and pigments (e.g. astaxanthin, phycocyanin, beta-carotene) that are synthesized by microalgae are also produced as health foods and for the cosmetics industry.

Microalgae are marketed as live concentrates, frozen, in dried form, or as purified extracts of specific secondary metabolites. Live microalgae concentrate has the highest nutritional value as feed for early developmental stages of aquaculture animals in hatcheries, especially if live food is a requirement for the species produced by the hatchery. However, shipping, handling, and logistics are more elaborate and costlier compared with frozen or dried microalgae. Frozen cultures may be revived if appropriate freezing media and procedures are used. They represent a good compromise between live concentrate and dried microalgae. Dried microalgae have a much lower weight per unit of nutrient contained, which is advantageous for shipping and handling. The trade-off is that the structural integrity of some nutrients is impaired during the drying process, which renders them less effective as a nutraceutical or hatchery food than living microalgae.

8.3.1 Aquaculture systems for production of microalgae

Microalgae are produced in extensive, semi-intensive, and intensive aquaculture systems. *Extensive systems* for microalgae aquaculture are very shallow ($\leq$ 0.5-m depth) ponds that have a very large surface area (up to 250 ha) and allow efficient capture of light. Such cultures achieve only moderate microalgae density (0.1–0.5 g/L dry weight) and ponds need to be harvested regularly to avoid a long stationary phase or deterioration of the culture (Paul and Borowitzka, 2019).

Extensive microalgae cultures are often located in hot, arid areas to ensure rapid growth at high temperature and light intensities. Water is drawn either from groundwater or from the ocean if the microalgae farm is located near the coast. Because of the significant extent of evaporation happening from extensive ponds, microalgae species cultured by this method are often extremely tolerant of wide ranges of salinity, including very high salinities. An example for a euryhaline microalgae species cultured extensively is *Dunaliella salina* (Chlorophyta). This microalga tolerates a salinity that is tenfold higher than seawater and approaches the maximum solvent capacity of water (300 g/kg). An example of extensive commercial *D. salina* aquaculture is the production of the natural pigment beta-carotene (provitamin A) in Australia (Paul and Borowitzka, 2019).

Semi-intensive culture of microalgae was pioneered in the 1950s in Taiwan to increase culture density to about 1g/L dry weight. For this type of culture either circular concrete ponds or rectangular raceways with a surface area of up to 500 m^2 are used. Semi-intensive systems represent the majority of commercial microalgae aquaculture systems currently in operation because they are most economical. Many different species are being cultured as monocultures in these systems, including *Arthrospira (Spirulina) spec., Chlorella spec., Nannochloropsis spec., Phaeodactylum spec., Scenedesmus spec.*, and *Haematococcus pluvialis*. Microalgae species selected for semi-intensive culture in open systems are environmentally highly resilient and can outgrow competitive phytoplankton species, resist pathogenic bacteria and fungi, and evade predatory zooplankton. An example for semi-intensive microalgae culture is the production of *Haematococcus pluvialis* (Chlorophyta) at the Kailua-Kona coast of Hawai'i. This aquaculture farm consists of open freshwater raceways to produce the secondary metabolite astaxanthin, which is a red carotenoid pigment with strong antioxidant properties. Astaxanthin is marketed as a nutraceutical.

Intensive aquaculture of microalgae uses closed systems that permit a maximum level of control over the culture conditions and, if proper water recirculating and treatment approaches are used, have the smallest environmental impact on nearby natural ecosystems. Such systems include axenic (bacteria-free, sterile) and non-axenic culture methods and they achieve high densities up to 10 g/L dry weight. So-called big bag systems and photobioreactors are the most common types of intensive microalgae aquaculture systems currently used. Big bag systems comprise a succession of increasingly larger volumes of indoor cultures starting with a 1L Erlenmeyer flask and culminating in 2000L indoor tanks. Variations of big bag systems are commonly used in hatcheries to produce live food for aquaculture animals.

Photobioreactors are more sophisticated and technically advanced systems that maximize the capture of light and culture density in a closed system. They consist of long tubes with a diameter of 20–30 cm that are arranged in either circular or linear arrays. Microalgae are continuously pumped through these tubes to keep them from settling while maximizing their exposure to light. The culture volume used in a typical photobioreactor is similar to that of the largest tanks in big bag systems (2000 L). An example for a large photobioreactor is the aquaculture plant operated by ALGOMED GmbH in Klötze, Germany. This plant utilizes water from an artesian aquifer to culture *Chlorella vulgaris* (Chlorophyta) in an array of glass tubes with an overall length of 500 km. The microalgae grown in this system are harvested by centrifugation and marketed in dried form as a nutraceutical.

In addition to more conventional photoautotrophic culture systems, heterotrophic intensive aquaculture of microalgae is also used. Microalgae are photoautotrophic organisms, meaning they are green plants with chloroplasts that enable them to utilize light energy for producing sugars from inorganic compounds to fuel their metabolism. However, some commercially cultured microalgae can also be grown heterotrophically, meaning that their energy metabolism does not rely on light under these conditions and photosynthesis is optional.

For heterotrophic culture of such microalgae, sugar must be included in the growth media to provide a source of metabolic energy. Examples of heterotrophically cultured microalgae include *Chlorella*, *Crypthecodinium*, *Aurantiochytrium*, *Schizochytrium*, and *Ulkenia* species (Guldhe *et al.*, 2017; Han *et al.*, 2019; Lopes da Silva *et al.*, 2019). Heterotrophic aquaculture of microalgae has the advantage that access to light is not limiting, which greatly reduces the spatial footprint and logistical complexity of such cultures. A main trade-off is that heterotrophic culture media require the use of nutrients (sugars) that also have nutritional value for other heterotrophic organisms (animals), which reduces resource use efficiency.

8.4 Ecological sustainability prospects

Plant aquaculture is generally more environmentally sustainable than animal aquaculture because of the lower position of plants in the trophic level transfer efficiency (TLTE) pyramid (Chapter 4). Less energy and less waste are associated with producing biomass at a lower trophic level. Moreover, some plant aquaculture is exclusively geared towards ecosystem restoration. Biofloc and turf scrubber technologies are examples of integrated multitrophic approaches for wastewater treatment that utilize aquatic plants (microalgae) in combination with bacteria and fungi (decomposers), to achieve more efficient bioremediation.

Less than 1% of the 40,000 species of macroalgae (and even fewer species of bacteria) have been explored for this purpose and genetic engineering of metabolic pathways in these microorganisms promises to improve wastewater treatment efficiency and detoxification. Furthermore, genetic engineering of microalgae has the potential to improve the yield of highly valued secondary metabolites and enable the production of new ones (pigments, nutraceuticals, etc.) on a smaller ecological footprint.

Along with filter feeding invertebrates like sponges and bivalves, seaweeds and other aquatic plants are referred to as extractive species because they remove nutrient-rich compounds from the water and incorporate them in their biomass. A major difference between extractive filter-feeding invertebrates and seaweed is that the animal filter feeders remove organic nutrients in the form of microalgae and small zooplankton from the water while aquatic plants remove mostly inorganic nutrients such as phosphates, nitrates, ammonia, and sulfates. Invertebrates are consumers as are all other animals while seaweeds and microalgae are primary producers along with all other green plants, that is, they occupy different positions in food webs (see Chapter 4). Therefore, seaweed farming is ecologically highly sustainable if practised properly, but it also requires considerable space as most seaweed aquaculture systems are extensive.

Seaweed polyculture has been practised with salmon and other fish species grown in semi-intensive marine cage culture and with crustaceans and bivalves produced by coastal mariculture and in saline pond culture systems. Such IMTA is generally more ecologically sustainable because it utilizes more diverse elements of trophic cycles (see Chapter 5). However, a major caveat in most IMTA systems is that production rates of consumers (especially carnivorous fish) and producers (green plants) are incompatible and not suitable for balancing trophic cycles.

As mentioned in Section 8.1.1, seaweed can be cultured by either bottom culture, line culture, or net culture. Bottom culture builds on the benthic lifestyle of seaweed, which naturally requires rocks or other solid substrates for attachment. For bottom culture, blasting rocky surfaces and temperate reefs to increase the surface area for kelp (*Laminaria spec.*, *Undaria spec.*) attachment is common practice and has large negative impacts on affected ecosystems (Pillay and Kutty, 2005). Therefore, bottom culture is less ecologically sustainable than line or net cultures. Moreover, bottom culture of seaweed is sometimes practised by attaching seaweed cuttings with rubber bands to coral branches and dropping them onto reefs. This practice is of concern in tropical areas where seaweed aquaculture is often situated in pristine coral reef areas.

Fortunately, most seaweed aquaculture is performed by either the line or net culture methods, which also have practical advantages for managing culture density, harvest, and monitoring of farmed seaweed. Ecologically responsible management of seaweed culture density is critical for providing sufficient light for benthic fauna that relies on photosynthetic symbionts to minimize the impact of aquaculture on reef resilience and biodiversity in the affected areas.

Another opportunity for increasing the ecological sustainability of seaweed aquaculture is improving the practice of fertilization, which is necessary for achieving economic competitiveness in the market. Fertilizer is often sprayed over the coastal marine habitat to improve the growth of seaweed. As much as 1 kg of fertilizer is used to produce 1 kg seaweed (Pillay and Kutty, 2005). This low efficiency of fertilizer utilization by seaweed is a result of tidal and coastal currents rapidly dispersing much of the applied fertilizer into other areas.

For pond culture of seaweed, animal manure and urea are also used in addition to inorganic fertilizer such as ammonium nitrate. For large-scale seaweed farms such fertilization threatens to cause imbalances in natural ecosystems and eutrophication. Large-scale fertilization is of particular concern when seaweed farms are located near intricately balanced ecosystems like coral reefs. An ecologically highly sustainable way of fertilization is the practice of immersing the seedlings in a solution containing ammonium nitrate and other nutrients, which promotes rapid growth, before transplantation into natural grow-out areas. Although this way of fertilization is temporally limited to the early growth phase of seedlings it does shorten the production cycle without introducing large amounts of nutrients into coastal waters.

Culture of seaweed in more intensive systems, for example, greenhouses or ponds with controlled water exchange instead of open ocean mariculture, provides means for minimizing the impact of fertilization on natural ecosystems. For greenhouses and well-managed ponds, waste-water collection and treatment of eutrophication, pests, fouling organisms, and pathogens is much more feasible than in open water coastal mariculture. However, significant economic incentives are necessary to render more ecologically sustainable, contained, intensive aquaculture systems like greenhouses economically viable.

Alternatively, open water mariculture of seaweed in intertidal and shallow subtidal zones of coastal areas has ecological benefits if nutrients resulting from urban, agricultural, or industrial waste can be assimilated in lieu of fertilization. Such bioremediation would have the dual gain of producing seaweed while improving water quality, which could be especially lucrative if seaweed is produced for non-food uses where food safety is not a concern.

Ecologically sustainable seaweed aquaculture utilizes local species and populations to prevent invasiveness into non-native habitats. Most of the cultured seaweed species are grown in their native range, which promotes the ecological sustainability of farming them. However, there are exceptions where invasive species of seaweed have been introduced into non-native habitat for aquaculture purposes. For example, the giant kelp (*Macrocystis pyrifera*) has been introduced from Mexico to China and the elkhorn sea moss (*Kappaphycus alvarezii*) has been translocated from the Philippines to China. An extremely invasive strain of the green macroalgae *Caulerpa taxifolia*, which had been artificially selected for increased environmental stress tolerance as an ornamental aquarium plant, has colonized large areas of the Mediterranean coast (Grewe *et al.*, 2007). This strain of *C. taxifolia* is native to the Caribbean, Indo-Pacific, and Red Sea but has successfully colonized coastal waters of California and Australia. It is now commonly dubbed the killer alga or death weed.

Artificial selection does not always have negative side effects but can be highly beneficial. For example, Japanese scientists have created a more heat-tolerant strain of wakame (NW-1) by cross-breeding and selecting different varieties of *Undaria pinnatifida*. This new wakame strain has increased performance for aquaculture in the face of global climate change (Niwa and Kobiyama, 2019).

Perhaps more than marine macroalgae, invasive freshwater aquatic plants are of substantial concern, especially the many ornamental species that have been introduced to non-native subtropical areas. An example of a highly invasive aquatic freshwater plant is the water hyacinth mentioned in Section 8.2 (Figure 8.4a, c), which is native to South America. This floating, fast-growing plant has invaded all continents except Antarctica and many isolated islands, including the most remote island chain in the world, the Hawaiian Islands. Highly invasive aquatic plants such as the water hyacinth have altered many ecosystems and managing the spread of such species, conservation of impacted native species, and preserving the balance of affected ecosystems comes at a significant cost.

Despite the challenges outlined here, farming plants for seafood production generates significantly more product (biomass) and less waste on a comparable ecological footprint than farming animals. Therefore, it is desirable to shift the future emphasis of aquaculture from animals, especially carnivores at high trophic levels, to plants. However, consumer preference is unlikely to shift massively towards a vegetarian diet and animal aquaculture also needs to be developed in parallel by advancing ecologically more sustainable approaches.

Some essential nutrients are underrepresented in current vegetarian diets and need to be provided in the form of dietary supplements. Research on producing such nutrients and supplements by alternative means that do not rely on animal production, along with research on adjusting the flavour and texture of plant-based foods towards the preference of consumers used to meat, will be key to curbing the demand for seafood produced from top-level carnivores such as salmon, tuna, and halibut towards organisms that occupy lower trophic levels.

Key conclusions

- Plant aquaculture contributes more than 96% of all wild and cultivated aquatic plants harvested and represents a very rapidly expanding branch of aquaculture.
- Marine seaweeds are the most common aquaculture commodity, contributing to over a quarter (in wet weight) of all aquaculture products harvested but accounting for less than 1% of resale value.
- Seaweeds (macroalgae) are primarily produced as seafood and marketed mostly in dried form. They have

high protein content and contain PUFAs, vitamins, and other compounds used as nutraceuticals.
- Non-seafood uses of seaweed as emulsifiers, pharmaceuticals, animal feeds, cosmetics, and biofuel add to its versatility.
- Seaweed production by extensive and semi-intensive aquaculture dominates but intensive production of microscopic life cycle stages in hatcheries and of all life cycle stages in greenhouses has become more common.
- The most common aquaculture approaches for seaweed grow-out are line and net culture methods.
- Aquatic freshwater macrophytes are cultivated for human consumption (e.g. watercress) and produced for bioremediation, soil conditioning, habitat restoration, and as ornamental species.
- Microalgae provide many of the same benefits as macroalgae when used for human consumption (as nutraceuticals) and they are marketed primarily in dried form.
- Microalgae are also cultured to produce pigments, pharmaceuticals, and other bioactive secondary metabolites; they are also manufactured for bioremediation purposes and biofuel production.
- Plant aquaculture represents an ecologically highly sustainable form of aquaculture but a shift from open to more closed culture methods would further increase its ecological sustainability by more efficient use and containment of fertilizer and by reducing spatial footprint and invasiveness.

References

Drew, K. M. (1949). 'Conchocelis-phase in the life-history of *Porphyra umbilicalis (L.) Kütz*', *Nature*, 164, pp. 748–749. https://doi.org/10.1038/164748a0

Food and Agriculture Organization of the United Nations (FAO) (2020). 'FishstatJ: fishery and aquaculture statistics. Global production by production source'. http://www.fao.org/fishery/statistics/software/fishstatj/env4.00.17 (Accessed December 2020).

Food and Agriculture Organization of the United Nations (FAO) (2020). *The state of world fisheries and aquaculture 2020*. Rome: FAO.

Food and Agriculture Organization of the United Nations (FAO) (2018). *The state of world fisheries and aquaculture 2018: meeting the sustainable development goals*. Rome: FAO.

Grewe, P. M., Patil, J. G., McGoldrick, D. J., Rothlisberg, P. C., Whyard, S., Hinds, L. A., Hardy, C. M., Vignarajan, S., and Thresher, R. E. (2007). 'Preventing genetic pollution and the establishment of feral populations: a molecular solution', in Bert, T. M. (ed.) *Ecological and genetic implications of aquaculture activities, methods and technologies in fish biology and fisheries*. Dordrecht: Springer, 103–114. https://doi.org/10.1007/978-1-4020-6148-6_6

Guldhe, A., Ansari, F. A., Singh, P., and Bux, F. (2017). 'Heterotrophic cultivation of microalgae using aquaculture wastewater: a biorefinery concept for biomass production and nutrient remediation', *Ecological Engineering*, 99, pp. 47–53. https://doi.org/10.1016/j.ecoleng.2016.11.013

Han, P., Lu, Q., Fan, L., and Zhou, W. (2019). 'A review on the use of microalgae for sustainable aquaculture. *Applied Sciences*, 9, pp. 2377. https://doi.org/10.3390/app9112377

Hwang, E. K., Choi, H. G., and Kim, J. K. (2020). 'Seaweed resources of Korea', *Botanica Marina*, 63, pp. 395–405. https://doi.org/10.1515/bot-2020-0007

Largo, D., Fukami, K., and Nishijima, T. (2004). 'Time-dependent attachment mechanism of bacterial pathogen during ice-ice infection in *Kappaphycus alvarezii (Gigartinales, Rhodophyta)*', *Journal of Applied Phycology*, 11, pp. 129–136. https://doi.org/10.1023/A:1008081513451

Liu, X., Bogaert, K., Engelen, A. H., Leliaert, F., Roleda, M. Y., and Clerck, O. D. (2017). 'Seaweed reproductive biology: environmental and genetic controls', *Botanica Marina*, 60, pp. 89–108. https://doi.org/10.1515/bot-2016-0091

Lopes da Silva, T., Moniz, P., Silva, C., Reis, A. (2019). 'The dark side of microalgae biotechnology: a heterotrophic biorefinery platform directed to ω-3 rich lipid production', *Microorganisms*, 7, pp. 670. https://doi.org/10.3390/microorganisms7120670

Mikami, K., Li, C., Irie, R., Hama, Y., 2019. A unique life cycle transition in the red seaweed *Pyropia yezoensis* depends on apospory. *Communications Biology*, 2, pp. 1–10. https://doi.org/10.1038/s42003-019-0549-5

Niwa, K. and Kobiyama, A. (2019). 'Development of a new cultivar with high yield and high-temperature tolerance by crossbreeding of *Undaria pinnatifida* (Laminariales, Phaeophyta)', *Aquaculture*, 506, pp. 30–34. https://doi.org/10.1016/j.aquaculture.2019.03.002

Paul, N. A. and Borowitzka, M. (2019). 'Seaweed and microalgae', in Lucas, J. S., Southgate, P. C., and Tucker, C. S. (eds.) *Aquaculture: farming aquatic animals and plants*, 3rd edn. Hoboken: John Wiley & Sons, Ltd, 313–337.

Pillay, T. V. R., Kutty, M. N. (2005). *Aquaculture: principles and practices*, 2nd edn. Oxford: Wiley-Blackwell.

CHAPTER 9

Aquaculture of sponges and cnidarians

'The marine environment in particular has attracted much attention as a source of natural product discovery due to its vast biodiversity, the variety of environmental conditions and natural compound production by marine plants, invertebrates, and their microbial communities.'

Brinkmann, Marker, and Kurtböke, *in: Diversity, vol. 9* (2017)

Porifera (sponges) and cnidarians (corals and jellyfish) represent primitive animal phyla that are, except for some jellyfish, not primarily cultured for human consumption. Most species of these phyla are marine and produced for a variety of non-food purposes by mariculture. These two phyla account for more than half of all biologically active, natural compounds that have been extracted from organisms. Nevertheless, the relative amount of these invertebrates produced by mariculture is small compared to other taxa, which is why they are often clustered together with other invertebrates as invertebrates not elsewhere included (nei) in annual production estimates (Figure 9.1).

9.1 Sponges

Sponges are the most primitive multicellular animals (metazoans). They are related to single-celled zooflagellates (protozoans). The scientific phylum name Porifera means pore bearers, which refers to the many pores that transverse their bodies from the outside to the interior cavity. Sponges are sessile animals that are permanently attached to benthic substrate, except for some early developmental stages. Different species range from a few mm to over a meter in size.

Sponges are classified into three classes, glass sponges (Hexactinellida), calcareous sponges (Calcarea), and demosponges (Demospongiae). Demosponges are most abundant and, along with glass sponges, have siliceous spicules. They can be distinguished from calcareous sponges by the resistance of their spicules to acid, which is useful for taxonomic classification (Figure 9.2).

There are at least 8500 described species of sponges and it is estimated that more than 10,000 extant species exist. Optimal culture conditions for these species can vary considerably and first need to be identified by corresponding basic research before attempting commercial scale aquaculture. A sponge farmer must have a good understanding of sponge biology to succeed in culturing these animals. Equipped with knowledge about the biology of the aquaculture species, aquaculturists can provide the optimal conditions for maximal growth and health of the farmed sponges.

9.1.1 Body plan and cell types

Sponges have a simple body plan consisting of one or more (depending on species) cavities (spongocoel) that are enclosed by a wall. Many sponges can be recognized by one or more oscules, which are openings enclosed by cells with large flagella through which water exits from the spongocoel (Figure 9.3). Each spongocoel has a single large opening (the osculum). In some species the spongocoel is a simple hollow cavity while in others it can be differentiated into an elaborate aquiferous system. Although sponges are metazoans, they possess very few cell types (Figure 9.4).

A Primer of Ecological Aquaculture. Dietmar Kültz, Oxford University Press. © Dietmar Kültz (2022). DOI: 10.1093/oso/9780198850229.003.0009

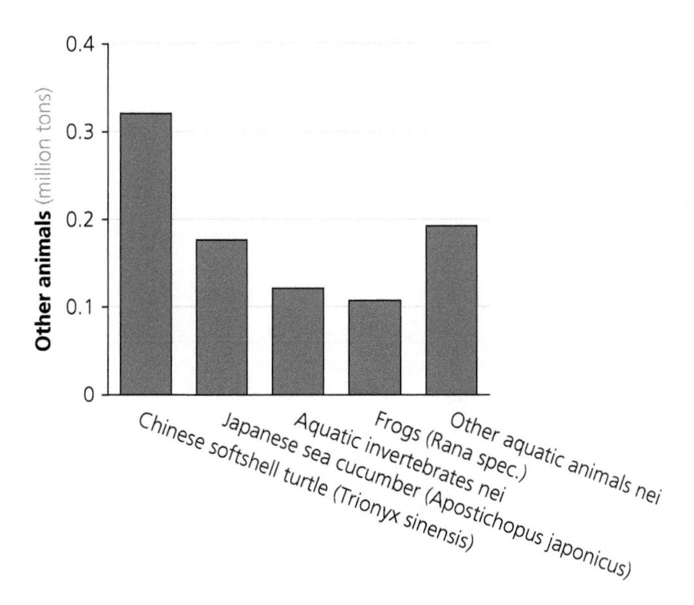

Figure 9.1 Annual production yields of some commercially important aquaculture animals other than fish, molluscs, and crustaceans. Sponges and cnidarians are included in the category aquatic invertebrates not elsewhere included (nei). Data source: (FAO, 2020).

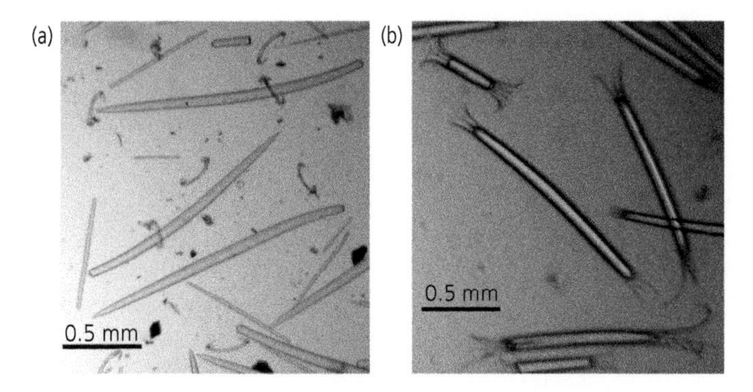

Figure 9.2 Spicules from **A)** a siliceous and **B)** a calcareous sponge collected at McMurdo Sound, Antarctica. Spicules were isolated by incubation in **A)** nitric acid and **B)** sodium hypochlorite (bleach). Photos by author.

Pinacocytes are cells that form the external surface (pinacoderm) of the sponge. Pore cells (porocytes) penetrate that surface to form small canals (pores, ostia) that connect the exterior with the spongocoel. The interior facing the spongocoel is covered with collar cells (choanocytes) that have large flagella on their apical surface, which are used to create a current that draws in water through the porocytes, catch bacteria and microalgae for feeding, and expel the water through the osculum.

Food particles, that is, phytoplankton (microalgae) and bacteria are transferred from choanocytes to the mesenchyme (mesohyl), which comprises the connective tissue, followed by engulfment and uptake by *amoebocytes*, a process called phagocytosis. Amoebocytes break down bacterial and microalgal cells and migrate throughout the mesenchyme to distribute the nutrients extracted from food as well as oxygen to use in aerobic metabolism. In addition, they discharge metabolic waste products and CO_2 or redistribute them to symbionts. Therefore, the water pumped into the spongocoel by the action of choanocytes is deprived of nutrients and oxygen and enriched with metabolic waste and CO_2

Figure 9.3 A typical calcareous sponge from Fiji (*Haliclona spec.*, Demospongiae) with the oscules visible at the top of each branch. Photo by author.

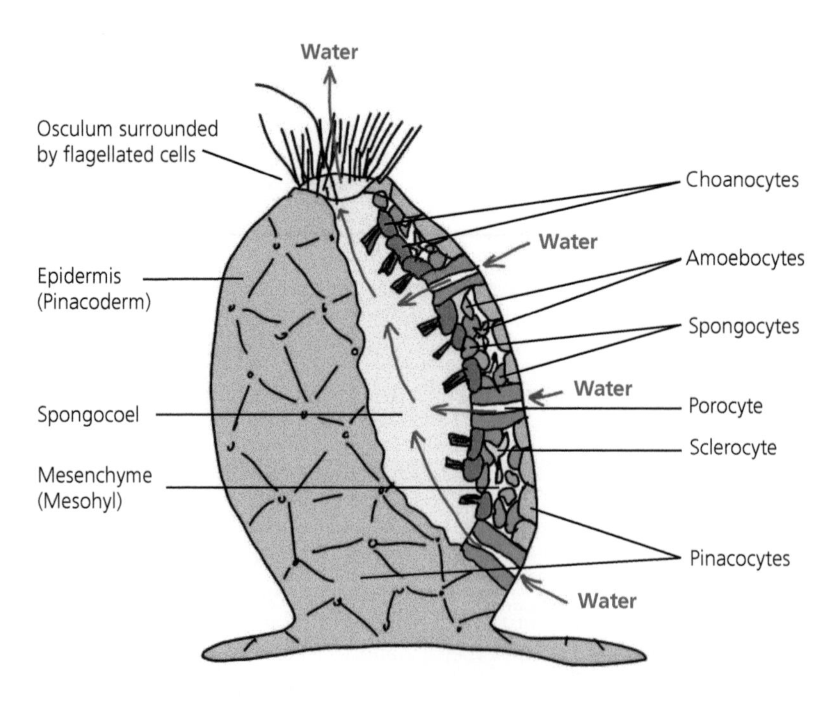

Figure 9.4 Body plan of porifera (sponges) illustrating the positioning of different cell types and the direction of waterflow. The right side of the animal has been cut open to illustrate the position of internal cell types. A single osculum and its corresponding spongocoel are shown.

before being expelled through the central opening (osculum). Some sponges can pump 24,000 L of water per kg biomass on a single day (Taylor *et al.*, 2007).

Sclerocytes excrete either $CaCO_3$ or SiO_2 to form the spicules, which are internal skeletal elements that provide structural support and enable sponges to grow very large. The connective tissue (mesenchyme or mesohyl) that holds all these different cell types together consists of glycoproteins and proteoglycans, the most prominent of which is spongin. These sticky compounds are secreted by mesenchymal cells called *spongocytes*.

9.1.2 Symbioses

Symbiosis *sensu largo* is a mutually beneficial relationship between at least two different species that live in close physical proximity for an extended period of time (Taylor *et al.*, 2007). True symbiotic associations between sponges and microbes are difficult to ascertain but there is clear evidence for high enrichment of specific microorganisms in certain sponges (Brinkmann *et al.*, 2017). Symbiotic relationships between sponges and a great variety of microorganisms, including cyanobacteria, heterotrophic bacteria, archaea, fungi, and microalgae have been documented (Brinkmann *et al.*, 2017). Some species of sponges represent integrated units of trophic microenvironments that are comprised of all three major trophic levels: consumers (sponge cells), decomposers (bacteria), and producers (photosynthetic microalgae). Other species are devoid of prominent symbionts, for example, commercially produced bath sponges do not typically contain photosynthetic symbionts (Duckworth, 2009).

The presence of symbionts is not only dictated by the species of sponge (genetics) but also by the environmental conditions. A great variety of different microbes is often present even within a single sponge,. However, the factors that control microbial diversity in sponges are still not well understood. A related unresolved question is how sponges distinguish between bacteria and microalgae as prey (food particles) or harmful pathogens versus maintaining them as symbionts. Microbial symbionts in sponges have been estimated to contribute up to half of the biomass of some sponges and reach densities of greater than 10^9 cells/cm^3 of sponge tissue (Taylor *et al.*, 2007). They are located either within the mesenchyme (mesohyl), on the outer epidermis (pinacoderm), or on the tissue lining the spongocoel (Figure 9.4).

Symbionts in the mesenchyme of some sponges even include anaerobic bacteria, which suggests that some areas within some sponges might be hypoxic or anoxic (Brinkmann *et al.*, 2017). Like other marine microbial communities, the microbial community of sponges is difficult to propagate in laboratory cultures. Only <1–5% of symbionts can survive outside the host in artificial media, which renders the analysis of symbionts and their functions a formidable research task.

9.1.3 Reproduction

Reproduction is one of the key physiological functions that must be managed well for any aquaculture organism. Sponges have multiple ways to reproduce. Most sponges are hermaphrodites, which means that a single individual or colony can produce both male (sperm) and female (eggs) gametes and reproduce sexually by either self- or cross-fertilization.

Sexual reproduction is currently researched as an alternative to the more common asexual sponge propagation in aquaculture. In this approach sponge aquaculture is divided into an initial hatchery phase that represents intensive aquaculture before open ocean grow-out during the extensive phase. Sexually produced offspring have the advantage of being more amenable to artificial selection of desirable traits such as growth rate (domestication). The trade-off is that the intensive hatchery phase adds significant costs for feeding and managing sponge larvae under controlled indoor conditions. Moreover, it takes significant time (several weeks) for juveniles to develop from fertilized eggs to a size that is comparable to asexual cuttings and suitable for transplantation to the coastal grow-out areas.

Sponges have multiple means of asexual reproduction, including the formation of *gemmules*. Gemmules are small packets of totipotent stem cells that form inside a sponge before being released and settling to the bottom to attach and grow into a new sponge. Gemmules can be dormant after being released during stressful environmental conditions and restart development after conditions improve. This form of reproduction is more typical for freshwater sponges than for marine species.

Budding is another asexual mechanism of sponge reproduction. This mechanism is more common in some species than others and consists of the formation of buds that separate from the parent when reaching a critical size. A third way of asexual sponge reproduction is *regeneration*, which occurs when a sponge is damaged or fragmented into pieces during a storm, strong wave action, or other environmental insults. Each of the fragments can grow into a new sponge because all necessary cell types are present even in very small fragments. This ability of sponges to reproduce from small fragments is commonly exploited for commercial sponge farming to produce offspring in the form of cuttings derived from a parent sponge.

9.1.4 Growth

Sponges have variable growth rates that depend on *temperature*. Tropical species grow faster than polar species. Nevertheless, cold-water sponges can have a lifespan of over a hundred years and some grow very large despite slow growth rates. Although growth is faster at higher temperatures, subtropical and temperate sponges used for aquaculture have higher survival rates and adapt faster when transplanted at lower water temperature (i.e. during the colder season). Lower water temperature may promote transplantation success by enhancing pinacoderm healing at the cuts, decreasing oxidative metabolism, and slowing the growth of fouling organisms and pathogens that explants are susceptible to (Duckworth, 2009).

In addition to temperature and other *abiotic parameters*, growth rate also depends on *biotic parameters* such as food, pathogens, and predators. Since sponges are filter feeders, water movement around them is key to providing nutrients in the form of microalgae (phytoplankton) and bacteria. Strong water currents generally promote sponge growth for this reason but only up to a threshold, above which damage may be inflicted and sponges must direct energy from growth towards damage repair and synthesis of skeletal structures (spicules, etc.) that strengthen the sponge to protect it from strong currents. Therefore, selecting a grow-out area with the optimal water current is critical for achieving rapid growth in sponge aquaculture.

Light is important for sponges with symbiotic microalgae and also encourages growth of fouling organisms that impair water movement through the porocytes. Since commercially cultivated bath sponges are devoid of photosynthetic symbionts (Duckworth, 2009), they are generally grown submersed at a depth of at least 5 m to minimize fouling from algal overgrowth, minimize damage from wave action, and maximize access to phytoplankton and bacteria that promote growth.

When using cutting as a method to produce offspring for sponge farming, it is important to find the optimal trade-off between producing a high number of explants and the time it takes for these explants to grow to harvest size. More explants will produce more sponges but, at the same time, take longer to heal the cuts and grow. The optimum number versus size of cuttings must be determined empirically for each species and each locality where it is grown.

A caveat of reproducing sponges by cutting is that the performance (growth, disease susceptibility, etc.) of explants (cuttings) can differ widely, even when comparing explants from the same individual. Reasons for these differences include dissimilar extents of damage from the cutting process, varying amounts of energy reserves in different pieces, varying health of different sponge areas, different cellular composition, and different size of cuttings. Large differences in physiological performance of sponge cuttings used for aquaculture are undesirable because production rates are rendered less predictable. Moreover, artificial selection for growth or other desirable traits is not feasible if large variation between cuttings from the same parent (genotype) is observed. Therefore, sexual reproduction and nursery of larvae by intensive aquaculture in hatcheries is sometimes explored if the added costs can be offset by the value of product (e.g. a bioactive metabolite commanding a high price).

9.1.5 Bath sponge farming

The use of sponges for sanitary purposes dates back several millennia. Ancient Egyptian and Greek civilizations collected marine sponges from the Mediterranean Sea by freediving and used their skeletons for bathing and washing (Duckworth, 2009). The term bath sponge refers to multiple, taxonomically diverse species that have in common a compact growth form and a sturdy but highly flexible skeletal consistency. For bath sponges, a spongin skeleton consisting of soft, durable, and elastic fibres and a spherical growth form are in highest demand.

Overfishing led to the depletion of bath sponge species in the Mediterranean and Atlantic more than a century ago. Because of overfishing and high demand for bath sponges by the sanitary and cosmetics industries, bath sponge aquaculture has grown significantly during the last few decades. Today most bath sponges are farmed in tropical and subtropical areas where high temperatures promote rapid growth. Some species of farmed tropical bath sponges can double and even triple in size within a year. The biggest obstacle for bath sponge farming are infectious diseases caused by pathogenic viruses, fungi, cyanobacteria, and bacteria that destroy both wild and farmed populations (Duckworth, 2009).

Commercial bath sponge farming is rather small in scale with the largest farms located in coastal waters surrounding Pohnpei Island in the Federated States of Micronesia. The main species of bath sponge farmed is *Coscinoderma matthewsi*, which grows rapidly and has many oscules on the surface, each approximately 1 cm in diameter.

Bath sponge aquaculture in Zanzibar supports the subsistence of families in local communities by producing local sponge species, which have yet to be taxonomically identified (Figure 9.5). Several small-scale pilot sponge farms have also been established in Australia. One farm is located near Masig Island in Torres Strait and focuses on the Pohnpei bath sponge. Another farm is operated by the indigenous community of Palm Island (North Queensland) and focuses on two species, *C. matthewsi* and *Rhopaloeides odorabile*. Because of the small, local scales of these sponge farms, the selection of local species, and the biology of sponges as filter feeders that do not require adding feeds or fertilizer to the ocean, they represent highly ecologically sustainable forms of mariculture.

Most commercial sponge aquaculture is extensive with explants grown off-bottom like beads on a string using the horizontal line method (Figure 9.5a). This arrangement allows sponge farmers to keep the sponges suspended and ensure maximal access to microbial food in the water by maintaining an optimal immersion depth. Sponges can also be grown inside a mesh or net lantern hung vertically from a horizontal line or floating rafts. Although horizontal lines are weighed down to sink to a depth of 5 m, they can be temporally brought to the surface by using floats, a process which allows sponge farmers to conveniently harvest sponges and deploy new sponge cuttings without the need for SCUBA equipment.

When first starting a sponge farm, a site selection considering the discussed environmental criteria is critical. Marine bath sponges have limited tolerance of low salinities and should not be farmed near an estuary or an area subject to large amounts of freshwater runoff. Broodstock for a species that is native to the selected site may be obtained from either wild stock, government nurseries, or another local farm.

Cutting the broodstock donor sponge into fragments must be done with a knife that is sharpened regularly since the spicules will dull the knife rapidly. A dull knife induces significantly more damage when tearing through the sponge tissue, which will increase the time and energy required for repair and healing and impair growth and survival. When handling sponges, it is also important to not keep them out of water for longer than a minute and not squeeze them, to avoid desiccation, stress, and injury.

Bath sponges are harvested when they reach a size of approximately 0.5–1 L volume (0.5–1 kg wet weight) by cutting the horizontal lines and filling the gap with a small, freshly cut explant. Harvested sponges are sun-dried on land for several hours to kill all cells. Unlike for seafood aquaculture products, the cellular tissue of bath sponges is discarded while the spongin skeleton is harvested and marketed.

Figure 9.5 Sponge aquaculture in Zanzibar. **A)** Extensive sponge mariculture using the horizontal line method. **B)** African bath sponge. Photos reproduced with kind permission from Christian Vaterlaus (marinecultures.org).

The process of cleaning the sponge skeleton involves letting the dead sponges rot in shallow containers filled with water for about a month. During this time bacterial decomposers will degrade most organic material leaving only the skeleton behind. The sponge is periodically cleaned by squeezing out water and rotten biomass until no spicules or cells are left, no strong odour is evident, and the skeleton returns quickly to its original shape after squeezing it. Rinsing with clean water removes any remaining sediment particles before the resulting natural bath sponges are washed using (ideally biodegradable) laundry detergent. After the final wash they are ready for marketing and can be stored in a dry place.

9.1.6 Bioactive metabolites

Only a small minority of sponge species have been studied regarding their metabolite profile. Sponge species yet to be explored represent a natural repertoire with enormous potential for the discovery of novel secondary metabolites. Secondary metabolites are compounds produced by organisms that are not essential for normal growth, development, and reproduction. These compounds include bioactive metabolites used to defend against predators and fouling organisms. Chemical defence mechanisms are crucial for sponges, which are sessile (immobile) organisms with a large, exposed surface area and are unable to escape from enemies by behavioural mechanisms.

The number of biologically active secondary metabolites produced from marine organisms has increased greatly during the last fifty years. Since 2008, more than a thousand new bioactive metabolites have been isolated annually from marine organisms reaching an overall number of almost 10,000 in 2010 (Mehbub *et al.*, 2014). A substantial portion of these compounds (almost 2500) were extracted from marine sponges (Figure 9.6a).

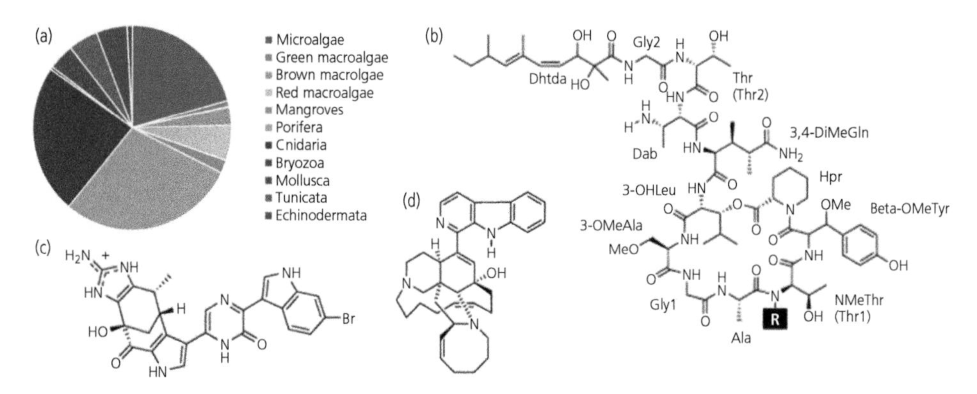

Figure 9.6 Biologically active secondary metabolites derived from sponges (Porifera), cnidarians, and other phyla. **A)** Contribution of different phyla to the number of bioactive metabolites harvested in 2010. Data source: (Mehbub *et al.*, 2014). **B)** Papuamide derived from *Theonella spec.* The R in the black box represents a methyl group (CH_3) for papuamide A and a hydrogen (H) for papuamide B. **C)** Dragmacidine F derived from *Halicortex spec.* **D)** Manzamine A derived from *Haliclona spec.* and *Pachypellina spec.*

The chemical nature of bioactive metabolites synthesized in sponges is very diverse and complex. They include alkaloids, terpenes, terpenoids, peptides, glycosides, sterols, steroids, and many other classes of organic chemicals (Figure 9.6b-d). Their spectrum of biological activities includes very potent anti-tumour, anti-viral, anti-bacterial, anti-inflammatory, anti-parasitic, anti-malarial, and insecticidal effects.

Bioactive metabolites are present in low concentrations of maximally a few mg per kg sponge mass (Duckworth, 2009). Moreover, their concentrations are highly variable and depend on the environmental context, food sources, and life history of the sponge. Often the production of these compounds is increased greatly upon stimulation by specific abiotic (salinity, pH, temperature, hypoxia) or biotic (fouling organisms, predators) stresses.

Tropical and Antarctic species of sponges appear to produce more bioactive metabolites than temperate and polar species in the Northern hemisphere (Brinkmann *et al.*, 2017). Because of the minute amounts of bioactive metabolites present in a single sponge, harvesting wild sponges to produce these compounds is not ecologically sustainable and aquaculture represents an attractive option.

Intriguingly, several of the bioactive metabolites isolated from sponges are not synthesized by sponge cells themselves but rather by symbiotic microorganisms residing within the sponge tissue. It is therefore critical to optimize the aquaculture conditions for the microbial symbiont species that harbours the metabolic machinery for biosynthesis of these compounds. Bacterial genera associated with sponges that are known to produce bioactive compounds include proteobacteria, actinobacteria, acidobacteria, fumicates, cyanobacteria, chloroflexi, poribacteria, bacteriodetes, and verrucomicrobia (Brinkmann *et al.*, 2017). Photosynthetic microalgae living in sponges include cyanobacteria, dinoflagellates, rhodophytes, chlorophytes, and diatoms (Lemloh *et al.*, 2009).

Aquaculture of sponges for the extraction of bioactive metabolites is currently only performed on a small scale. Extensive grow-out systems, like those used for bath sponges, and intensive aquaculture systems in indoor facilities are used to about equal extent. The higher cost associated with intensive, small-scale aquaculture is offset by the high value of the compounds produced.

The environmental conditions can be controlled in intensive aquaculture systems and, if optimal conditions, such as a proper diet, have been determined and can be maintained, then the growth rates of sponges can be very high, resulting in a several-fold increase in biomass within a year. For siliceous species it is important to include a source of silicon in the diet. Intensive aquaculture systems are also used for producing ornamental sponges for the aquarium trade, much as they are used for intensive aquaculture of ornamental cnidarians (Section 9.2.2).

Small-scale experimental sponge aquaculture often serves to identify and characterize rather than mass-produce novel bioactive compounds. Compounds are then extracted from the sponges by using sophisticated purification methods, for example, chromatographic separation after cutting and homogenizing sponge tissue. Once novel secondary metabolites have been identified, methods for large scale ex vivo chemical synthesis of some of the structurally simpler compounds have been developed to increase the scale of their production.

9.1.7 Ecological sustainability prospects

Sponge aquaculture is environmentally friendly because of its small scale and the low position of sponges in the trophic pyramid, which minimizes the ecological footprint required to provide resources for feeding and growing them. As filter feeders they consume tiny (<10 μm) bacteria and phytoplankton (microalgae) and, therefore, counteract eutrophication (algal blooms). For this reason, sponges are commonly classified as 'extractive species'.

Sponge aquaculture reduces the pressure on harvesting wild sponges, which has decimated wild populations of bath sponges in the North Atlantic and Mediterranean. Moreover, the use of local species for aquaculture has the added benefit of restocking natural ecosystems as a side effect of extensive grow-out. Sponge farms can be used as components of biofilters to counteract eutrophication and control microbial population growth. Dissolved organic carbon (DOC) present in wastewater and produced via the metabolism of sponge cells can be converted into symbiont biomass for sponges with bacterial and algal symbionts, the extent of which depends on which species of sponges are being used and the species and density of microbial symbionts present in sponge tissue.

Therefore, the value of sponges for bioremediation depends on the species composition and density of their microbial symbiont community, which is influenced by the environmental and culture conditions. Some sponge symbiotic communities can be regarded as mini ecosystems with an internal trophic cycle that includes all three main levels: consumers, decomposers, and producers. If symbiont density is very high and the metabolic capacity of bacterial decomposers, photosynthetic microalgae, and heterotrophic sponge cells are well matched, then sponges require little food to sustain themselves, for example, they can thrive in highly oligotrophic marine environments.

However, in oligotrophic waters they grow very slowly. In theory, an increased amount of food provided in the form of eutrophic water that is high in DOC stimulates sponge metabolism to assimilate more nutrients into sponge biomass, which would have the dual benefit of increasing growth rates and treating nutrient-enriched wastewater. In praxis, however, it is not trivial to control the conditions in such a way that an optimal balance between all three trophic levels within a symbiotic sponge can be achieved, especially in a dynamic environment.

Sponges cultured in intensive indoor systems where conditions are better controlled than in extensive aquaculture produce large amounts of nitrogenous waste. This waste excretion indicates that either symbionts are absent, or their metabolic capacity cannot match that of the heterotrophic sponge cells when food provided in the form of bacteria and phytoplankton is plentiful. Consequently, under these conditions (and just as for other animals) wastewater is produced, rather than treated, by the sponges and regular water changes or treatment are necessary to maintain optimal culture conditions. Combining sponge cultures with algal turf scrubbers or large bacterial biofilters provides a means for removing the animal waste by conversion into microbial biomass. This biomass can be harvested and used as fertilizer, soil conditioner, biofuel, or animal feed. Alternatively, sponge waste can be removed by integrating sponge and seaweed mariculture. In such an integrated multi trophic aquaculture system, macroalgae utilize ammonia and inorganic nutrients that are derived from sponge organic waste by decomposing bacteria.

The main aspect of ecological concern for extensive sponge aquaculture is the cleaning step when sponges are left to rot in tanks filled with ocean water and later cleaned using laundry detergent. These steps produce concentrated wastewater that, when disposed of into the environment, requires ecosystem services. The added demand for ecosystem decomposer capacity could alter the balance of fragile local ecosystems and drive ecosystem succession onto an undesirable trajectory. Since sponge cleaning is performed in containment (i.e. small tanks), it is possible to use filtration devices for wastewater treatment that include mechanical, biological, and (if necessary) chemical filter stages before disposal. The drawback of using such filters is mostly of economic nature as their acquisition and maintenance adds to the cost of production. Economic incentives are critical to encourage their use.

Cell culture represents a promising alternative for the production of bioactive metabolites. However, not a single cell line has been generated to date from any aquatic invertebrate (except for very few hybrid cell lines) and more basic research is needed to generate such tools. Even though immortalization of sponge cells to generate cell lines has not yet been achieved, sponge cell primary culture is possible. Since primary cultures have a limited lifespan, their proliferative capacity is insufficient to allow them to be used as 'bioreactors' for producing large amounts of bioactive metabolites.

Some bioactive metabolites are synthesized by microbial symbionts, which suggests culturing these microorganisms instead of sponge cells to produce bioactive metabolites in bioreactors. However, because less than 5% of marine microbes is currently amenable to laboratory culture, basic research to improve cultivation approaches is also necessary for microbial symbionts.

Another area of research focuses on genetic engineering of microorganisms that can be easily cultured to express heterologous (foreign) biosynthetic gene clusters that are responsible for bioactive metabolite synthesis in sponge symbionts. Synthesis of complex bioactive metabolites in laboratory bioreactors, as opposed to live sponges, has great potential for improving production efficiency while minimizing the ecological footprint.

9.2 Cnidarians (jellyfish and corals)

Cnidaria is an animal phylum with a more complex body plan than the phylum Porifera. These animals have a true body cavity, a coelenteron, which is functionally more advanced than the spongocoel of porifera because it can be closed off completely from the environment by a mouth. Although cnidarians are more complex than sponges, they are more primitive than most other animal phyla, which have a coelom. The coelenteron of cnidarians lacks a distinctive cell layer called the mesothelium, which lines the coelom of more advanced animal phyla that are clustered together as the Coelomata.

Cnidarians also have a radial symmetry and only a single opening of their body cavity, which further distinguishes them from Coelomata that have bilateral symmetry (Bilateria) and two body cavity openings (mouth and anus). The phylum Cnidaria consists of four classes, three of which are jellyfish (hydrozoa, scyphozoa, and cubozoa) and the other consisting of corals and anemones (anthozoa). The scientific name of the class anthozoa means 'flower animals', which is an apt description of their flowerlike appearance. The purpose of jellyfish aquaculture is very different from that of corals, which is why they are considered separately in the following sections.

9.2.1 Jellyfish aquaculture

The primary objective of jellyfish aquaculture is seafood production. All species of jellyfish consumed as seafood are from the order Rhizostomae of the class Scyphozoa. Most Scyphozoa have a dual-stage life cycle consisting of a polyp stage that is sessile and reproduces asexually to generate a medusoid stage. The medusoid stage is mobile and reproduces sexually to generate larvae that develop into polyps. The medusoid stage, which essentially represents a large, inverted polyp, is used as seafood.

Eating jellyfish has been practised for centuries in Asia, where its consumption is associated with health benefits. The main part that is consumed is the bell of the jellyfish, which is rich in collagen-like proteins, although the tentacles of some species are also sometimes used. Jellyfish are also in demand as ornamental species for aquaria, but commercial-scale production has not been established for any ornamental species. Some species, for example, the moon jellyfish (*Aurelia aurita*) are cultured in research laboratories on a small, experimental scale.

Aquaculture of the flame jellyfish, *Rhopilema esculentum*, was developed in China as a stock enhancement effort to supplement commercial capture fishery harvests after the turn of the millennium (Purcell *et al.*, 2013). Ephyra larvae (juvenile stage of the medusae) are produced and grown to a size of at least 1 cm in a hatchery before being released into the ocean. Medusae reach harvest size (approximately 30 cm) within two to three months before being recaptured. Recapture rates vary between 1% and 3%, which is significant when considering that hundreds of millions of ephyra larvae are being released annually.

Captive grow-out of the medusoid stage of ornamental jellyfish in intensive aquaculture systems or smaller scale marine aquaria requires a special tank design. Since the medusoid stage of jellyfish is mobile and adapted to utilize water currents, such currents need to be mimicked well in the culture tanks. Jellyfish feed on live zooplankton, which can be provided in the form of brine shrimp (*Artemia salina*), a convenient source of live zooplankton that is commonly used along with copepods and rotifers in aquaculture hatcheries for both marine and freshwater species (Chapter 11). Some captive jellyfish species can also be fed frozen zooplankton mixtures.

Markets for jellyfish as seafood are centred in Asia. Expanding these markets by consumer education could potentially help mitigate population blooms of edible species of jellyfish (e.g. the purple-striped jelly, *Pelagia noctiluca*) while providing animal protein and other nutrients in the form of seafood. Jellyfish blooms cause significant economic losses due to human health conditions resulting from jellyfish stinging, impacts on tourism and recreation, fish mortality, clogging of mariculture cages, and obstructing seawater intake for desalination and nuclear power plants (Purcell *et al.*, 2013). However, since jellyfish blooms are periodic in nature, capture fishery harvest of such species would be highly stochastic. Aquaculture provides a means to buffer the impact of fluctuating capture fishery harvest and help stabilize demand.

9.2.2 Aquaculture of ornamental corals

Coral aquaculture has increased greatly since the turn of the millennium due to 1) enhanced incentives for producing corals in captivity versus harvesting them from their native habitat and 2) improvements and reduced costs of technologies for intensive mariculture and ornamental marine aquaria.

Both soft corals (order Alcyonacea) and hard corals (order Scleractinia) are being produced to supply ornamentals for the aquarium trade (Figure 9.7). Coral aquaculture is performed by extensive grow-out in natural habitat and in intensive recirculating aquaculture systems (RAS). For both approaches, asexual reproduction is utilized to propagate cuttings from parent broodstock.

Like sponges, corals employ multiple mechanisms of reproduction including asexual budding, polyp bail-out (stress-induced polyp detachment from colonies), fragmentation, and production of planula larvae. Corals also produce gametes for optional sexual reproduction. Some coral species form separate male and female individuals, while other species are hermaphrodites that produce both types of gametes within the same individual (Gulko, 1999).

For aquaculture, asexual fragmentation provides the easiest, fastest, and most convenient way of coral propagation. Depending on the species, a small piece (usually a few cm long) is broken off from the broodstock parent colony, which is the donor. The donor is then allowed to regenerate such that it can be used repeatedly for obtaining cuttings. No more than 20% of donor colony is removed by fragmentation into cuttings to promote rapid regeneration of the broodstock. Cuttings are mounted onto heavy supports that keep them stationary during grow-out either by gluing calcareous stony corals to the supports or by attaching leathery soft corals to the substrate with suitable strings. The grow-out period is three to six months for soft corals and six to twelve months for stony corals.

Extensive grow-out is practised in the cultured species' native areas. Site selection for placing the coral cuttings attached to supports is critical and

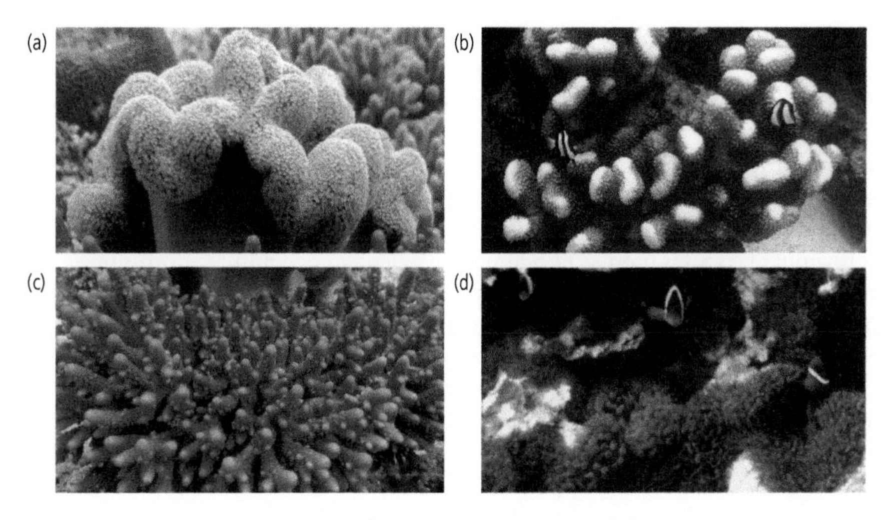

Figure 9.7 Corals are produced by aquaculture to supply ornamentals for the aquarium trade. Examples include soft corals (order Alcyonacea) such as **A)** *Sarcophyton spec.*, and stony corals (order Scleractinia) such as **B)** *Pocillopora spec.* and **C)** *Acropora spec.* In contrast to corals, anemones (order Actiniaria) are still exclusively supplied by capture fisheries from their native habitat. An example is **D)** the bubble tip anemone (*Entacmaea quadricolor*), which is a preferred symbiont species for anemonefish (e.g. *Amphiprion barberi*). Photos by author.

needs to consider multiple criteria, including depth of immersion, limited wave action to promote nutrient supply but prevent mechanical damage, and minimal obstruction of wild coral reef habitat, especially if only few species are being cultured.

Depth of immersion is critical for optimizing coral health and growth, which depends greatly on light intensity. For most species, the preferred depth of grow-out is 1–2 m. If the depth is too shallow, then solar radiation will be too intense, and UV-B and UV-C will damage the coral polyps. Many shallow water corals have molecular defence mechanisms against UV-induced damage, but those mechanisms are only effective at lower doses of UV radiation. These corals express fluorescent proteins that convert high-energy, short-wavelength UV radiation to less-damaging, longer-wavelength visible light.

Coral fluorescence can be observed when using a blue light for night snorkelling in shallow water tropical reefs, an activity that has become popular because of the spectacular appearance of fluorescent corals. However, it should be cautioned that when shining blue light onto corals and activating the fluorescence defence mechanism then the coral may incur damage. Therefore, it is in the interest of reef preservation to not undertake this activity and refrain from developing it into a mass tourist attraction.

While it is important to limit the amount of harmful short-wavelength light it is also critical to provide the corals with sufficient levels of longer wavelength visible light to promote their growth and health. Therefore, immersion of corals in depths >2 m is suboptimal for most ornamental species as the photosynthetic, symbiotic microalgae that are present in most cultured corals do not function optimally, which affects the overall physiological (functional) performance of their coral host.

Symbiotic microalgae of corals are less taxonomically diverse than those of sponges. They belong to the phylum dinoflagellates and are termed zooxanthellae. The most common zooxanthellae in corals belong to the aptly named genus *Symbiodinium* (e.g. *Symbiodinium microadriaticum*). During their symbiotic stage zooxanthellae reside mostly in the endoderm (the inner cell layer) of corals but some are also present in the coral ectoderm (the outer cell layer). They can be directly transmitted to new corals during coral reproduction. However, zooxanthellae also have a free-living stage, which can be transmitted indirectly by colonizing a new coral host. The free-living stage is released from coral tissue when fish that feed on corals (e.g. parrotfish and pufferfish) digest the coral cells but are unable to digest the zooxanthellae, which have a tough cell wall that is missing from animal cells. These photosynthetic symbionts are responsible for most of the carbon fixation in coral reefs, which is estimated at 2.5 kg/m^2 reef area annually (Gulko, 1999).

Intensive grow-out of corals is used as an alternative to extensive aquaculture. In intensive systems it is possible to control lighting and optimize not only the intensity but also the wavelength of light. Lighting technology has improved greatly during the past few decades, which has enabled intensive aquaculture of many coral species. In addition to lighting, optimal water flow rates that supply nutrients but do not damage corals or reposition them, and optimal feeding can be maintained. Raising corals under permanently optimal conditions maximizes their growth and health, which boosts their value for the aquarium trade.

9.2.3 Coral aquaculture for ecological conservation

Both extensive and intensive aquaculture produce corals for conservation purposes, that is, to grow corals for restocking reefs that have been subject to coral die-offs. Coral die-offs are referred to as bleaching events because the corals' living components, the polyps harbouring pigmented symbionts, perish to leave only their bleach-white calcareous exoskeletons behind. Coral bleaching severely affects the ecological balance and global marine biodiversity as approximately 25% of all marine species are found in or near coral reefs. Coral bleaching is reinforced by stresses such as pollution, global warming, sedimentation, ocean acidification, and proliferation of predators, for example, the crown-of-thorns starfish (*Acanthaster planci*).

Half of the world's coral reefs were severely stressed in 2016 and about a third of Australia's northern Great Barrier Reef shallow water corals and half of the corals in the Seychelles were killed between 2014 to 2017. Estimates indicate that almost three-fourths of all coral reefs will disappear by 2030 if the anthropogenic impact on coral reefs, that is, stresses caused by humans, are not alleviated (NOAA, 2021). Disappearance of coral reefs will have dramatic consequences for wave action and protection from storms in tropical coastal areas and for the sustenance of more than half a billion people worldwide whose livelihoods depend intimately on ecosystem services provided by coral reefs. Consequently, aquaculture efforts directed at reviving reef areas that have been depleted by coral bleaching are essential.

Coral restoration efforts are happening in many areas off the coasts of Australia, Africa, Asia, the Americas, and islands in the Pacific and Indian Oceans; these are exemplified by conservation efforts in the Florida Keys, which have experienced a nearly 90% decline of corals in just a few decades. Most coral conservation approaches are based on transplantation of fragments derived from nearby healthy reef areas. However, this approach has several severe shortcomings. First, restocking efforts can only be successful if the environmental conditions in the restocking area are supportive of healthy coral reefs. If these conditions continue to be stressful for corals, then restocking will not solve the problem.

Moreover, since restocking based on transplantation of coral fragments relies on healthy donors, such donors must be obtained from nearby local populations. If local populations have greatly declined (e.g. the Florida Keys), then it may be difficult to find healthy donors; even if they are available, it would be questionable to compromise their health by obtaining fragments for massive restocking efforts. Alternatively, use of non-local donors most likely will alter the gene pool. Thirdly, restocking of depleted areas with fragments derived from donors reduces genetic diversity since fragments from the same donor are clonal. Concerns associated with stock enhancement (Section 6.5) do apply. On the one hand, as many donors as possible should be utilized to establish a high genetic diversity and robust gene pool that promotes resilience in depleted areas. On the other hand, as few donors as possible should be compromised by cutting off fragments to sustain the health of donor reef habitat.

Many of these shortcomings associated with coral transplantation can be addressed by intensive hatchery aquaculture. RAS for growing coral cuttings as ornamental species can be utilized for restocking purposes if local species are cultured. Development of intensive RAS coral hatcheries not only supports restocking and conservation efforts, but also benefits the discovery and production of bioactive metabolites present in corals. Corals and other cnidarians are second only to sponges in the variety of bioactive metabolites that they produce (Figure 9.6).

Sexual reproduction of corals in hatcheries is preferable over asexual fragmentation to generate a high genetic diversity of populations and increase the number of individuals used for restocking. To achieve sexual reproduction of coral species in captivity a significant investment in research and development efforts is needed, especially for those species that are in danger of extinction. An alternative option, which would maximize genetic diversity, is the collection of a fraction of planula larvae produced during synchronized mass spawning events in healthy coral reefs. Millions of larvae can be collected this way but successful hatchery culture of these early embryonic stages into juvenile corals that are useful for restocking also requires more research.

In theory, corals and their symbionts could be genetically engineered to increase their environmental resilience and keep pace with the rate of anthropogenic acceleration of climate change. However, predicting the effects of such 'accelerated artificial evolution' in an ecosystem context and selecting which species should be engineered is exceedingly difficult. This approach may represent a desperate ecosystem scale experiment that may be attempted if the current rate of environmental pollution and climate change cannot be reversed or at least slowed down, and coral reef ecosystems are in acute danger of global extinction. Since the outcome of any true scientific experiment is uncertain, it is preferable to manage the environmental conditions in such a way that current ecosystems and biodiversity can be maintained on a healthy trajectory of succession, rather than resorting to strategies that artificially adapt living organisms and ecosystems to degrading environments.

Key conclusions

- The animal phyla Porifera (sponges) and Cnidaria (corals and jellyfish) include species that are used for mariculture aimed at a variety of non-seafood purposes, in addition to jellyfish seafood production.
- Sponges are the most primitive multicellular animals with a simple body plan that consists of only a handful of cell types. Sponges often harbour microbial symbionts (bacteria and microalgae).
- Sponges can reproduce sexually and by using diverse mechanisms of asexual reproduction, one of which (fragmentation) is commonly used for their propagation in aquaculture.
- Optimal conditions are necessary for maximizing sponge production; temperature and food are particularly important factors. Sponges generally grow more rapidly at higher temperature, but transplantation of fragmented cuttings is best performed at lower temperature to minimize disease.
- Bath sponges are farmed on a small but commercial scale to obtain their spongin skeleton in tropical and subtropical areas using extensive mariculture based mostly on the horizontal line method.
- Sponges are farmed on a small scale in intensive and extensive aquaculture systems to extract a very wide variety of exotic bioactive metabolites that are used to improve human health.
- Sponge aquaculture is highly ecologically sustainable with sponge symbionts contributing to organic waste recycling but can be further improved by treating wastewater generated from cleaning bath sponge skeletons and biofiltration of nitrogenous waste from intensive sponge culture.
- Cnidarians are more complex than sponges and include jellyfish, corals, and anemones. Jellyfish have dual stage life cycles and corals, like sponges, can reproduce sexually and asexually.
- Jellyfish aquaculture provides seafood by hatchery production of juvenile stages, which are used to restock and supplement natural jellyfish populations that are harvested by capture fishery.
- Both extensive and intensive aquaculture of ornamental corals and anemones has increased greatly since the turn of the millennium due to increased incentives, demand, and better technologies.
- Both extensive and intensive aquaculture of corals utilizes transplantation of fragments from donors to initiate cultures; even though hatchery production of

sexually generated larvae would be in many ways preferable, more research is needed to render this approach commercially feasible.

- Coral aquaculture, although small in scale, has great potential for coral reef conservation if it is combined with effective environmental and habitat restoration efforts.

References

Brinkmann, C. M., Marker, A., and Kurtböke, D. İ. (2017). 'An overview on marine sponge-symbiotic bacteria as unexhausted sources for natural product discovery', *Diversity*, 9, p. 40. https://doi.org/10.3390/d9040040

Duckworth, A. (2009). 'Farming sponges to supply bioactive metabolites and bath sponges: a review', *Marine Biotechnology*, 11, 669–679. https://doi.org/10.1007/s10126-009-9213-2

Gulko, D. (1999). *Hawaiian coral reef ecology*. Honolulu: Mutual Publishing.

Lemloh, M.-L., Fromont, J., Brümmer, F., and Usher, K. M. (2009). 'Diversity and abundance of photosynthetic sponges in temperate Western Australia', *BMC Ecology*, 9, p. 4. https://doi.org/10.1186/1472-6785-9-4

Mehbub, M. F., Lei, J., Franco, C., and Zhang, W. (2014). 'Marine sponge derived natural products between 2001 and 2010: trends and opportunities for discovery of bioactives', *Marine Drugs*, 12, pp. 4539–4577. https://doi.org/10.3390/md12084539

NOAA. (2021). 'Coral Reefs'. Available at https://coast.noaa.gov/states/fast-facts/coral-reefs.html (Accessed 3.23.21).

Purcell, J., Baxter, E., and Fuentes, V. (2013). 'Jellyfish as products and problems of aquaculture', in Allan, G. and Burnell, G. (eds.) *Advances in aquaculture hatchery technology*. Cambridge: Woodhead Publishing, pp. 404–430. https://doi.org/10.1533/9780857097460.2.404

Taylor, M. W., Radax, R., Steger, D., and Wagner, M. (2007). 'Sponge-associated microorganisms: evolution, ecology, and biotechnological potential', *Microbiology and Molecular Biology Reviews*, 71, pp. 295–347. https://doi.org/10.1128/MMBR.00040-06

CHAPTER 10

Mollusc aquaculture

'I am not sure where it originated, but oyster spat and other types of invertebrate larvae, as well as the early life stages of finfish, are often referred to by aquaculturists as seed. To my mind, seeds come from plants, not animals . . .'

Stickney, *Aquaculture* (2017)

Molluscs are a diverse animal phylum that includes almost 90,000 extant species, which are grouped into at least seven classes: shell-less molluscs (Aplacophora), tusk shells (Scaphopoda), chitons (Polyplacophora), limpet chitons (Monoplacophora), snails (Gastropoda), bivalves (Bivalvia), and cephalopods (squids and octopi, Cephalopoda). Molluscs have a bilaterally symmetric body plan and a true body cavity, a coelom, that is, they belong to the Bilateria and Coelomata.

Overall, mollusc aquaculture accounts for about 22% of animal aquaculture (Figure 7.5), but it is much more prevalent in marine settings. In 2018, molluscs farmed as seafood accounted for approximately 60% of animal mariculture but only for 0.4% of animal freshwater (FW) aquaculture (FAO, 2020). These production estimates do not include pearl mussels, which are produced in FW and by mariculture and have a high economic value. Even though the economic value of cephalopods traded as seafood exceeded that of other molluscs in 2018, most cephalopods are caught by capture fisheries, which is why they are not considered here. Bivalves are the largest commodity of mollusc aquaculture. This class of molluscs includes the commercially important bivalve orders Ostreoida (oysters), Mytiloida (mussels), Venerida and Cardiida (clams), Arcida (ark clams), Pectinida (scallops), and Pterida (pearl oysters). Snails (class Gastropoda) are also produced by aquaculture, most notably several species of abalone (family Haliotidae, order Vetigastropoda). However, snail aquaculture is practised on a much smaller scale than bivalve aquaculture.

Most mollusc aquaculture is extensive although early developmental stages are also produced in intensive hatchery systems. China is the leading producer country, accounting for more than 75% of global mollusc biomass produced in aquaculture. Several other Asian and Southern European countries, the US, Canada, and New Zealand also produce significant amounts of molluscs.

10.1 Oysters

Several species of cupped oysters (genus *Crassostrea*) accounted for a combined aquaculture production of almost 6 million tons in 2018 (Figure 10.1). The great majority of this is attributed to the Japanese cupped oyster (*Crassostrea gigas*) (Figure 10.2a). In addition, flat oysters (genus *Ostrea*) such as the European oyster (*Ostrea edulis*) are also used in aquaculture. Of all oyster species, *C. gigas* is best suited for aquaculture because it is larger, environmentally more resilient, and more disease resistant than other oysters. It also grows faster. Oyster consumption as seafood has a very long history in Europe, Asia, and Africa (Chapter 2). Immigrants from these continents brought this tradition with them to North America.

California is a good example for how *C. gigas* replaced other species as the primary choice for oyster production. In California, oysters became very popular during the nineteenth century, which led to

A Primer of Ecological Aquaculture. Dietmar Kültz, Oxford University Press. © Dietmar Kültz (2022). DOI: 10.1093/oso/9780198850229.003.0010

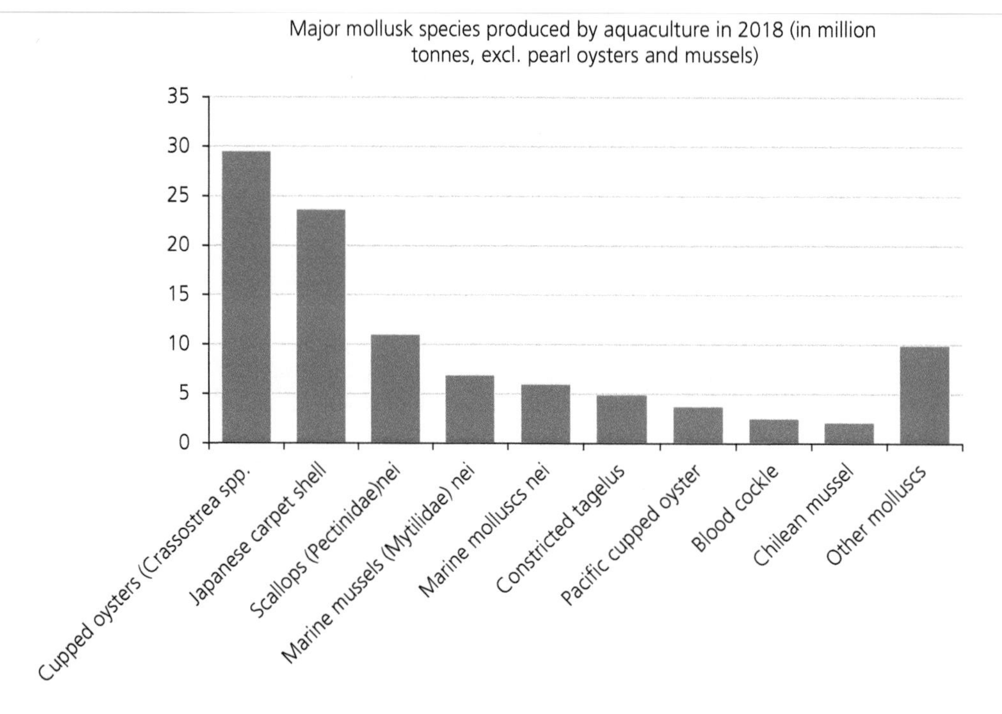

Figure 10.1 Annual 2018 production estimates for different taxa of molluscs. nei = not elsewhere included. Data source: (FAO, 2020).

overfishing and severe depletion of the oyster beds consisting of the native Olympia oyster (*Ostrea conchaphila*) by 1851. At that time, the California gold rush led to such a large increase in oyster demand and price that seaway shipments of Eastern oysters (*Crassostrea virginica*) became economical and profitable.

After the Transcontinental Railroad had opened in 1875, Eastern oyster larvae were transported from the US east coast to the west coast by train and transplanted into San Francisco Bay and other bays along the Pacific coast. In the late nineteenth century, over a ton of *C. virginica* larvae were annually imported and transplanted on the Pacific coast and over a thousand tons of Eastern oysters were produced. The grow-out of Eastern oysters on the Pacific coast did not last long as stocks suffered massive mortality of unknown cause in the early twentieth century, which led to a halt in Eastern oyster larvae imports.

Imports of *C. gigas* from Asia to the US started in the mid-twentieth century to fill the void left by the mortality of Eastern oysters and the overfishing of native Olympia oysters. *C. gigas* emerged as the most suitable species for aquaculture worldwide as it withstands brackish water upwards of 16 g/kg salinity, resists temperatures slightly below 0 °C, grows at temperatures of 10–30 °C, performs well in turbid water, and has higher resistance against pathogens than other oyster species.

However, *C. gigas* production by extensive grow-out in Northern California and the Pacific Northwest was fully reliant on imports of larvae and spat from Asia. The ocean water temperature in these areas is lower than the 20 °C required for *C. gigas* reproduction, which led to very infrequent spawning. To address this challenge, oyster hatcheries opened on the US west coast in the 1980s during the 'blue revolution' (Chapter 2).

10.1.1 Oyster development and hatchery production of larvae

Oyster development incudes several larval and juvenile stages that are typical for bivalves. After fertilization, the egg quickly undergoes multiple

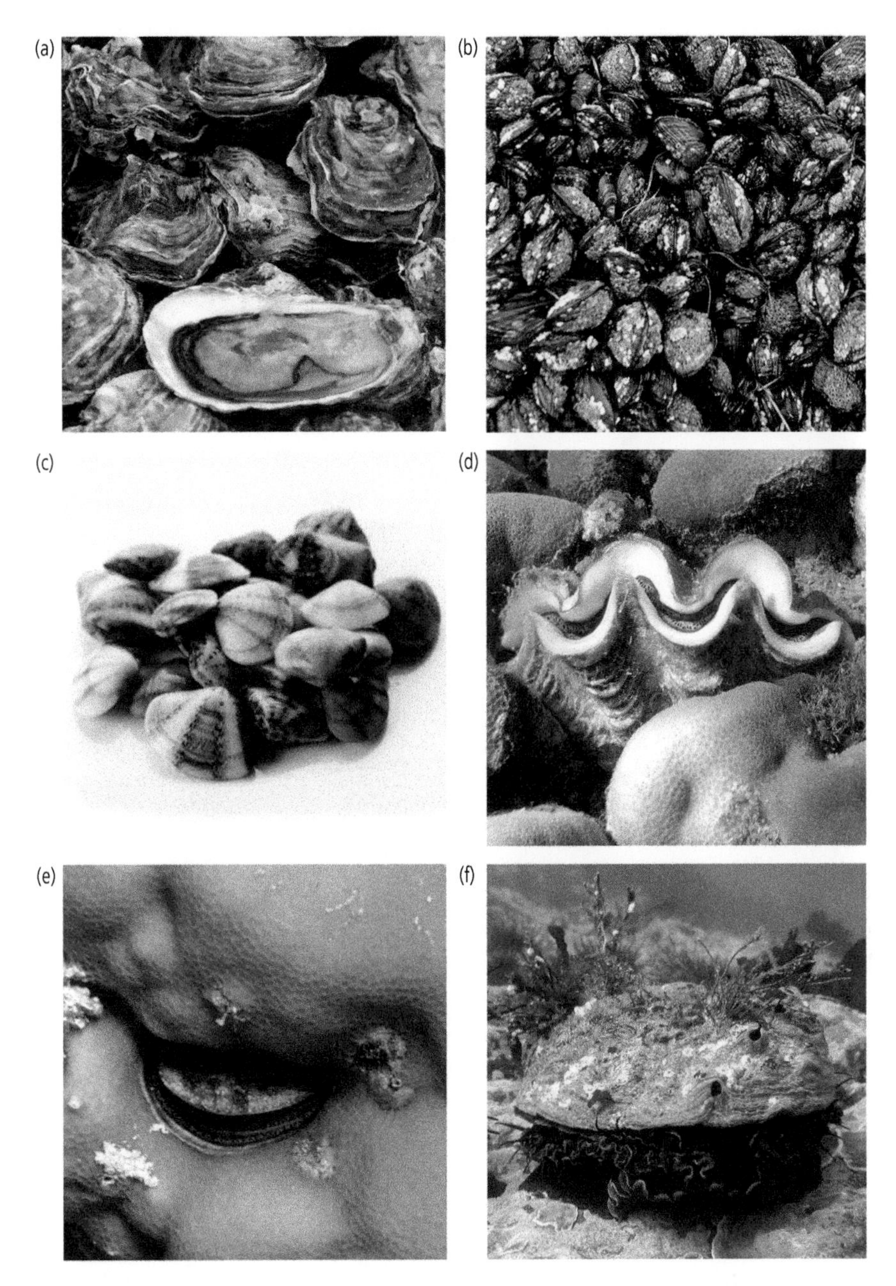

Figure 10.2 Molluscs used for aquaculture. **A)** Japanese cupped oyster, photo by Ben Stern (UnSplash). **B)** California mussel (*Mytilus californianus*), photo by author. **C)** Manila clam (*Venerupis philippinarum*), photo reproduced from Wikimedia (National Institute of Korean Language) under CC BY-SA 2.0 KR. **D)** Giant clam (*Tridacna spec.*), photo by author. **E)** Coral clam (a scallop, *Pedum spondyloidum*) anchored into a lobe coral, photo by author. **F)** Red abalone (*Haliotis rufescens*), photo by Athena Maguire (CDFW, CC-BY 2.0).

rounds of cell division to successively pass through the morula, blastula, and gastrula stages and form a free-swimming trochophore larva within less than a day. Within a week, the trochophore transforms into a veliger larva, which contains a light shell and is still free-swimming (Figure 10.3).

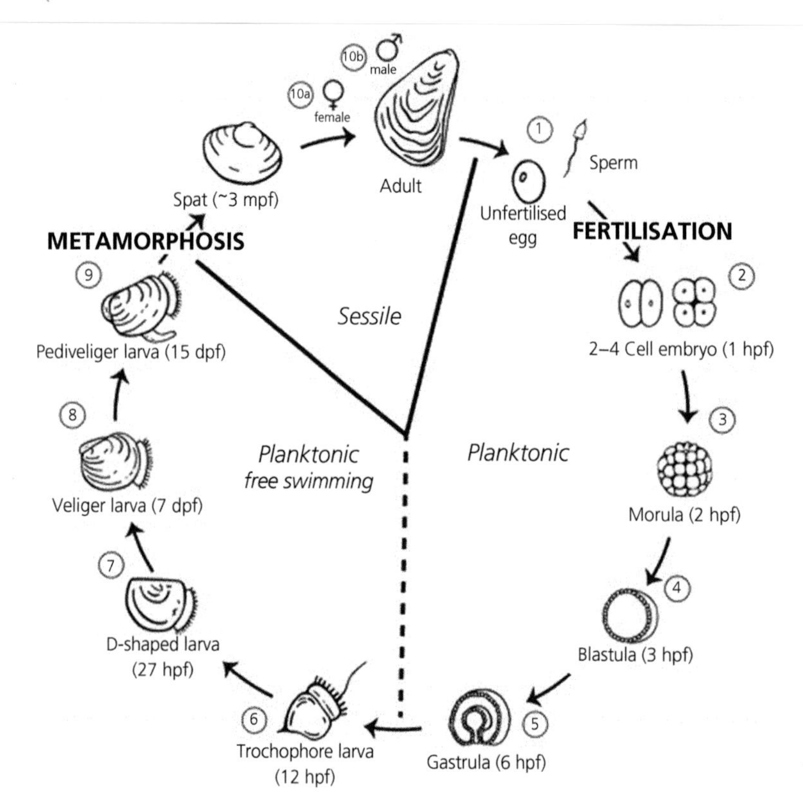

Figure 10.3 Typical life cycle of bivalves illustrated by using oysters as an example. Note that attachment to substrate after metamorphosis varies among different taxa. hpf = hours post fertilization, dpf = days post fertilization. Figure reproduced from Wikimedia under CC-BY-SA 4.0 Licence.

In cupped oysters (*Crassostrea spec.*) fertilization and development of larval stages is external while in flat oysters (*Ostrea spec.*) fertilization takes place in the pallial cavity near the gills of females and larvae develop internally until expelled at the veliger stage. The veliger larva develops a small foot and at the age of approximately two weeks transforms into a pediveliger larva, which becomes heavier as the shell grows and within another week attaches to a suitable substrate. Once attached, the juvenile oyster is referred to as spat. The substrate for oyster attachment is called cultch. In oyster aquaculture unused oyster half shells are recycled as a preferred cultch. Alternative substrates include synthetic fibres, wood, palm threads, cement tiles, and even (before being embargoed) asbestos tiles.

The phrase 'spat on cultch' is common in oyster aquaculture, as this is the form in which juvenile oysters are translocated from intensive hatchery systems to extensive coastal grow-out areas. Cultchless spat can be produced by attachment to very small chips of mollusc shells, which are fully engulfed by the growing oysters. They have better access to oxygen and food than spat on cultch and can be grown out by placing them into wire mesh cages or plastic mesh nets. Traditionally, spat was not raised in hatcheries but collected using attachment substrates placed in spawning areas at times of oyster reproduction, that is, during spat fall. These substrates were then transferred to the most suitable grow-out areas. Although this way of obtaining spat is still practised today on many oyster farms, the importance of intensive hatchery production of spat is increasing.

Intensive hatchery production of oyster spat starts with the selection of broodstock males and females that are conditioned for two to six weeks in flow-through trays while being fed microalgae. The optimal sex ratio is 30% males (1.5–2 years old) and 70% females (2.5 years and older). *C. gigas* are protandrous animals, which means they are sequential hermaphrodites that change sex from males to females as they grow older. Each female will produce between 50 and 100 million eggs. The optimal temperature for broodstock conditioning is about 20 °C and the optimal salinity is 33 g/kg.

Oyster spawning is induced by a short environmental stress, for example, a fifteen- to thirty-minute heat shock in the form of a temperature increase to 30 °C or a salinity shock in the form of a decrease from 33 g/kg to 10 g/kg in small tanks.

Other stresses that have been used to induce oyster spawning include pH shock by adding ammonium hydroxide and oxidative stress by adding hydrogen peroxide. The stress causes the synchronized release of female and male gametes, that is, the broodstock animals expel unfertilized eggs and sperm at the same time into the water. If spawning cannot be induced, then eggs and sperm can be extracted from broodstock animals. It is important to select genetically diverse broodstock for mixing gametes to minimize inbreeding and genetic dilution of stocks. The gametes are mixed to maximize fertilization, which is facilitated by a small tank volume and use of blender techniques. The ratio of sperm to eggs is critical for maximizing fertilization while minimizing polyspermy.

Feeding with microalgae starts one day after fertilization to support rapid development of the trochophore larvae and progression through the veliger and pediveliger stages. The substrate for spat is often bundled together into mesh bags to facilitate transport. Before translocation, spat is usually grown for several additional weeks in nursery tanks and growth trays to increase survival after transplantation to natural habitat.

10.1.2 Extensive oyster grow-out

Extensive grow-out of oysters occurs for a period of one to three years in sheltered bays in coastal areas that are rich in nutrients. On-bottom culture is the traditional method of oyster grow-out, which is still common in some areas. For this approach, the seafloor is often pre-treated to remove predators before transplantation of spat on cultch, with cultch in this form of oyster culture being most commonly oyster half-shells.

Off-bottom culture uses wooden sticks or metal poles that are driven into the seafloor to hold horizontal bars, ropes, cages, or other structures containing the attached oysters. These structures are maintained at a constant distance above the seafloor but exposed to variable depth due to the tides. In intertidal areas, off-bottom cultures are often exposed to air for a short period during low tide, which helps to reduce the activity of fouling organisms and pathogens. Off-bottom cultures are less susceptible to predation than on-bottom cultures since many of the predators are bottom dwellers. Another advantage of off-bottom cultures is that larger amounts of oysters can be cultured in a comparable area as they are surrounded by water on all sides and have better access to oxygen and nutrients delivered by the current. A disadvantage of off-bottom cultures relative to on-bottom cultures is that they obstruct the area above the seafloor and require more resources.

The most environmentally friendly form of extensive oyster grow-out are suspension cultures. These are similar to off-bottom cultures but the distance of oysters to the seafloor changes with the tides while the depth of immersion remains constant. Long lines or floats like those discussed for sponges are used (Chapter 9). Except for anchoring the long lines and floats, no physical obstruction of the seafloor is necessary, and relocation or rotation (fallowing) of the grow-out areas is possible.

Commercial scale harvesting is performed in a variety of ways, which depend on the type of culture system used. Cultchless oysters can be harvested by hand using tongs. Smaller clusters of suspension culture and off-bottom cultured oysters can also be harvested manually. For heavier clusters, cranes are used to detach the oysters and lift them into collection trays on boats. Dredging the seafloor is a common method for harvesting oysters grown with the on-bottom method. In California it takes between one and two years to produce a 10-cm *C. gigas* oyster while it may take up to four years to produce a Sydney rock oyster (*Crassostrea commercialis*) in Australia. Extensive grow-out of other oyster species in other areas also takes between one and four years.

After harvest, oysters are depurated in clean seawater that is often treated with UV, chlorine, or ozone to accelerate the depuration process and remove bacteria that are hazardous for humans. Oyster depuration is legally required in many countries, especially if human pathogens have been detected during mandated regular monitoring in

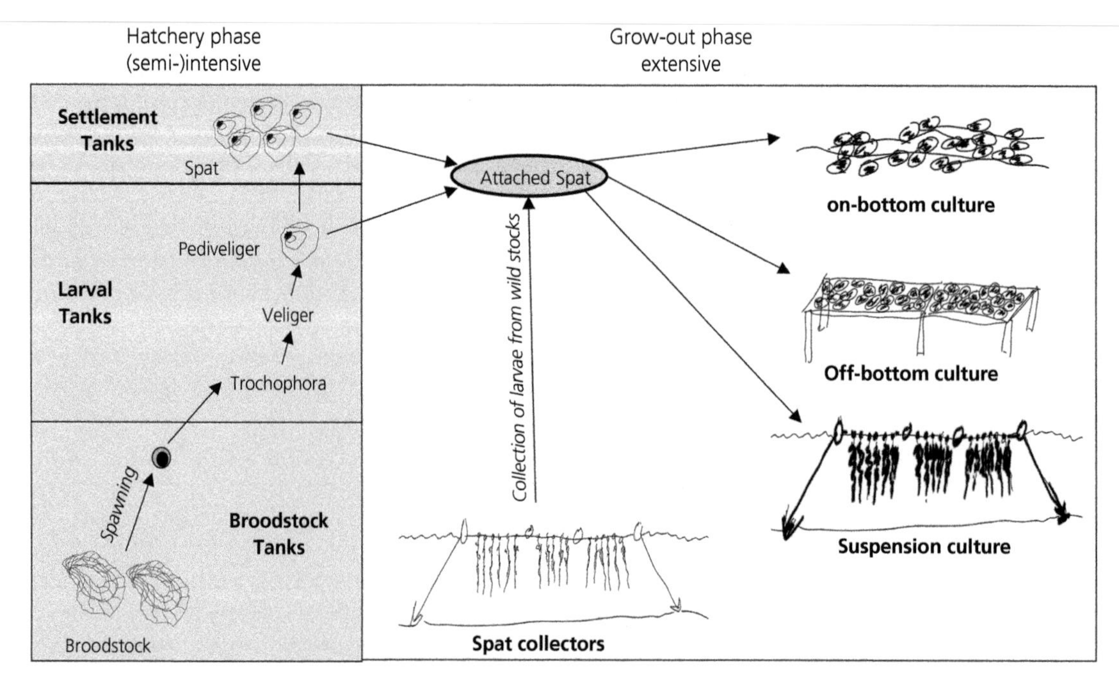

Figure 10.4 Production cycle of the Japanese cupped oyster (*Crassostrea gigas*) illustrating several culture methods. Intensive hatchery production of spat (grey shading) is contrasted with extensive grow-out in natural habitat (larger box with white background).

oyster grow-out areas. Oysters are commonly processed and marketed shackled, that is, they are opened and left attached on a half-shell. The other half of the shell that is removed can be recycled into cultch. Figure 10.4 shows the whole production cycle for oyster aquaculture.

10.2 Mussels

Mussels have two similarly sized, elongated shells and attach to the substrate using their byssal threads, which are very strong, silk-like proteinaceous fibres. Two genera of mussels are commercially important as aquaculture species, *Mytilus spec.* (Figure 10.2b) and *Perna spec.* In Northern Europe, the blue mussel (*M. edulis*) is the primary species cultured. This species is also cultured in China. Its congeners are distributed in Southern Europe (Mediterranean mussel, *M. galloprovincialis*) and Eastern Asia (Korean black mussel, *M. crassitesta*). In the US, *M. galloprovincialis* is preferred for aquaculture since it grows faster and is more disease-resistant than the native congener, *M. trossulus*. The green mussel (*P. viridis*) is cultured in Southeast Asia while small-scale culture of its congeners has been attempted in Venezuela (brown mussel, *P. perna*), India (brown mussel, *P. indica*), and New Zealand (green-lipped mussel, *P. canaliculus*).

Mussel development is very similar to that of oysters, and spawning methods have been developed for the species mentioned here. However, almost all mussel aquaculture is entirely extensive and relies on natural spat-fall. All three grow-out approaches discussed earlier for oyster aquaculture are also used for mussels (on-bottom, off-bottom, and suspension cultures).

When using rafts, vertical ropes are periodically turned over and moved from the centre to the edge of the rafts to encourage even growth. Mussels are harvested after one to two years when they have reached a length of 5–8 cm. The harvest method depends on the culture approach and can be by hand picking, using a crane, or dredging. When using suspension culture, a single line suspended from a raft can weigh almost 100 kg when mussels have reached harvest size. Since mussel shells weigh less than oyster shells, the weight of edible soft tissue in mussels is as high as one-third of total body weight. Mussels are marketed in fresh/live, frozen, canned, smoked, or pickled forms.

10.3 Clams

The Japanese carpet shell (*Venerupis philippinarum*, Figure 10.2c) is the second most produced aquaculture species of mollusc with over 4 million tons harvested annually, second only to the *C. gigas* (Figure 10.1). Other clams that are produced in significant quantities by aquaculture are blood cockles (*Anadara spec.*) and the North American hard clam (*Mercenaria mercenaria*). Australia and Oceania engage in experimental aquaculture of giant clam species (*Tridacna spec.*), which is of particular interest for commercial farming because of its very large size and as ornamental species in the aquarium trade (Figure 10.2d).

Many species of clams are cultured extensively by collecting spat from natural spat-fall in wild habitat and translocating it to suitable intertidal grow-out areas in sandy, shallow bays or as a combination of intensive hatchery production of juveniles, which are then grown-out extensively in bays. Hatchery spawning and propagation of embryonic stages is similar to methods developed for oysters. Intensive recirculating aquaculture systems (RAS) are also used for intensive grow-out of some clams. For example, the hard clam (*M. mercenaria*) is cultured in raceways using recirculating seawater. Harvesting of clams is usually done by hand or by using rakes fitted with a collection net.

In some areas of Asia, coastal fishponds with controlled water flows to and from the ocean are used for clam grow-out. These ponds are often prepared by mechanically and chemically eradicating predators and by fertilization to encourage growth of microalgae that serve as a food source. Since the water in these ponds is exchanged with the ocean every few days, they represent a source of coastal pollution, the extent of which depends on the amount of fertilizer and chemical predator deterrents used.

The Pacific geoduck (*Panopea abrupta*) is distributed widely throughout the Pacific Northwest of the US and British Columbia. This clam buries deep (more than half a meter) into the sediment but remains connected to the surface by a very large siphon, through which it sucks in water for filter feeding. A stock enhancement aquaculture programme that utilizes intensive hatcheries has been developed for production of geoduck larvae in the US. Growth of this species is relatively slow, and harvest is usually done not before six years of grow-out. For this reason, and because harvesting requires manual digging, commercial-scale aquaculture of geoduck for seafood production is challenging.

10.4 Scallops

Most scallops are free-living bivalves that can flap their hinged valves (shells) to move around in spurts. The hinge of scallops is usually more prominent than that of other bivalves. This morphological feature is utilized for aquaculture: holes drilled through the hinge area allow scallops to be attached to cords like beads on a string for suspension culture. Alternatively, cages and on-bottom cultures are employed for scallop aquaculture. Some scallops are sessile and have unique adaptations that protect them from predators, for example, the coral clam (*Pedum spondyloidum*), which is buried deep inside pore corals inhabiting shallow coral reefs (Figure 10.2e).

Over a million tons of scallops are produced by aquaculture annually. Farmed scallop species include the Farrer's scallop (*Chlamys farreri*), bay scallop (*Agropecten irradians*), and Yesso/Ezo scallop (*Mizuhopecten yessoensis*) in Asia, the Iceland scallop (*Chlamys islandica*), the deep-sea scallop (*Placopecten magellanicus*) and bay scallop (*A. irradians*) in North America, the Southern Australian scallop (*Pecten fumatus*) in Australia, and the Great scallop (*Pecten maximus*) in Europe.

Scallop aquaculture is very similar to oyster aquaculture. Early-life stages up to the spat larval stage are commonly produced in a hatchery. During the end of the cold season broodstock animals are collected from either wild populations or stocks from extensive aquaculture and acclimated to a higher temperature and a diet of live microalgae over several weeks. To induce spawning, broodstock is kept emersed (out of the water) for up to two hours to simulate a low tide and then heat-shocked by submersion into seawater that has an elevated temperature. Within a few minutes of submersion into warmer water the gametes (eggs and sperm) are being released synchronously over a period of an hour. Several million eggs can be released by a single female, and fertilization rate is high if eggs are mixed well with synchronously released sperm.

Scallop development includes the same larvae stages (trochophore, veliger, pediveliger) as oysters. The fittest larvae are selected by utilizing their attraction to light. A light source is placed above the water, which separates strong from weak larvae that are unable to reach the water surface. Development is optimal when feeding the larvae with live microalgae. Unlike heavy half-oyster shells, low-weight substrates (palm threads or synthetic fibres) are chosen for attachment of pediveliger larvae for scallop aquaculture to minimize the weight and cost during transplantation. These substrates are kept in nets that initially have a small mesh size and are later (as the spat grows) substituted with larger mesh size nets to maximize water circulation. The spat reaches a size of 1–2 cm in about three months and is then ready to be transplanted for extensive grow-out.

Extensive grow-out of scallops occurs in bays with a sandy or rocky bottom in a depth of about 20 m. The seafloor in grow-out areas is often pre-treated to remove predators, which include echinoderms and crustaceans. This practice significantly increases survival and growth of transplanted spat at the expense of significant destruction of habitat for wild flora and fauna. Often, experimental pre-stocking efforts utilizing a small batch of the hatchery-produced spat are undertaken to test for survival over a period of a month, after which growers select the most promising area for grow-out.

After transplanting the spat to the selected grow-out area at a density of ten to twenty juveniles per m^2 they can be left to grow on their own for approximately two years before they are harvested. After harvest, most scallops are shucked, that is, their adductor muscles (and sometimes gonads) are collected and transported to markets in frozen form.

10.5 Marine pearl oysters and FW pearl mussels

Many mollusc species generate pearls, although only under rare circumstances. For most species, pearls are small, irregular, and rather dull. Sometimes pearl formation occurs in species produced as seafood, which represents a problem. For example, *Mytilus* mussels produced in the Pacific Northwest sometimes form pearls in their mantle tissue, presumably in response to infection by parasitic nematodes (Pillay and Kutty, 2005).

Unlike in most molluscs, regular pearl formation in FW pearl mussels (family Unionidae, river mussels) and marine pearl oysters (family Pteriidae, feather oysters) occurs and the appearance of pearls is more attractive and predictable than those occasional artifacts formed by other mollusc species. Aquaculture of such pearl mussels is highly profitable and not subject to food safety concerns or regulations. Most commercial pearl aquaculture is centred in Oceania and China. China is the leading producer of FW pearls while marine pearls are produced primarily by Pacific Island countries. Aquaculture of pearl mussels started around 1916 in Japan after a technique for implantation of pearl nuclei into marine pearl oysters had been invented.

10.5.1 Composition of bivalve shells and pearls

Mollusc shells consist of three layers. The outer layer is the thin periostracum (cuticula), which consists of proteinaceous organic material including conchiolin and other glycoproteins, proteoglycans, and carbohydrates (Figure 10.5). These materials are produced and secreted by the epithelial cells of the mantle tissue. These cells also secrete calcium carbonate either as calcite or aragonite to form the second (middle) and third (inner) layers of mollusc shells.

The second layer is the chalk-like ostracum (prism layer). It is composed of prism-like calcium carbonate crystals that are between 0.5 and 1 μm in size. The size of these prisms matches the range of wavelengths for visible light (400–700 nm). The calcium carbonate prisms are embedded in conchiolin, which serves as a binding material. Conchiolin's

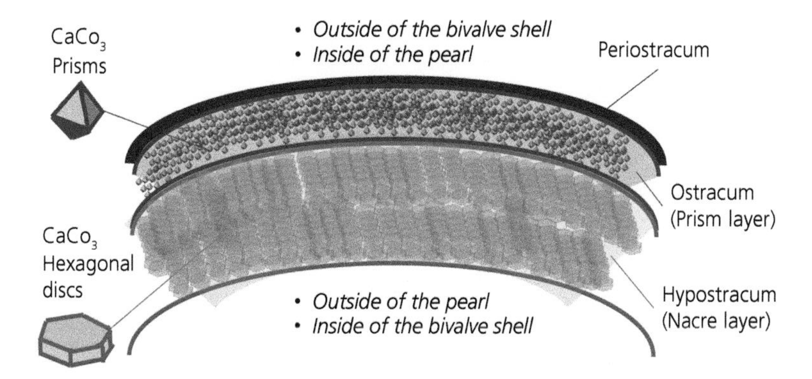

Figure 10.5 Composition of a bivalve shell. A cross-section through a small piece of shell is depicted. The inside layer of a mollusc shell is equivalent to the outside of a pearl. This layer always faces the bivalve's soft tissues and is called the nacre layer (hypostracum). It consists of tiny calcium carbonate crystals that form stacks of thin hexagonal discs (blue). The middle layer of a bivalve shell is the ostracum or prism layer, which consists of tiny calcium carbonate prisms (purple). The outer layer of the shell and innermost layer of pearls is the periostracum (coat layer, black). The periostracum is thinner than the other two layers.

glue-like properties result from quinone-tanning, which is an oxidative polymerization process that takes place after synthesis and secretion of these proteins by mantle epithelial cells.

The third layer of mollusc shells is the hypostracum (laminar, nacreous layer). It consists of nacre, which is made up to 95% of hexagonal aragonite platelets that have a diameter of 5–20 μm. Like the size of prisms in the ostracum, the thickness of these hexagons falls neatly within the range of wavelengths for visible light. Individual aragonite hexagons are arranged in laminar parallel stacks that are held together by conchiolin. Nacre's characteristic iridescent properties are a result of the matching between the thickness of calcium carbonate crystals in the ostracum and hypostracum layers and the range of wavelengths of visible light.

Iridescence represents a change in local colour when the angle of light hitting the nacre surface, and consequently the surfaces of the many prisms and hexagons in the ostracum and hypostracum, are altered. Part of the light is reflected by each hexagonal layer, which generates a dynamic pastel shade colour pattern across the nacre surface. Like the outer bivalve shell nacre, the nacre of pearls becomes thicker, and the pearls increase in size as the bivalves grow.

10.5.2 FW pearls

The large Chinese triangle shell (*Hyriopsis cumingii*) accounts for almost all commercial FW pearl production (Figure 10.6a). In these mussels, pearl formation is artificially induced by inflicting a small injury (cut) into a recipient mussel's mantle tissue and placing a piece of mantle tissue from another individual (the donor) into that wound. The mantle cells from the donor tissue remain alive and reorganize to form a small sac-like structure, the pearl sac (Box 10.1) (Lucas, 2019).

Box 10.1 How do natural pearls form?

When foreign particles intrude into pearl mussels or pearl oysters and damage the mantle tissues in such a way that small pieces of tissue get severed, then it is likely that these small mantle fragments are responsible for the formation of a pearl sac, that is, a tissue that engulfs the intruder such as a sharp object or a parasite to isolate it from the mussel or oyster tissue (Southgate and Lucas, 2008). Sometimes a pearl sac may form even if the intruding agent has already been removed, in which case the pearl does not have a foreign object as a nucleus but is virtually entirely made of nacre (Lucas, 2019). Nacre is the material that forms the inner layer of mollusc shells and outer layer of pearls. This iridescent nacreous layer is also referred to as the mother of pearl.

Using a piece of mantle from a Chinese triangle shell donor mussel that is different from the recipient mussel apparently promotes the segregation of the pearl sac as a piece of mantle tissue that remains distinct from the rest of the recipient mussel's mantle tissue. The donor mantle tissue forming the pearl sac secretes calcium carbonate into the saccular cavity to form the pearl, which is a piece of inside-out mussel shell that is comprised mainly of the third (inner) layer. Thus, pearls from FW mussels are made almost entirely of nacre. Another advantage of FW pearl aquaculture is that a single triangle shell can produce up to 40 pearls simultaneously. Furthermore, Chinese triangle shells may be reused to grow another batch of pearls after harvesting pearls from the pearl sacs.

FW pearl mussels are grown out extensively in large lakes in suspension culture, either by keeping them in net baskets or by drilling a hole in the hinge region and hanging them on cords like beads on a string. It takes about two to six years for FW pearls to reach market size. FW pearls from a single species (*H. cumingii*) can have a wide range of shapes, sizes, and colours.

Figure 10.6 Tahiti black pearl produced by the black-lipped pearl oyster (*Pinctada margaritifera*) mounted on a ring. Photo by author.

10.5.3 Marine pearls

Several species of marine pearl oysters belonging to the two genera *Pinctada* and *Pteria* are used for commercial pearl production. Akoya oysters (*Pinctada fucata*) occur throughout the Indo-Pacific region and in the Red Sea. Their pearls are white or cream-colored with rose, cream, silver, or ivory overtones and are produced by extensive aquaculture in Japan and Southeast Asia. Black-lip oysters (*Pinctada margaritifera*) occur in the Gulf of Persia and the Indo-Pacific region and are produced largely in French Polynesia under the name Tahiti black pearls. Their pearls have a unique dark metal purplish or dark metal greenish colour (Figure 10.6).

Gold-lip/white-lip oysters (*Pinctada maxima*) are the largest pearl oysters in the South Pacific. This species also forms the largest pearls, which are golden to silver-white in colour. They are mostly produced in the Philippines, Indonesia, and Australia. The Sharks Bay oyster (*Pinctada albina*) is another species produced in Australia. The Gulf pearl oyster (*Pinctada radiata*) is distributed throughout the Red Sea and Indo-Pacific. Its range was expanded into the Mediterranean Sea after construction of the Suez Canal and by introduction for aquaculture in Southern Europe. A species of oyster used for pearl production in North and Central America is the Mazatlan oyster (*Pinctada mazatlanica*). The Pacific wing-oyster (*Pteria sterna*) is another North American species, which occurs in the Gulf of California and is produced by extensive aquaculture in Mexico.

Implantation of a nucleus distinguishes pearl production in marine oysters from FW mussel pearl production. Nuclei are made of small (about 0.5 cm diameter) beads manufactured from mollusc shells, that is, they also consist of calcium carbonate. This bead nucleus is implanted into a small pocket that is cut into the recipient oyster gonad. A piece of mantle tissue from a different donor oyster is inserted into the pocket along with the bead nucleus. Cells from this donor tissue form a pearl sac that encapsulates the bead nucleus. The donor cells lining the bead nucleus along the inner margin of the pearl sac then secrete calcium carbonate to deposit nacre onto the surface of the bead nucleus (Lucas, 2019).

Surgical implantation of polished shell beads and donor mantle tissue requires highly skilled technicians that perform this process rapidly and with minimal injury to the oyster. Efficient bead implantation ensures that the level of stress is kept minimal, and the animals resume normal growth and metabolism rapidly after returning them from the laboratory to the grow-out area.

As in other sectors of aquaculture, intensive hatchery production of juvenile stages is increasing for several species of pearl oysters. Hatchery production relieves pressure on wild populations and provides an opportunity for artificial selection of desirable traits such as pearl size and oyster disease resistance. These advantages of intensive hatchery systems come at the expensive of increased costs for technological aids and training of skilled personnel.

Captive spawning of broodstock animals is in principle like the procedures developed for other oysters (Section 10.1) but the exact conditions need to be optimized for each species. Mortality of trochophore larvae of pearl oysters can be very high and larvae need to be monitored carefully within the initial twenty-four-hour post-fertilization period to avoid the proliferation of infectious pathogens that thrive in water containing a high load of decomposing organic material.

Stocking densities must be managed carefully during larval development to maximize health and growth. The time of transplantation from hatcheries to open water grow-out cages is often determined as a compromise between losses resulting from predation and biofouling that affect younger spat more severely than older animals versus the higher costs associated with prolonged holding of spat in a hatchery (Bondad-Reantaso et al., 2007).

Extensive grow-out of pearl oysters is commonly performed using anchored rafts. Vertical lines are suspended from the rafts to hold the oysters in mesh and net cages of various shapes. On-bottom culture is sometimes practised due to legal restrictions that limit the use of floating structures visible at the water surface to avoid 'visual pollution'. However, on-bottom culture is very rare for pearl oysters. Grow-out areas are sheltered bays that are rich in nutrients. Since grow-out is extensive and relies on natural phytoplankton as a food source, maintenance is low and limited to periodic monitoring for health status and removal of biofouling organisms.

10.5.4 Pearl harvest

Oyster pearls are harvested after a growing period of one to three years. The period of growth corresponds to the thickness of nacre, which is an important determinant of the pearl value. Very large pearls containing a small nucleus are most valuable if they are free of blemishes. The thickness of nacre for commercial pearls is routinely determined using X ray machines. The minimal accepted nacre thickness for commercial pearls is usually 1 mm to ensure durability. When the mussels are ready for harvest, they are opened, and the pearls are removed. It is possible to reimplant a new bead nucleus and return the oysters to the grow-out area.

Pearls are cleaned and sterilized by incubating them in bleach solution and sorted based on quality criteria that determine their grade. The highest grade of 'Presidential' is conferred if the nacre is very thick, the shape is perfectly spherical to the eye, the iridescence generates a supreme lustre, and the surface is smooth and >95% free of blemishes. Less perfect pearls are given successively lower grades (AAA, AA, A). Pearls from marine oysters excel at matching in colour, size, and shape, which is why many of them are sold as strands for necklaces. Because of the implanted bead nucleus, they are generally larger than FW pearls and reach up to 2 cm in diameter. Marine pearls generally have a better lustre, greater smoothness, and are more symmetrical than FW pearls. Whether one values FW pearls over marine pearls, or vice versa, is a matter of personal preference.

10.6 Abalone

Abalone are large herbivorous marine gastropods represented by more than fifty species, all within the genus *Haliotis*, that are generally larger in temperate areas than in tropical and subtropical climate zones. Their nacre is very shiny and iridescent and comparable in that regard to pearl-producing bivalves. Consequently and because of the large size and thickness, abalone shells are also used as souvenirs and for making jewellery items.

Abalone, like other gastropods, rarely produce pearls and, if they do, the pearls are highly irregular in shape. Therefore, abalone aquaculture is primarily focused on producing seafood, although abalone pearl farms have been set up on an experimental scale in New Zealand, South Korea, and the US. When all species of abalone (*Haliotis spec.*) are combined, then they contribute the most of any group of gastropods to total global gastropod aquaculture. In 2017 over 160,000 tons of abalone were harvested from aquaculture farms worldwide, 140,000 tons of which were produced in China alone (Cook, 2019).

Commercial abalone aquaculture in China has been increased to top that of other countries after the turn of the millennium. Other commercially important aquaculture gastropods besides abalone include the Chinese mystery snail (*Cipangopaludina chinensis*) with more than 100,000 tons produced by FW aquaculture in 2013 (Lucas, 2019), and top shell snails (*Trochus spec.*) (Pillay and Kutty, 2005). Abalone may well be the most highly priced molluscan seafood. Their large foot muscle can be sliced into steaks and represents a delicacy in Japan, China, Korea, South Africa, the US, Canada, Mexico, Chile, Australia, and New Zealand.

The largest and commercially most important species is the red abalone (*H. rufescens*), which is native to the Pacific Coast of North America (Figure 10.2f). This species can reach a size of 20 cm. In California, aquaculture grow-out of *H. rufescens* in raceway or tank systems takes between two to five years to reach market size, which is 5–10 cm shell length. Other species of abalone farmed in aquaculture systems include the disk abalone (*H. discus hannai*) and Siebold's abalone (*H. gigantea*) in Japan and Korea and the variously coloured abalone (*H. diversicolor*) in Southeast Asia and Australia.

10.6.1 Hatchery production of abalone larvae

Commercial hatchery production of juvenile abalone is becoming more common as wild stocks of many species have been depleted by overfishing. For example, in recent years, the perlemoen or South African abalone (*Haliotis midae*) has been heavily depleted by illegal fishing (FAO, 2020). In California, populations of all native abalone species have been decimated by overfishing and a disease (withering syndrome) in the 1980s and 1990s, which led to a legal ban on all commercial abalone fisheries and has also limited recreational fishing since 1997. Therefore, hatchery-based stock enhancement programmes have been implemented to promote the recovery of wild abalone populations.

Intensive hatchery production starts with the selection of broodstock animals from either wild habitat or an existing aquaculture farm. Spawning is induced in designated tanks by controlling water temperature and imposing an environmental stress, for example, oxidative, UV, or salinity shock, that increases reactive oxygen species (ROS) and stimulates synchronized release of female and male gametes (eggs and sperm). Gametes are mixed to maximize fertilization success and fertilized eggs are collected, washed, and incubated at a density of about 15 eggs per mL until the hatching of free-swimming trochophore larvae. At this stage, the fittest juveniles that swim actively towards the water surface are collected and transferred to larval rearing tanks while those that are weaker settle to the bottom of the spawning tank.

The trochophore larvae start to develop a shell within about a day and transform into veliger larvae, which develop a foot that is characteristic of the pediveliger larvae. The pediveliger are ready to settle onto corrugated plastic substrates within a few days. After settlement, post-larvae develop into juveniles, which are grown for one to several months until the first respiratory pore appears, and the shell length reaches 0.5 cm. Respiratory pores are holes in the shell that are characteristic for abalone to facilitate breathing via their gills. At that stage the abalone juveniles are dislodged from their attachment substrate and transferred to nursing substrates positioned inside larger rearing tanks. Chemical anaesthesia is often used to dislodge the juveniles without physically damaging their foot. Alcohol, ethylcarbamate, and ethylaminopionate are among the anaesthetics used for this purpose.

Abalone larvae are fed suitable mixtures of microalgae and bacteria grown as biofilms on substrate. When the juveniles reach about 0.5 cm in length, they need to be weaned off microalgae to feed exclusively on macroalgae (seaweed) or an artificial pellet diet. Juveniles are grown for about half a year in rearing tanks to reach a size of 2–3 cm length. At this stage, they can be transferred to either extensive or intensive grow-out systems.

The hatchery conditions must be managed carefully to ensure optimal development, growth, and health. Most species of abalone have a narrow salinity tolerance and require ocean salinity (32–35 g/kg) for development and growth. Depending on the length of exposure, the lower salinity tolerance limit for most species is 20–25 g/kg. All hatcheries and farming facilities should therefore be located away from estuarine, brackish water areas.

10.6.2 Abalone grow-out

Grow-out of abalone is done using intensive RAS, semi-intensive flow-through raceway or tank systems, and extensive farming in open coastal areas or tide pools. When intensive systems are used it is important to limit light intensity as too much light inhibits the growth of abalone, which are nocturnal molluscs that seek shelter in low-light areas during the day, for example, in rock crevices and under rock ledges.

Many semi-intensive flow-through systems used for abalone aquaculture have the potential to be reconfigured as RAS such that waste can be collected and recycled. Unfortunately, the trend in abalone production points in the opposite direction. Semi-intensive open systems appear to be preferred over closed RAS because they are more economical from a producer perspective. Current economic incentives are insufficient to favour ecologically sustainable aquaculture over cheaper alternatives.

Extensive abalone grow-out may consist of release and recapture of animals in wild habitat, which is essentially a form of stock enhancement and sea ranching. Another form of extensive abalone grow-out utilizes marine cage culture on floating rafts and seaweed as a food source, sometimes in an integrated multitrophic aquaculture (IMTA) system along with seaweed aquaculture.

Depending on species and culture area, abalone grow out for at least two years to reach harvest size (typically 6–8 cm length). Abalone harvesting from cages is performed by emptying them while recapture of released abalone is more laborious and time-consuming and done by divers. Triploid abalone that show improved growth and reduced fertility have been successfully developed by researchers, but these have not yet been utilized on a commercial scale.

10.7 Ecological sustainability of mollusc aquaculture

Mollusc aquaculture can be highly ecologically sustainable if proper management and technologies are applied since bivalves are filter feeders that represent a low trophic level. The remainder of this chapter discusses possibilities for increasing the ecological sustainability of mollusc aquaculture. This can be achieved by taking better advantage of technologies for recycling animal waste, minimizing invasive species problems, employing aquaculture for species conservation, and increasing biosecurity (Chapter 18) and food safety.

10.7.1 Improving the ecological sustainability of mollusc aquaculture

There are many opportunities for improving the ecological sustainability of commercial mollusc aquaculture farms. Like sponges and cnidarians, bivalves are low trophic level filter feeders and most bivalve aquaculture grow-out is extensive. Bivalves use their gills to remove microscopically small plankton from the water and counteract eutrophication, which is why they are considered extractive species in IMTA systems.

However, bivalves used for aquaculture do not have photosynthetic symbionts and produce nitrogenous waste as all animals do. They excrete waste in the form of dissolved ammonia and soluble and particulate organic matter (e.g. faeces). When bivalves are cultured at high density, then the amount of this organic waste is also high and culture areas may be over-burdened regarding their waste assimilation capacity. Such overburdening of ecosystem service capacity can cause anoxia and an upsurge of anaerobic bacterial populations in the sediment. Mass mortalities, which have occurred in the past, compound this problem severely as the decomposing oyster biomass adds to the environmental burden.

To prevent these undesirable ecological consequences, it is best to employ suspension cultures that can be moved to different locations to allow for fallow periods that promote ecosystem recovery and sustenance. The downside of suspension cultures is that they obstruct a significant part of the upper water column and are perceived as 'visual pollution' of coastal areas. Ecological concerns associated with mollusc aquaculture also include the spread of pathogens and diseases that result from high culture densities and crowded conditions in extensive and intensive aquaculture farms.

On-bottom culture is the least ecologically sustainable form of mollusc aquaculture. It is less productive than off-bottom and suspension cultures and commonly uses dredging to remove predators before transplanting and for harvesting oysters. Mollusc predators include bottom-feeders such as sea stars, large bottom-feeding fish, snails (oyster drills), crustaceans, flat worms, and other worms. In addition to dredging or mopping the seafloor before the application of spat on cultch, applying quicklime (calcium oxide) to oyster beds kills these predators and most other organisms. Therefore, even though raft and line suspension cultures introduce a component of 'visual pollution' and obstruct the upper water column more than on-bottom culture, they are more ecologically sustainable and should be encouraged.

Cultchless cultures are especially environmentally friendly as they require less space and do not use environmentally questionable attachment substrates such as asbestos strips, which have been used as a substrate for spat attachment in the past despite rising concerns about environmental contamination and food safety (Pillay and Kutty, 2005). For abalone, artificial diets have been developed that substitute for seaweed, but they contain fish meal and fish oil (Cook, 2019), which is ecologically unsustainable. Alternative, plant-based diets with a smaller ecological footprint are necessary and currently being developed, but the rationale for shifting abalone feeds from seaweed to pellets made from fish meal and fish oil is ecologically unsound.

10.7.2 Invasiveness of aquaculture molluscs

Another concern associated with mollusc aquaculture pertains to the introduction of non-native species that have the potential to become invasive and profoundly alter ecosystems. Transplantation of larvae and broodstock animals to aquaculture farms outside their native distribution range is often tempting if non-native species show better growth and higher disease resistance than native species.

A good example of an invasive aquaculture species is *C. gigas*, which is larger, grows more rapidly, has better disease resistance, and is more environmentally resilient than other species of oysters. This species has been transplanted from its native distribution range in Eastern Asia to facilitate aquaculture in North America, Europe, and Australia. In Australia, *C. gigas* is now threatening to replace the Sydney rock oyster, which itself is a seafood species that is in high demand and produced by aquaculture.

The most prominent example of an invasive bivalve may be the zebra mussel (*Dreissena polymorpha*), which is native to the Caspian Sea in Asia but has invaded large parts of North America with devastating ecological impacts in the Great Lakes region. The zebra mussel is not an aquaculture species and has not been transplanted via aquaculture activities but rather with ballast water from ships. Nevertheless, it illustrates the grave consequences that invasive species can have on ecosystems and represents a warning that should discourage the use of non-native species for aquaculture. An example of a highly invasive aquaculture mussel is *Mytilus galloprovincialis*, which has invaded multiple locations around the world, including the east and west coasts of North America and South Africa.

Transplantation of exotic species (e.g. *C. gigas*) is believed to account for the occurrence of exotic diseases among native molluscs and the associated risks are well recognized. Such effects add to pressure on native species from direct displacement by farmed molluscs. Thus, biosecurity measures are important not only for preventing disease outbreaks in aquaculture farms but also the spread of pathogens to adjacent wild ecosystems (Chapter 18). Many other mollusc species have been transplanted to non-native habitat to establish aquaculture farms. Bay scallops have been introduced from the US to China in 1982 and now represent a significant portion of Chinese aquaculture production, indicating that they have flourished in their new, non-native habitat. Manila clams have been introduced from Asia to North America and Europe. Chile experienced the introduction of the red abalone in 1977 and the ezo abalone in 1982. Both species have since flourished there.

If transplantation is performed, then at least several precautionary steps should be implemented to minimize potential negative consequences. Translocation of live organisms should always be preceded by proper assessment of the health status of the organisms. The translocated organisms should be free of pathogens or vectors for the spread of diseases. Moreover, appropriate quarantine procedures should be implemented to ensure that no infectious disease is present before transplantation. Although these measures have merit, they are difficult to enforce in praxis and, even if all are accounted for, it is still possible that unintended ecological consequences result from competition of introduced species with local species. Therefore, it is preferable from an ecological sustainability perspective to utilize local species for aquaculture.

10.7.3 Food safety considerations

To prevent biofouling of bivalves during extensive grow-out they are periodically cleaned. Cleaning with pressure hoses and mechanical scrubbers or exposure to air represent more environmentally friendly methods than chemical methods that use bleach or coating of equipment with tar, synthetic organic chemicals, toxic paints, or heavy metals to deter the growth of biofouling organisms. In addition to being toxic to the environment, these harsh chemical treatments also harm the farmed molluscs and can be bioaccumulated in bivalve tissue with potential food safety risks.

Two other food safety concerns with bivalves are the bioaccumulation of toxins and pathogenic bacteria that cause human diseases. Pathogenic bacteria can be killed by cooking oyster meat before consumption. However, they will cause disease when oysters or other bivalves are consumed raw, which is common. Pasteurization and cooking do not prevent poisoning from bioaccumulated toxins. The toxins produced by certain species of cyanobacteria and microalgae forming large blooms (e.g. red tides caused by dinoflagellates) are durable and cause paralytic shellfish poisoning and amnesic shellfish poisoning in humans (e.g. dinoflagellate toxins produced by the genus *Goniaulax*). The resulting economic losses and decreased consumer confidence have severely affected the oyster aquaculture industry in the past, and subsequently triggered significant corrective improvements.

Regular monitoring of oyster grow-out areas for counts of coliform and other pathogenic bacteria and concentrations of toxic microorganisms, clear demarcation of farmed and wild oyster harvest areas, and the introduction of routine depuration after harvest are significant improvements made to increase mollusc seafood safety. Such efforts must be enhanced in the future because virtually all marine aquaculture molluscs are grown out in nearshore habitats characterized by high productivity, which makes them susceptible to coastal pollution and other anthropogenic impacts.

10.7.4 The role of mollusc aquaculture for species conservation

Mollusc aquaculture contributes significantly to ecosystem restoration and species conservation. It has been used for restocking many populations of endangered molluscs, including several species of abalone, oysters, and pearl mussels. Using local species for extensive aquaculture is ecologically most sustainable as it avoids the introduction of non-native species that may significantly alter local ecosystems.

Many FW pearl mussel species are endangered, which is why countries importing pearls from FW mussels require them to be marked as 'freshwater cultured pearls' to designate that they have been produced by aquaculture and to discourage poaching from the wild. Over 100 species of FW mussels were considered declining and have been included on the International Union for the Conservation of Nature (IUCN) red list in 2007. Moreover, in Fall 2021, the US Fish and Wildlife Service has declared eight species of FW mussels as extinct, which were once inhabiting the south-eastern part of the US (Musselman, 2021).

The European FW pearl mussel (*Margaritifera margaritifera*) is classified by IUCN as endangered and its congener, the Moroccan FW mussel *Margaritifera marocana* as 'the most threatened bivalve on earth' (IUCN, 2017). The distribution area of *M. margaritifera* is Central Europe, where its population has declined by 95%. Aquaculture of FW pearl mussels contributes to relieve pressure on wild populations and restore them in habitats that they had been depleted from by overfishing or pollution. Aquaculture of *M. margaritifera* is sometimes practised as polyculture with migratory salmonids (*Salmo salar*, *S. trutta*), which takes advantage of its parasitic trochophore and veliger larval stages that attach to the gills of salmon and brown trout.

FW pearl mussel larvae are distributed across a large riverine habitat by letting them attach to fish gills before releasing the fish for sea ranching. The mussel larvae mature into a pediveliger that detaches from the gills and burrows into the riverbed on the downstream journey of the fish to the ocean. This smart mussel conservation strategy does not harm salmonid production since the parasitic stage of FW pearl mussels is only short-lived and does not cause permanent gill damage to the fish.

A future prospect for conserving pearl mussels and oysters, for preventing poaching, and for preserving natural resources is the development of mantle tissue cultures that can produce pearl sacs and pearls in vitro. This prospect requires significant future research investments as it has proven difficult to generate immortalized cell lines for aquatic invertebrates. However, the existence of primary cultures for many species of marine invertebrates and of cell lines and tissue culture systems for many mammalian and insect species clearly attests that this goal is not unattainable.

Key conclusions

- Mollusc aquaculture accounts for 22% of all animal aquaculture and for 60% of animal mariculture.
- The majority of commercially important molluscs belong to the class Bivalvia, but some species of the class Gastropoda are also used for aquaculture, notably abalone.
- Bivalves develop by successive progression through the trochophore, veliger, and pediveliger larval stages. After attachment, the pediveliger metamorphoses into spat.
- Mollusc aquaculture is divided into two major phases: 1) spat collection and juvenile development and 2) grow-out to market size. Both phases can be intensive or extensive aquaculture, but often the first phase is performed in intensive hatcheries while grow-out is using extensive open ocean systems.
- Three extensive grow-out systems are used for mollusc aquaculture: on-bottom culture, off-bottom culture on poles or racks, and suspension culture using anchored rafts or long horizontal lines.
- Oysters (order Ostreoida) contribute most to mollusc aquaculture production, the Japanese cupped oyster (*Crassostrea gigas*) being the species with the highest seafood yield.
- Mussels (order Mytiloida) used for aquaculture are represented by marine species from two genera: *Mytilus spec.* and *Perna spec.*
- Unlike oysters and mussels, clams (orders Venerida, Cardiida, Arcida) prefer soft sediment substrate into which they burrow themselves. The Japanese carpet shell (*Venerupis philippinarum*) is the second most produced aquaculture species of mollusc.
- Scallops are free-living bivalves that include several species of commercial importance in Asia, Europe, North America, and Australia.
- Aquaculture of pearl mussels (orders Unionida and Pteriida) produces jewellery and is used for re-stocking and species conservation. Commercially important species include the FW triangle shell (*Hyriopsis cumingii)* and marine pearl oysters of the genus *Pinctada*.
- Abalone are aquatic gastropods produced by aquaculture to restock declining natural populations, to process their large foot muscle as seafood, and to produce jewellery.
- Although aquaculture bivalves are low trophic level filter feeders and considered 'extractive species', they are heterotrophs that generate organic waste. Of the many mollusc aquaculture approaches some are much more ecologically sustainable than others, but their utilization requires economic incentives.

References

Bondad-Reantaso, M. G., McGladdery, S. E., and Berthe, F. C. J. (2007). 'Pearl oyster health management: a manual', *FAO Fisheries Technical Paper*, 503, pp. 1–120.

Cook, P. (2019). 'Worldwide abalone production statistics', *Journal of Shellfish Research*, 38, pp. 401–404. https://doi.org/10.2983/035.038.0222

Cook, P. A. (2019). 'Abalone', in Lucas, J. S., Southgate, P. C., and Tucker, C. S. (eds.), *Aquaculture*. Oxford: John Wiley & Sons, Ltd, pp.573–585. https://doi.org/10.1002/9781118687932.ch23

Food and Agriculture Organization of the United Nations (FAO). (2020). *The state of world fisheries and aquaculture 2020*. Rome: Food and Agriculture Organization of the United Nations.

International Union for Conservation of Nature (IUCN). (2017). 'Breeding the most threatened bivalve on earth: *Margaritifera marocana*'. IUCN [online]. https://www.iucn.org/news/species/201711/breeding-most-threatened-bivalve-earth-margaritifera-marocana (accessed 29 March 2021).

Lucas, J. S. (2019). 'Bivalve molluscs', in Lucas, J. S., Southgate, P. C., and Tucker, C. S. (eds.) *Aquaculture*. Oxford: John Wiley & Sons, Ltd, pp. 549–571. https://doi.org/10.1002/9781118687932.ch23

Musselman, A. (2021). 'Why were so many species of mussels just declared extinct?' *Sierra* [online]. https://www.sierraclub.org/sierra/why-were-so-many-species-mussels-just-declared-extinct (accessed 26 October 2021).

Pillay, T. V. R. and Kutty, M. N. (2005). *Aquaculture: principles and practices*, 2nd edn. Oxford: Wiley-Blackwell.

Southgate, P. and Lucas, J. (eds.) (2008). *The pearl oyster*. Oxford: Elsevier Science.

Stickney, R. R. (2017). *Aquaculture: an introductory text*, 3rd edn. Oxford: CABI.

CHAPTER 11

Crustacean aquaculture

'Integrated systems that make use of ecosystems without degrading the resource base on which they depend will be more sustainable and have positive contributions to the surrounding ecosystems and socio-economy.'

Fitzgerald, 'Silvofisheries' in: *Ecological aquaculture: the evolution of a blue revolution* (2007)

Crustacean aquaculture expanded unsustainably during the second half of the twentieth century at the cost of damaging the ecological balance of coastal wetlands (Hopkins *et al.*, 1995). Significant efforts are now made to develop approaches for ecologically sustainable crustacean aquaculture based on knowledge gained during this initial uncontrolled era of expansion. Ecologically sustainable aquaculture requires significant economic and societal incentives to offset the cost of technologies, professional development, and management practices that minimize environmental impacts while maintaining high productivity. Such incentives must match (or exceed) the economic incentives of exploiting ecosystem services. This principle is exemplified by aquaculture of shrimps and other crustaceans, which traditionally relies heavily on exploitation of coastal wetlands.

Crustaceans are invertebrates that comprise a subphylum of the phylum Euarthropoda (arthropods). This phylum consists of true segment-legged animals, which besides Crustacea also includes the subphyla Chelicerata (spiders, horseshoe crabs), Myriapoda (centipedes, millipedes), Hexapoda (insects), and the extinct Trilobita. The four main classes of crustacea are Branchiopoda (gill-legged), Ostracoda (two-shelled), Maxillopoda (jaw-footed), and Malacostraca (soft-shelled) (Figure 11.1). Although some Branchiopoda (water fleas and brine shrimp) and Maxillopoda (copepods) are important live food species for aquaculture, all commercially produced seafood species of crustaceans belong to a single order (Decapoda) of the class Malacostraca. The reason for this focus is simple: decapods grow much larger than the many other taxa of crustaceans.

Like other Malacostraca, Decapoda have five thoracic and six abdominal segments, each of which bears a pair of appendages. Appendages of the head segments are antennae, mandibles, maxillae, and maxillipeds. The five thoracic segments each contain a pair of walking legs (pereiopods). These ten walking legs are highly prominent in decapods. The five abdominal segments each contain a pair of swimming legs (pleopods). The last (sixth) abdominal segment has a pair of uropods as appendages.

Crustaceans undergo multiple rounds of moulting, the frequency of which decreases during their life cycle. Moulting refers to the process of shedding and replacing the exoskeleton to permit growth. This unique aspect of crustacean biology must be considered for aquaculture of crustacean species by adjusting feeding, handling, and water quality accordingly. Moulting causes a stepwise growth process that is characteristic for crustaceans but contrasts to the continuous growth of other aquaculture organisms.

11.1 Overview of crustacean aquaculture

In 2018, 9.4 million tons of crustaceans were produced by aquaculture world-wide (FAO, 2020). This amount is less than the 54.3 million tons of fish

A Primer of Ecological Aquaculture. Dietmar Kültz, Oxford University Press. © Dietmar Kültz (2022). DOI: 10.1093/oso/9780198850229.003.0011

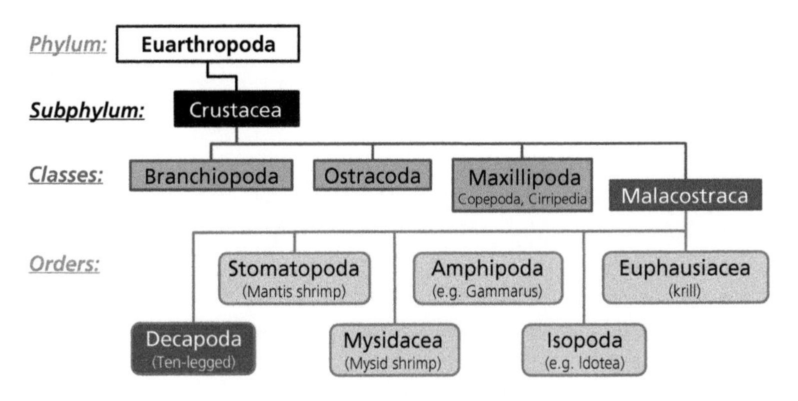

Figure 11.1 Simplified taxonomic classification of the subphylum Crustacea. All commercially produced crustacean seafood species belong to the order Decapoda within the class Malacostraca. However, there are many other taxa of Crustacea that consist of smaller species, some of which are important live food species for commercial aquaculture.

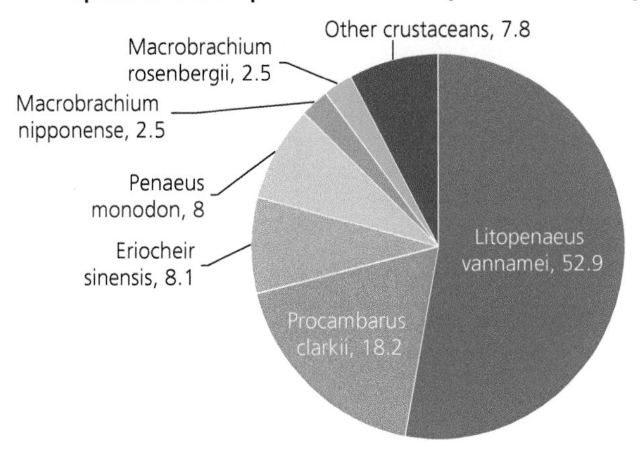

Figure 11.2 Major species of crustaceans produced by aquaculture world-wide. Numbers indicate percent contribution to total crustacean aquaculture production in 2018 (9.4 million tons overall). Data source: (FAO, 2020).

and 17.7 million tons of molluscs produced by aquaculture but considerably more than the aquaculture production of other animal phyla. The economic value of aquaculture crustaceans produced in 2018 ($US 69.3 billion) exceeded that of molluscs ($US 34.6 billion) and was almost half as that of fish ($US 139.7 billion) (FAO, 2020).

A single marine species, the Whiteleg shrimp (*Litopenaeus vannamei*) accounts for more than half of all crustacean aquaculture production. In total, marine shrimps and prawns account for approximately 75% of all crustacean aquaculture production. The remainder comprises freshwater crayfish dominated by a single crayfish species (*Procambarus clarkii*) (Figure 11.2).

The majority of world aquaculture production of crustaceans (90%) is located in Asia with almost all of the remainder produced in the Americas and less than 0.2% in Africa and Oceania (FAO, 2020). Technologies and methods currently exist to produce commercially important aquaculture crustaceans in intensive recirculating system and in ecologically more sustainable polycultures with mangroves or rice. However, encouraging their wide-spread adoption for commercial aquaculture requires more research and incentivization.

11.2 Whiteleg shrimp and other marine shrimps

Crustacean aquaculture is primarily performed as semi-intensive or extensive coastal mariculture of Whiteleg shrimp (*L. vannamei*, Figure 11.3a) and other marine shrimps, sometimes in polyculture with fish. Intensive recirculating aquaculture systems (RAS) are also increasingly used for *L. vannamei* aquaculture. Their benefits for ecological sustainability are discussed in Section 11.7.

Figure 11.3 Major species of crustaceans used for aquaculture. **A)** Whiteleg shrimp (*Litopenaeus vannamei*), photo by John Cameron (Unsplash). **B)** Chinese mitten crab (*Eriocheir sinensis*), photo from Wikimedia (CC BY-SA 4.0). **C)** Giant tiger shrimp (*Penaeus monodon*), photo by CSIRO, Wikimedia (CC BY-SA 3.0). **D)** Red swamp crayfish (*Procambrus clarkii*), photo by Sidney Pearce (Unsplash).

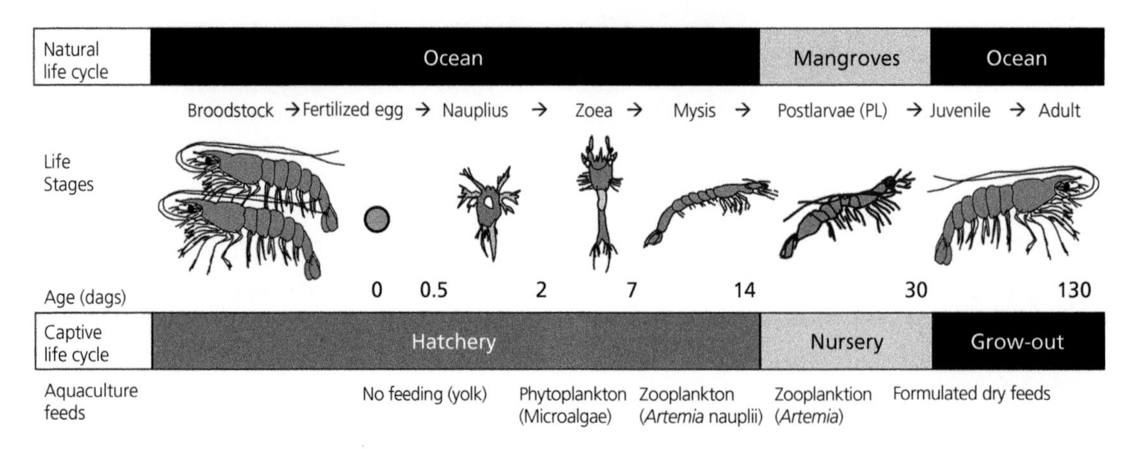

Figure 11.4 Life cycle of Whiteleg shrimp (*Litopenaeus vannamei*). The top bar shows the natural habitat, while the bottom bar depicts the corresponding aquaculture system for each stage.

Shrimps have a migratory life cycle with most developmental and the adult phases spent in the open ocean. In contrast, the post-larvae (PL) and part of the juvenile phases utilize coastal brackish areas such as mangroves, lagoons, bays, and estuaries (Figure 11.4). Because of this lifestyle that utilizes variable habitat, shrimps are environmentally highly tolerant, especially regarding salinity and temperature. Commercial shrimp aquaculture became possible after reproduction of shrimp in captivity was achieved. Japanese researchers first accomplished captive reproduction for the Kuruma shrimp (*Marsupenaeus japonicus*) in 1933 (Fujinaga, 1967).

Even though more than 2000 shrimp species have been described and many dozens of them are used as commercially important seafood in China and other parts of Asia, only six species besides

L. vannamei are utilized on a larger scale for shrimp aquaculture. They include species from the family Penaeidae (genera *Penaeus, Litopenaeus,* and *Marsupenaeus*). Some shrimp species are farmed extensively by sea ranching, which combines attributes of both aquaculture and capture fisheries. Shrimp broodstock for producing larval stages to be released for sea ranching are obtained in two ways. First, broodstock animals can be captured in their native spawning habitat or during migration to their spawning grounds. Although this method is inexpensive, it may threaten wild populations and negatively impact population structure. Alternatively, broodstock can be collected from aquaculture farms. This method has a lower impact on natural population size but is limited to species that mature and reproduce well in captivity and has the potential drawback of genetically diluting wild stocks (Chapter 6).

Historically, the giant tiger shrimp (*Penaeus monodon*) and Kuruma shrimp (*Marsupenaeus japonicus*) have been predominantly used as seafood in Japan and Southeast Asia. Nevertheless, Whiteleg shrimp is now the dominant species produced in Asia and elsewhere in the world. The giant tiger shrimp (*Penaeus monodon*) is also sometimes referred to as giant tiger prawn but in this book the term prawn is not used for marine species or those that spawn in the ocean. The common name for giant tiger shrimp was derived from easily recognizable traits, namely its large size and the yellow-black banding pattern of its abdominal segments (Figure 11.3c). This species can reach almost 40 cm, which makes it attractive for aquaculture. However, because reproduction in captivity is difficult, aquaculture of giant tiger shrimp relies on wild broodstock, which is not always reliable. Broodstock shortages often exist, which is why giant tiger shrimp is exclusively produced in its native distribution range in Asia. Developing a reliable method for captive breeding and improving hatchery survival of this species in intensive RAS represents a significant research opportunity.

The Whiteleg shrimp is the most farmed species of shrimp. Before the 1980s, this species was distributed only along the Pacific coast of the Americas from Mexico to Peru. However, because of massive utilization for aquaculture and escapes into wild habitat it is now considered a ubiquitous species with a worldwide distribution range in tropical and subtropical areas. Several reasons led to the dominance of Whiteleg shrimp in crustacean aquaculture, including easier reproduction in captivity, highest hatchery survivals, better tolerance of high-density culture, better use of dry feeds, lower protein requirements in feeds, and high environmental tolerance compared to other shrimp species. Moreover, Whiteleg shrimp grows rapidly at a more uniform rate than other shrimps and is most amenable to genetic improvement of stocks by selective breeding. Captive stocks that are either specific pathogen free (SPF) or specific pathogen resistant (SPR) have been generated and are now used for aquaculture. Furthermore, this species tolerates unilateral eyestalk ablation well. Eyestalk ablation is a common procedure used in crustacean aquaculture to increase the fecundity of broodstock.

11.2.1 Shrimp hatcheries and nurseries

Division of shrimp aquaculture into hatchery, nursery, and grow-out stages supports ecologically responsible intensification of shrimp aquaculture by reducing space and water required while increasing stocking density. In addition, intensification by utilizing these three production stages maximizes shrimp survival and growth. Shrimp aquaculture is most efficient when it is divided into these three stages as they mimic the conditions experienced during the natural life cycle (Figure 11.4).

Shrimp hatcheries principally use two different approaches of culture: the community tank method and the Galveston method. The community tank method does not separate different larval stages into separate tanks and utilizes fertilization to induce the growth of planktonic food organisms in shrimp tanks. Under these conditions, survival is relatively low, often less than 50%, but management costs are also low. Hatcheries employing the community tank method are usually small. Larger hatcheries commonly use the Galveston method, which utilizes diverse plankton grown in separate tanks for feeding shrimp larvae and often separates nauplius and zoea larvae from mysis larvae in different tanks. Under these conditions, which require more elaborate management, survival is as high as 90% and feeding can be better customized, improving growth rate (Figure 11.4).

Lack of feeding during the nauplius stages maximizes water quality. Nauplii do not require external food and survive exclusively on yolk for two to three days before transforming into zoea larvae. They develop in three to six sub-stages, which are delineated by intermittent moulting. Moulting of nauplii and the start of feeding represents a critical period during which mortality can be high. Zoea larvae (three sub-stages) feed on small phytoplankton (microalgae) and transform into mysis larvae after three to six days. Mysis larvae (three sub-stages) feed on small zooplankton and transform into PL after three to five days. Shrimp hatcheries often use brine shrimp (*Artemia salina*, Section 11.6) nauplii to feed mysis larvae. The PL stage is fed with adult brine shrimp and weaned off live food to consume dry feed pellets that are formulated to optimize growth and health.

Formulated dry feeds of shrimp are complex and consist of major nutrients (i.e. proteins, polyunsaturated fatty acids (PUFAs) and other fatty acids, and carbohydrates). They also include many micronutrients such as vitamins, minerals, and cofactors. One aspect of formula dry feeds currently under scrutiny is the use of significant amounts of fish meal and fish oil, which reduces the ecological sustainability of such feeds (Box 17.3). Eliminating these ingredients by using alternative sources of PUFAs and other essential micronutrients will make shrimp aquaculture more ecologically sustainable. The use of formula dry feeds reduces the potential of pathogen introduction via live food.

Hatcheries are increasingly utilized for shrimp aquaculture, but shrimp nurseries are less common and PL are also transferred directly from hatcheries to grow-out ponds. When using nursery ponds, the PL are grown for two to three weeks to reach a size of about 2 cm before being transferred to grow-out systems. For example, in northwest Mexico PL are either directly transferred to grow-out ponds when they reach a weight of 6–10 mg or, alternatively, they are first reared in nursery ponds and then transferred to grow-out ponds when they have reached a weight of 80–200 mg.

Shrimp nurseries represent an intermediate stage that mimics coastal nursery grounds for PL. Nursery tanks or ponds are used to maximize survival and health during acclimation from hatcheries to grow-out systems. The optimal environmental conditions change during larval development as different developmental stages experience different salinity, temperature, pH, and other abiotic and biotic parameters (Figure 11.4). Using intermediate nursery systems to grow PL and acclimate them to the conditions of grow-out systems affords the opportunity to mimic coastal brackish areas for optimal shrimp development. Including a nursery stage in the shrimp production cycle decreases stress and disease susceptibility while increasing acclimation efficiency, productivity, and economic gain.

11.2.2 Shrimp grow-out

During the final grow-out stage juveniles grow to adult shrimp until they reach market size. Traditional extensive and semi-intensive systems for shrimp grow-out heavily rely on ecosystem services. They require large coastal areas for ponds and high exchange rates of water in sensitive tidal and mangrove habitat. Semi-intensive pond systems that utilize large amounts of feeds and fertilizer are especially taxing on the environment.

Shrimp maturation takes place during the grow-out stage. Although Whiteleg shrimp can reach a maximum size of 23 cm in length, market size is often less than half of that. Shrimp are commonly harvested at four to five months of age to prevent redirection of energy from muscle growth to gonadal development and avoid a resulting reduction of meat quality. Outdoor ponds are most used for shrimp grow-out, but raceways and greenhouse tanks are also used. The latter are either open or closed systems. Open systems require large amounts of clean water and depend on ecosystem services for wastewater treatment. Fully or partly closed RAS demand a greater up-front investment and technical management skills than traditional semi-intensive pond aquaculture or open raceway and tank systems. However, closed systems facilitate division of developmental stages in separate hatcheries, nurseries, and grow-out systems, which maximizes growth and survival.

Additional advantages of shrimp aquaculture closed systems include higher feed utilization, reduced generation of waste, and greatly reduced risk of diseases by limiting introduction of pathogens and predators. RAS require minimal water to replace evaporative losses, water sequestered in shrimp biomass, and other minor losses. They afford a high level of biosecurity by minimizing pathogen introduction, employing sterilization devices such as UV irradiation, ozone, or chlorine treatment sumps, and utilizing SPF shrimp stocks. RAS require minimal water to replace evaporative losses, water sequestered in shrimp biomass, and other minor losses.

Intensive RAS units for shrimp aquaculture may use bioflocs, which are aggregates consisting of particulate waste, decomposing bacteria, and other decomposers (Box 17.4). Shrimp are detritivores that can readily utilize bioflocs to derive nutrients from bacterial biomass. Faecal matter excreted by shrimp serves as a substrate for recolonization of bioflocs by bacterial decomposers, which can be reutilized as a food source for shrimp. This repeated utilization of bioflocs for supplemental feeding renders nutrient recycling very efficient, reduces feed costs, improves feed efficiency, and reduces the environmental impacts of shrimp production.

To minimize transport stress, hatcheries, nurseries, and grow-out systems are ideally near each other. However, in praxis, this is often not the case and shrimp are transported for many hours. During transport, they are kept at high density in sealed plastic bags topped off with pure oxygen to maximize oxygen saturation for an extend time. The plastic bags are placed inside Styrofoam boxes or canvas barrels to maintain a constant temperature. Shrimp farmers in north-west Mexico transport PL in large plastic containers of 1 m^3 or larger volume on trucks while providing aeration via air stones. The transport temperature is kept low (18 °C) to minimize metabolism, movement, stress, and oxygen consumption and maximize survival. PL are 1–2 cm in length when they are stocked into grow-out systems. They are typically grown for four months in these systems until they reach market size, when they are harvested and processed for distribution to markets.

11.3 Euryhaline crabs

The Chinese mitten crab (*Eriocheir sinensis*, Figure 11.3b) is a coastal species that accounts for approximately 8% of all crustacean aquaculture production. The mud crab *(Scylla serata)* is another coastal species used for aquaculture. Like Whiteleg shrimp and other marine shrimps, these crabs are euryhaline, that is, they have a wide salinity tolerance range for inhabiting fresh water and brackish water and they are coastal species.

Chinese mitten crabs have a catadromous life cycle, meaning that they mate in marine or brackish water, after which gravid females migrate to freshwater. Gravid females that have mated are ready to fertilize eggs, which then develop into larvae that hatch in freshwater. The larvae develop into juveniles that continue to grow as adults until they reach a final size of approximately 10 cm in diameter. Chinese mitten crabs are mostly produced in China and Korea at the centre of their native distribution range. Consumption of these crabs as seafood has a rich cultural tradition in this part of Asia.

11.3.1 Mitten crab hatcheries and nurseries

Like shrimps, crab broodstock can be obtained from either wild populations or from other aquaculture farms. Broodstock should be free of pathogens and is kept in concrete tanks or earthen ponds. Rearing is similar to that of shrimp broodstock, but bricks, rocks, or other shelters are often added at the bottom of the tanks or ponds.

Chinese mitten crabs are opportunistic omnivores that prefer to feed on invertebrates and mature best when fed live molluscs, small fish, and smaller crustaceans. Optimal conditions for mating besides diet include a brackish salinity of half-strength seawater (17 g/kg) and 10 °C water temperature. Fertilized eggs are attached to the abdomen of females and held in place by the pleopods. Hatching of larvae takes place in freshwater after two to four months of embryonic development, which can be accelerated by increasing the temperature.

Selection for the most robust larvae is commonly performed at the zoea stage by collecting only larvae that show strong phototactic swimming behaviour and can be caught at the water surface when using a light at night. Smaller larval stages are fed a mixture of phytoplankton and small zooplankton, while later stages are fed exclusively larger zooplankton. A net is added to tanks to provide substrate and prevent cannibalism when zoea larvae transform into megalopa larvae.

Optimal salinity for larval development is around 20 g/kg but a wide range is tolerated. Before moulting of megalopa larvae to transform into juvenile crabs they are weaned off live food, which is gradually replaced by other feeds, including meat derived from other crustaceans, molluscs, and fish, but also dry pelleted feeds. The entire phase of larval growth after hatching typically takes between three to four weeks. Most of Chinese mitten crab hatcheries that produce megalopa larvae for distribution to grow-out areas are in the Yangtze River Delta area in China.

The nursery phase of Chinese mitten crab aquaculture consists of two stages. The first stage takes approximately one month until juvenile crabs reach a weight of 0.2 g. It is performed in small ponds. The second stage takes about six months and is performed in larger ponds or rice fields that have been modified to grow crabs to a weight of several grams. The water level in crab nurseries is gradually raised from 25 cm to 75 cm during the culture period. Crabs feed on plankton, macrophytes, and benthic invertebrates that naturally occur in nurseries. Supplemental feeding with crop plants, minced fish, and dry formula feeds is common.

11.3.2 Mitten crab grow-out

Grow-out of Chinese mitten crabs takes place in ponds, net enclosures, or rice fields. Crabs are obtained from nurseries when they weigh 5–10 g and then grown out to market size of 80–200 g wet weight. For pond culture, extensive pre-treatment of ponds is performed before stocking with juvenile crabs from nurseries. Pond pre-treatment includes disinfection by liming, repopulation with aquatic plants and invertebrates that serve as natural food sources, and fertilization.

As an alternative to pond grow-out, net enclosures are placed into larger lakes in areas of slow water flow and good water quality at a depth of 1–2 m. Supplemental feeding and pre-stocking of mud snails maximizes growth in net enclosures. When rice fields are used for grow-out of *E. sinensis*, then they are modified in similar ways as for nursery production (e.g. by fencing and adding channels) except that perimeter channels are deeper (1–2 m). Yields are similar for pond, net-enclosure lake, and rice paddy grow-out systems and range between 0.2–1 ton per hectare surface area annually.

11.3.3 Mud crab aquaculture

Another important aquaculture crab species is the mud crab, which has a more southerly distribution range than the Chinese mitten crab and a higher thermal optimum. It is economically important in tropical countries of the Indo-Pacific region. Like Chinese mitten crabs, mud crabs tolerate a wide range of salinities (5–33 g/kg) and are opportunistic omnivores that preferably feed on a carnivorous diet. However, unlike Chinese mitten crabs, mud crabs do not occur in freshwater.

The fecundity of mud crabs can be very high with one female crab producing up to two million eggs. As the embryos develop, egg colour changes from bright orange to greyish or black before hatching occurs. Hatching takes place after two to three weeks, depending on temperature. After hatching, larval development of mud crabs proceeds through zoea and megalopa stages before transforming into juvenile crabs within three to four weeks.

Grow-out of mud crabs is usually performed by semi-intensive aquaculture in estuarine and tidal mud flats. These areas are brackish with a salinity that fluctuates around half that of seawater. Maximal feeding and growth occur at a temperature of 25 °C. To reduce food waste and minimize cannibalism, crabs are fed after dusk or before dawn in adequate amounts that are distributed evenly. Grow-out of mud crabs to market size (200–400 g) takes a similar amount of time as that of shrimps (four to five months).

11.4 Freshwater prawns

The giant freshwater prawn (*Macrobrachium rosenbergii*) has a short larval cycle, is readily amenable to reproduction in captivity throughout the year, and attains a large size. The maximum length is 25 cm for males and 15 cm for females. Therefore, this species is attractive for commercial aquaculture. It contributed to 2.5% of all crustacean aquaculture in 2018 (Figure 11.2). The same contribution to aquaculture was made by a congeneric species, the oriental river prawn (*Macrobrachium nipponense*). Both species are native to Asia but *M. nipponense* is a more temperate species and much smaller (max. 9 cm length) than *M. rosenbergii*, which is a subtropical/ tropical species.

The main advantages for aquaculture of *M. nipponense* over *M. rosenbergii* are its low temperature hardiness and completion of the full life cycle in freshwater. Even though *M. rosenbergii* is considered a freshwater prawn, the adults are found in both freshwater and low salinity brackish water and larval development requires brackish water (12 g/kg), which complicates aquaculture of *M. rosenbergii* relative to *M. nipponense*. Dimorphic *M. rosenbergii* broodstock is used for reproduction in hatcheries. Mature males are much larger than females and have a greatly enlarged pair of second pereopods. Alternatively, gravid females can be collected from wild habitat, but this increases the risk of introducing diseases.

11.4.1 Prawn hatcheries

Optimal reproduction and early development require high temperatures (28 °C for *M. rosenbergii*). Larvae hatch three weeks after fertilization when eggs have turned from orange to black or greyish colour. Like for shrimp aquaculture, feeding of early developmental stages is either performed by fertilizing tanks and growing phytoplankton and zooplankton communities directly in the larval rearing tanks, or by growing food organisms separately and periodically adding them to the larval tanks, which are supplied with clear, filtered water. The former method is referred to as the green water method and roughly corresponds to the community method of shrimp aquaculture. The latter is the clear water method roughly corresponding to the Galveston method of shrimp aquaculture.

Various combinations of these two principal methods have been practiced in both open flow-through systems and closed RAS. When using RAS, then disinfection procedures (UV irradiation, ozonation, chlorination/de-chlorination) can be incorporated to greatly reduce the risk of infectious disease outbreaks during susceptible stages of development. Moreover, water consumption and waste disposal are minimized, and larval stocking densities are higher in RAS.

Freshwater prawns develop by succession through the same planktonic developmental stages as marine shrimps (nauplii, zoea, and mysis) and feeding of larvae is similar to that described above for shrimp. Under optimal larval rearing conditions, the PL stage is reached three to four weeks after hatching. At that time, prawns are just under 1 cm long and acclimated to freshwater before being transferred to either nurseries or grow-out facilities. The survival rate of *Macrobrachium* species is approximately 50% for most hatcheries.

11.4.2 Prawn grow-out

Grow-out of freshwater prawns is performed either in batch culture or continuous culture. In batch culture, stocking is synchronized, and the entire stock is grown for six months or longer such that most prawns reach market size. Ponds, which are still more commonly used than RAS, are then drained and all prawns harvested. This system is simpler than continuous culture but has the drawback that prawns grow at uneven rates and a significant portion will not be of market size when ponds are drained.

In continuous culture systems stocking can be done multiple times a year and harvests are performed using dip nets and seines in regular intervals without ever draining the grow-out ponds. The advantage of continuous culture is that the biggest prawns that have attained market size (>30 g for *M. rosenbergii*) can be selectively harvested starting four months after stocking and continuously size-graded in short intervals thereafter.

Feeding of freshwater prawns is comparable to that of Whiteleg shrimp as *Macrobrachium* and

Litopenaeus species have similarly low protein requirements of about 25%. A feed conversion ratio of 2 and a growth rate of up to 2 cm per month can be achieved if specifically formulated diets are used. Since prawns and shrimps are benthic omnivores, they have been used in polycultures with herbivorous freshwater fishes such as tilapia and carps.

Extensive aquaculture is common for freshwater prawns. However, yields (0.5–3 tons per hectare pond surface area annually) are lower than for semi-intensive or intensive aquaculture (up to 10 tons per hectare tank surface area annually). The prospects for managing ecological sustainability are highest for intensive RAS.

11.5 Freshwater crayfish

Three genera of crayfish are primarily used for aquaculture, *Procambrus* (family Cambridae), *Astacus* (family Astacidae), and *Cherax* (family Parastacidae). *Procambrus spec.* are native to North America, *Astacus spec.* to Europe, and *Cherax spec.* to Australia and Oceania. The most cultured freshwater crustacean is the red swamp crayfish (*Procambarus clarkii*, Figure 11.3d). This species is native to the south-eastern US (Louisiana) and of considerable economic and cultural importance in this region. Because of its superior potential for aquaculture, it has been exported to Asia, Europe, and Africa.

Historically, crayfish consumption in Louisiana (called crawfish or crawdads in that region) became a cultural tradition when French Acadians migrated from Nova Scotia to the Bayou area around New Orleans in the mid-1700s. Red swamp crayfish are abundant in this region and became part of the Cajun cuisine, Cajun being an evolved version of the word Acadian. Today, Louisiana accounts for approximately 85% of the US crayfish production and consumption. Aquaculture, which started in the 1960s, now contributes more to crayfish production than capture fisheries.

Crayfish are easy to culture with minimal management, lack planktonic larval stages, grow rapidly, tolerate many environmental stresses well, and are opportunistic omnivores that consume a wide variety of inexpensive food. They are mostly produced by extensive aquaculture in outdoor earthen ponds. Intensive hatchery production of juveniles is sometimes combined with extensive grow-out in outdoor ponds. Pond grow-out represents semi-intensive rather than extensive aquaculture if pond fertilization, liming, planting of macrophytes, stocking of food organisms such as molluscs, and water level management are performed. Ponds are surrounded by levees and often fenced to prevent escapes. Water levels are maintained according to developmental state of the crayfish and seasonal temperature (0.5–1 m). The boundary between extensive and semi-intensive aquaculture of crayfish is rather fluid and depends on the extent of pond management and stocking density.

Crayfish monoculture and polyculture with rice are both common. In monoculture, ponds are prepared and stocked with juvenile crayfish obtained from hatcheries in April. Juveniles will grow to market size (10 cm in length) in as little as three months and can be harvested starting in summer. Some juveniles may take a year to reach market size and selective harvesting is common. In August, ponds are stocked with broodstock crayfish before ponds are gradually drained. Pond draining increases water temperature and induces mating and retreat of crayfish into burrows. Burrows are sealed with clay to prevent evaporation and provide cooling and protection from predators. During autumn and winter, fertilized eggs develop into larvae, which remain attached to the female by means of glair, a viscous protein mixture secreted by the females, and are secured by the female's pleopods.

A single female crayfish can produce up to 300 eggs. Larvae hatch when temperature starts to increase, and females emerge from the burrows in spring. Juveniles are released after they have moulted twice. Adult crayfish are harvested from the ponds at that time and the juveniles are left for grow-out to market size until summer or longer. Many crayfish species can reproduce year-round, and the main limiting factor is temperature. Therefore, females can produce multiple clutches of offspring in hatcheries under controlled environmental conditions. Hatchery survival of juveniles is higher than under natural conditions, which increases yield when stocking ponds with hatchery-raised juveniles that have reached a size of 1–2 cm.

Rice paddies are used for polyculture with crayfish alone or with combinations of crayfish and herbivorous fish that do not prey on crayfish juveniles, for example, silver carp (*Hypophthalmichthys molitrix*). Modification of rice paddies is necessary for accommodating this form of polyculture. Fences that minimize escapes, water channels, and structures that provide shelter for crayfish must be added while the use of insecticides that kill aquatic arthropods must be avoided. The increase in water level from spring to summer must be coordinated to meet the demands of both the growing rice plants and crayfish. Pond draining in late summer must be coordinated such that both rice and market size crayfish can be harvested sequentially while allowing broodstock and juveniles that have not yet reached market size to retreat into burrows. The costs of these additional management tasks are offset by the advantages of crop and market diversification that are achieved by this polyculture system.

Harvesting constitutes a significant portion of the overall cost of crayfish aquaculture. Crayfish yields of monoculture systems are typically twice as high (6000 kg/ha/y) as for polyculture systems. Traditional harvesting is labour intensive and based on manually setting and collecting baited traps. Harvesting by using motorized boats or electrofishing increases efficiency greatly. For electrofishing, trawl nets are being used. Manual and motorized boat harvesting uses traps made of wire mesh with funnel openings that allow entry but not exit of crayfish. Traps are baited with decaying meat or manufactured bait that emits a strong odour to attract crayfish.

Global demand for crayfish has increased significantly in recent years, particularly in China, because of successful marketing as 'small lobsters' that are sold as a delicacy in restaurants. Besides being sold live as fresh seafood, crayfish are also marketed frozen whole hard-shelled, soft-shelled, or as frozen meat. When sold live, they are often packaged into onion bags and transported at low temperature to minimize metabolism, damage, and mortality.

11.6 Brine shrimp as live food for aquaculture

Non-decapod crustaceans are too small for commercial seafood production. Nevertheless, copepods and branchiopods are important live food organisms for aquaculture. Copepods, daphnia, and brine shrimp (*Artemia salina* and *A. franciscana*, Figure 11.5) are used in hatcheries to feed larval stages of decapods and fish. Brine shrimp is particularly valuable because this species can form spores (cysts) that represent an environmentally highly tolerant, dehydrated, quiescent state of embryonic development before hatching. In the spore state, brine shrimp eggs can be preserved inside sealed cans for several years.

Figure 11.5 Brine shrimp (*Artemia salina*) are small branchiopod crustaceans that are commonly used as live food in aquaculture hatcheries. Photo by Hans Hillewaert, Wikimedia (CC-BY-SA 4.0).

The encysted spore state of brine shrimp is characterized by extremely slow, virtually undetectable rates of metabolism and very long anoxia tolerance. This state is called diapause, which is a period of suspended development that serves to endure unfavourable environmental conditions. Not only is diapause very pronounced in brine shrimp but exit from diapause can be triggered very easily by immersing Artemia spores in water. Immersion initiates rapid rehydration, resumption of metabolic activity, and hatching of nauplius larvae after twenty-four hours.

Although daphnia and several other branchiopods also undergo diapause and form spores (e.g. ephippia), triggering exit from diapause in these other species is not as easy as for brine shrimp. The convenient way to control the brine shrimp life cycle by utilizing its spores makes this branchiopod crustacean a superior live food organism for aquaculture hatcheries.

Brine shrimp hatching rate is generally highest in freshwater or at low salinity (5 g/kg), presumably because those conditions mimic rainfall in natural habitat (i.e. dried out riverbeds or shores of salt lakes). For aquaculture, the spores are often decapsulated prior to inducing hatching by treatment in a bleach (sodium hypochlorite) solution and rinsing in freshwater. Decapsulation removes the hard outer shell (capsule) of the spores and accelerates hatching without harming the embryos.

Spore capsules can be removed by using negative phototaxis of nauplii larvae to concentrate them at the bottom of the hatching tank while netting off the floating spore capsules from the water surface. Removing spore capsules from brine shrimp cultures before using them as live food prevents diseases and digestive disorders resulting from spore capsule ingestion by larvae of aquaculture animals.

Brine shrimp are extremely euryhaline. They tolerate salinities ranging from 10 g/kg to saturation (340 g/kg) indefinitely. They even survive freshwater for many hours, which makes them versatile live food organisms for mariculture and freshwater aquaculture. After brine shrimp nauplii have resorbed their yolk, they transform into zoea larvae that start feeding on phytoplankton, which is commonly produced in aquaculture hatcheries.

Traditionally, brine shrimp spores were harvested in two areas of the US: the San Francisco South Bay in California, and the Great Salt Lake in Utah. The main species occurring there is *A. franciscana*. The San Francisco South Bay site is no longer a major contributor but brine shrimp spore harvest from the Great Salt Lake in Utah is still a viable industry. Because of increased demand, brine shrimp have been exported for producing and harvesting spores to many non-native areas in Asia, Australia, and Europe. In these areas they have the potential to become invasive species.

An important consideration when using brine shrimp as live food for aquaculture is that they are naturally low on PUFAs. Therefore, PUFAs must be actively enriched in brine shrimp by feeding them corresponding diets. PUFA-enriched brine shrimp are preferable as live food since they better meet the dietary PUFA requirements of aquaculture animals.

11.7 Ecological impact and sustainability prospects

Crustacean aquaculture, in particular semi-intensive shrimp farming, has faced serious environmental pollution concerns that have paralleled the steep increase in productivity since the 1980s. Because shrimp aquaculture contributes noticeably to global aquaculture production it has been under heavy scrutiny for the unsustainable exploitation of coastal wetlands. As a result of assessing the environmental impacts of uncontrolled shrimp aquaculture development, the utilization of coastal wetlands for aquaculture is now more stringently regulated in many countries.

11.7.1 Wastewater containment and treatment

Shrimp farms have reduced water usage and flow-through rates of natural water to better conserve this precious resource. Efforts to increase the ecological sustainability of crustacean aquaculture are apparent across the past decade but there is still much room for further improvement. Environmental impact resulting from the destruction of wetlands for shrimp pond construction can be further minimized by encouraging intensive aquaculture that utilizes RAS.

Pond fertilization in semi-intensive aquaculture represents input of nutrients that requires balancing the recycling of corresponding amounts of animal waste by ecosystem services. Eutrophication of estuaries by nutrient-rich shrimp pond effluents has caused rapid deterioration of wetland ecosystems near semi-intensive shrimp farms. Therefore, more intensive aquaculture systems that fully utilize available technologies and provide opportunities for professional development of employees to minimize the use and cost of fertilizer should be better incentivized. Indoor greenhouse RAS of shrimp is already successful, but mostly targets local niche markets that offset the higher cost by higher prices. However, RAS is currently much less common than outdoor pond-based extensive or semi-intensive crustacean aquaculture and requires better economic incentivization to improve the ecological sustainability of crustacean aquaculture.

Build-up and removal of organic sludge at the bottom of semi-intensive shrimp grow-out ponds represents another environmental impact that is of significant concern. If not removed, such sludge will become anoxic, which causes chemical reduction of the organic matter it consists of and accumulation of anaerobic bacteria that convert organic matter into compounds that are toxic to animals. Such chemically reduced compounds include hydrogen sulfide (H_2S), nitrite, and reduced (ferrous) iron. Several of these compounds promote pathogenic bacteria that cause shrimp diseases, for example, *Vibrio parahaemolyticus*, the pathogen responsible for acute hepatopancreatic necrosis disease (AHPND) also referred to as early mortality syndrome (EMS, Section 18.4.1).

The common practice of flushing out accumulated sludge from shrimp ponds after harvest into wild coastal ecosystems puts considerable strain on ecosystem service capacity in delicate coastal wetlands. In addition, pre-treatment of ponds before restocking to kill pathogens and predators, for example, by applying piscicides or liming (increase in pH and water hardness), adds additional strain on ecosystem health. Limiting the impact of these practices by minimizing seepage of water from treated ponds into groundwater or natural surface water increases ecological sustainability.

11.7.2 Recycling of non-seafood by-products and feed waste

Ecological sustainability of crustacean aquaculture can be increased by minimizing waste disposal via better utilization of non-seafood by-products. For example, chitin found in crustacean exoskeletons can be used as a potential source of antimicrobial substances. Chitin derivatives such as chitosan have been applied for wastewater treatment and the production of agrochemicals, cosmetics, and pharmaceuticals. Pigments (e.g. astacene, astaxanthin β-carotene, canthaxanthin, lutein and zeaxanthin) present in crustacean waste can be extracted during wastewater treatment and used as antioxidants for biomedical applications.

Although aquaculture crustaceans feed preferentially on a carnivorous diet during juvenile stages, they are benthic omnivores as adults. They can function as detritivores that recycle bacterial decomposer biomass into animal biomass. Detritivores consume bacterial and other microbial decomposers that recycle organic waste by growing on faecal pellets, rotting plant material, animal cadavers, and food particle remains. Therefore, the presence of detritivores increases the recycling capacity of ecosystems.

Crustaceans and other detritivores (e.g. sea cucumbers) are added to integrated multi trophic aquaculture (IMTA) systems to more efficiently use particulate matter and energy that are added to aquaculture systems in the form of feeds and fertilizer. However, crustaceans are animals that generate ammonia and organic waste, which increases the demand for bacterial decomposers to break down these additional waste compounds. Balancing the different components in IMTA systems to achieve maximally integrated circuits that are self-sustained is only possible in closed circulation systems that have a large biological filter component consisting of microbial biomass. In open-water IMTA systems this microbial biomass is provided by ecosystem services.

11.7.3 Containing invasiveness and the spread of pathogens

North American crayfish species have been introduced to Europe and Asia. Because of their high environmental and disease resilience they have strong potential for becoming invasive species. Many other aquaculture crustaceans have been distributed widely to non-indigenous habitats. Kuruma shrimp, which is indigenous to Japan, was introduced to large parts of coastal Asia, Southern Europe, and the Americas after it captured the interest of aquaculture in the second half of the twentieth century. The Chinese mitten crab is an invasive species in the US and Europe. Whiteleg shrimp has been distributed from its original range in Western Central America to tropical and subtropical areas world-wide. The giant freshwater prawn has been distributed from Southeast Asia to tropical and subtropical areas on all continents (Pillay and Kutty, 2005). These are just a few examples of prominent aquaculture crustaceans that already have (or may become) invasive species in many parts of the world.

Sometime invasiveness may be promoted intentionally (e.g. by restocking crayfish in Scandinavian lakes), while other times it may result from unintended aquaculture escapes. In both cases the result may be replacement of indigenous species with non-indigenous species that occupy similar ecological niches. The consequences of invasions of new species into ecosystems are often unclear until non-indigenous species have established themselves and management options are very limited, expensive, and/or prone to significant side effects. Most importantly, ecosystems are dynamic systems that are always in a state of succession. Therefore, efforts to conserve or even revert to a certain status quo may be futile, or sometimes even harmful. Proactive prevention of invasiveness by limiting the anthropogenic distribution of species to non-native habitat is the best management strategy.

Another important ecological aspect of crustacean aquaculture is the potential for spreading diseases that affect wild stocks and competition for resources by non-indigenous aquaculture species. Aquaculture crustaceans have caused devastating outbreaks of the crayfish plague and resulted in wide-spread eradication of the indigenous noble crayfish (*Astacus astacus*) in Europe. This disease is caused by the fungal pathogen *Aphanomyces astaci*, which was introduced to Europe along with red swamp crayfish (*P. clarkii*) from North America during the nineteenth century. Unlike European *A. astacus*, North American species such as *P. clarkii* and the signal crayfish (*Pacifastacus leniusculus*) are resistant to this fungal pathogen. *P. leniusculus* is now produced in aquaculture hatcheries to restock lakes in Scandinavia. The goal of restocking is to refill the ecological niche that used to be occupied by *A. astacus* before the crayfish plague occurred.

Key conclusions

- Crustaceans are aquatic arthropods that are taxonomically highly diverse, but only few species from the order Decapoda are used for commercial seafood production. In addition, small crustaceans from the order Branchiopoda are used as live food in aquaculture hatcheries.
- Aquaculture of crustaceans is mostly based on semi-intensive pond culture that utilizes coastal wetlands and has been developed unsustainably during the initial period of rapid growth.
- Crustaceans grow in steps, rather than continuously, by moulting (i.e. by replacing their exoskeleton).
- Crustacean aquaculture ranks second after fish in terms of value and third after fish and molluscs in terms of production yield.
- The Whiteleg shrimp (*L. vannamei*) accounts for more than half of all crustacean aquaculture. Other important species include euryhaline crabs (*E. sinensis* and *S. serrata*), freshwater prawns (*M. rosenbergii* and *M. nipponense*), and freshwater crayfish (*P. clarkii*).
- Shrimps and prawns develop by succession through planktonic nauplius, zoea, and mysis larval stages, while crab development includes planktonic zoea and megalopa larval stages. However, crayfish lack any planktonic larval stages, which simplifies their culture.
- Hatcheries operating in intensive production mode and nurseries producing juveniles are increasingly utilized for aquaculture of crustaceans before transferring juveniles to grow-out systems.
- Opportunities for increasing the ecological sustainability of crustacean aquaculture include utilization of RAS for grow-out, reduction of water used by semi-intensive pond aquaculture, treatment of waste and sludge

generated in aquaculture ponds instead of disposal into wild habitat, increased utilization of by-products, improved IMTA systems, increased biosecurity, and utilization of indigenous species.

References

Food and Agriculture Organization of the United Nations (FAO). (2020). *The State of World Fisheries and Aquaculture 2020*. Rome: Food and Agriculture Organization of the United Nations.

Fitzgerald, W. (2007). 'Silvofisheries: integrated mangrove forest aquaculture systems', in Costa-Pierce, B. A. (ed.) *Ecological aquaculture: the evolution of a blue revolution*. Oxford: Blackwell Science, pp. 161–262. https://doi.org/10.1002/9780470995051.ch8

Fujinaga, M. (1967). 'Kuruma shrimp (*Penaeus japonicus*) cultivation in Japan', in Mistakidis, M. N. (ed.) *Proceedings of the World Scientific Conference on the Biology and Culture of Shrimps and Prawns*. Rome: Food and Agriculture Organization of the United Nations. p. 811–832.

Hopkins, J. S., Sandifer, P. A., DeVoe, M. R., Holland, A. F., Browdy, C. L., and Stokes, A. D. (1995). 'Environmental impacts of shrimp farming with special reference to the situation in the continental United States', *Estuaries*, 18, pp. 25–42. https://doi.org/10.2307/1352281

Pillay, T. V. R. and Kutty, M. N. (2005). Aquaculture: principles and practices, 2nd edn. Oxford: Wiley-Blackwell.

CHAPTER 12

Ornamental fishes

'The hobby aquarium industry, in its public education effects, can have an incalculable positive effect on the need for public understanding of biology and ecology ... However, as practiced today, there are enormous losses of organisms in the commercial aquarium trade. The suffering of the animals is deplorable, and there exists the very real possibility that intensive collection will deplete the environment and upset the balance of natural communities.'

Adey and Loveland, *Dynamic aquaria: building living ecosystems*, (2007)

Aquaculture of ornamental fishes and aquatic invertebrates is rapidly expanding owing to the large market and multi-billion-dollar trade value of ornamental species. Intensive, ecologically sustainable recirculating aquaculture systems (RAS) for aquatic ornamentals represent a strong alternative to capture fisheries. Aquaculture has the added potential benefit of improving the welfare of animals that are being traded. However, detailed production yields are not as well documented for the ornamental aquaculture sector as are those from the aquaculture farms that produce seafood (FAO, 2020).

Even though the exact number of species cultured as ornamental aquatics is unknown, it ranges in the thousands. Lack of data on capture fisheries, aquaculture, and trade of ornamental aquatics is particularly eminent for marine fishes and invertebrates. Better regulatory oversight and reporting by governments and the Convention on International Trade in Endangered Species of Wild Fauna and Flora (CITES) is deemed critical for improving the ecological sustainability of this expanding sector (Biondo and Burki, 2020).

Unlike the production of seafood, aquaculture of ornamental species is not a necessity but creates a luxury product. Nevertheless, it can have a positive socio-economic impact on developing countries in tropical and subtropical areas of the world where most ornamental fishes naturally occur. Such positive impact will be long lasting if ecological sustainability is considered properly during initial development.

12.1 Economic value of ornamental fishes

The enormous value of aquatic ornamentals is best illustrated by the following comparison: 1 kg of ornamental marine fishes produced for the aquarium trade may be valued as high as US $1800, whereas 1 kg of a fish species produced as seafood may a have a hundredfold lower value. The precious nature of ornamental fishes is not only evident in their high price but also in their appearance. They have been dubbed 'living jewels' due to their colourful and attractive appearance (Jayasankar, 1998).

This comparison illustrates why aquaculture farms producing ornamental species are often smaller in size than aquaculture systems producing seafood. However, ornamental aquaculture farms are numerous and support the production of diverse species. For example, in Florida alone there are hundreds of aquaculture farms that produce many species of ornamental fishes, invertebrates, and aquatic plants (Tlusty, 2002). Likewise, there are about 100 ornamental fish farms within an area of only 150 ha in Singapore's Agrotechnology

A Primer of Ecological Aquaculture. Dietmar Kültz, Oxford University Press. © Dietmar Kültz (2022). DOI: 10.1093/oso/9780198850229.003.0012

Park (Mutia *et al.*, 2007). The smaller size of most ornamental aquaculture farms means that fewer resources are required, less water is used, less waste is produced, and less space is needed. For all these reasons, ornamental aquaculture is well-suited for utilizing RAS.

Despite the lack of reliable estimates of ornamental aquaculture yields, some trends can be deduced from the sparse data available and previous analyses of ornamental fish trade value. Estimates indicate that as many as 150 million specimens of marine reef fishes, most of them obtained by capture fisheries, are traded annually (Biondo and Burki, 2020). The total number of ornamental fishes traded annually across the world is 1.5 billion, the majority (90%) being freshwater (FW) fishes, while the remaining 10% are marine reef fishes. The estimated value of the global ornamental fish trade in 2007 was US $45 billion and the annual growth rate is 8% (Mutia *et al.*, 2007). This strong market drives the rapid development of ornamental fish aquaculture.

The growth of ornamental fish aquaculture is fostered by an increasing number of households with aquaria in many parts of the world. In addition, hotels, restaurants, and other commercial businesses are expected to increasingly utilize aquaria for decorative purposes. Moreover, small-scale aquarium systems for aquatic ornamental species are progressively used at research institutions worldwide to discover solutions for biomedical, environmental, and other challenges. For example, the bicolour damselfish *Stegastes partitus* is a model for neurofibromatosis type I (Schmale, Gibbs, and Campbell, 2002). This growing market represents a strong economic driver for ornamental aquaculture. The major export countries of marine ornamental species are in the Coral Triangle of Southeast Asia. FW ornamental fishes are exported from Asia, South America, Africa, and Australia. Major importers of ornamental aquatic species are the EU, the US, Japan, and Australia. Large parts of Asia also represent a rapidly growing market.

Estimates of the number of ornamental fishes kept in aquaria in the US are 160 million, the majority (140 million) being FW fishes (Biondo and Burki, 2020). In the US, approximately 10% of households keep ornamental FW fishes, compared to 0.8% that keep marine fishes (Tlusty, 2002). In the UK, about 14% of homes keep ornamental fishes (Jayasankar, 1998), while in Australia 12–14% of homes have aquaria (Corfield *et al.*, 2008). Aquaria with ornamental fish are likely to have stress-relieving effects and benefits for human health but unambiguous scientific evidence supporting this claim is still sparse (Clements *et al.*, 2019). Although most home aquaria use FW, the development of suitable lighting and filtration technology has sparked an increase in the number of households with marine aquaria.

The technological advances making marine aquaria more accessible to hobby aquarists have increased the demand for ornamental marine fishes over the past decades—a trend that is likely to continue. Between 1991 and 2011 the number of marine reef fish species traded as ornamentals increased from 200 to over 2300 in the US alone (Biondo and Burki, 2020). Almost half of all marine ornamental fish species imported into the EU in 2018 were not evaluated by the International Union for the Conservation of Nature (IUCN) Red List. This illustrates the increasing threat that is posed by the commercial trade of ornamental fishes to coral reef ecosystems. Aquaculture can contribute to alleviating this threat by providing a sustainable alternative to destructive capture fisheries practices, such as cyanide poisoning or dynamite fishing.

12.2 Aquaria as microcosms

Modern aquaria systems are technology intensive and require significant experience and management skills. Despite the skills needed, the small scale of home aquaria and technological innovations of recent decades have made the management of aquaria both cost- and time efficient. In that sense, hobby aquarists and the aquarium industry are contributing tremendous innovation and knowledge required for future development of aquaria systems. Furthermore, this innovation and knowledge represents an invaluable source of inspiration for the development of larger, commercial-scale RAS.

Although scaling up from small-scale home aquaria microcosms to larger-scale RAS mesocosms presents challenges, it represents a sustainable and eco-friendly solution to environmental issues associated with the growth of commercial aquaculture. Research efforts that translate progress in managing small aquaria microcosms as closed systems to the operation of larger-scale RAS continue to be critical if the increased demand scenario outlined in Section 4.2 mandates expansion of intensive commercial aquaculture.

Aquaria represent closed-system microcosms (Chapter 4) with a balanced input and output of matter and energy. All microcosms are dynamic and characterized by a constant throughput (flow) of matter and energy that is required to maintain them. Aquarists now have powerful and highly efficient technologies at their disposal to manage the flow of matter and energy and the ecological balance in aquaria.

Optimal composition and quantity of feeds, lighting, and filter technologies assist in matching the amount of food and fish in the system with the capacity to recycle animal and food waste. Many elements present in fish food (e.g. sulfur, selenium, phosphorus, sodium, potassium, calcium, magnesium, nitrogen) need to be balanced and properly recycled to maintain an aquarium microcosm in ecological equilibrium. Figure 12.1 shows an (albeit oversimplified) schematic illustration of inputs, outputs, and filtration steps.

One aspect of food for ornamental fishes that impacts the ecological sustainability of ornamental aquaculture is the use of fish meal and fish oil. Feeds for ornamental fishes, including the widely used flake foods, use fish meal as a major ingredient. Therefore, more ecologically sustainable ways to produce ornamental fish feeds are needed. Alternatives include the production of essential feed ingredients such as polyunsaturated fatty acids (PUFAs) and certain amino acids in bioreactors that utilize microbes engineered for supplying these feed ingredients.

If all three trophic levels (producers, consumers, decomposers) are matched appropriately in their capacity to process matter and energy flows, then aquaria require very small amounts of clean water, as they generate no wastewater (Figure 12.1). All waste (output) can be captured in a concentrated form (mechanical filter sludge and exhausted chemical resins) that is amenable to environmentally benign disposal (e.g. deep geological storage) or processing (e.g. fermentation). Mechanical filtration using extra-fine filter materials may diminish beneficial decomposing bacteria along with potential pathogens in the water and their benefits and drawbacks need to be weighed on a case-by-case basis before use. Similar considerations apply to pathogen control via water treatments such as UV radiation, ozonation, reverse osmosis, and chlorination/de-chlorination in closed RAS.

Matching the three trophic levels requires a large primary producer (plant) and bacterial decomposer biomass per unit biomass of fish in the aquarium, particularly in marine systems. Many aquarists aim to keep as many ornamental animals as possible, which requires large mechanical, biological, and chemical filters. When stocking aquaria with fish one must consider the trophic level transfer efficiency (TLTE, Figure 4.4), which is generally 10–20% to keep trophic cycles balanced (Section 4.6). Adding aquatic plants to aquaria provides nutrient recycling capacity from primary producers and helps keep algal blooms that fulfil the same role to a minimum.

This principle is also pursued in aquaponics by combining fish aquaculture with the production of agricultural crops (e.g. brassica species). The footprint required for decomposers and producers to achieve a balanced aquarium microcosm is often underestimated, which then necessitates regular water changes to maintain water quality. If wastewater must be exchanged for clean water, then completion of the nutrient cycle is left to external sources, that is, ecosystem services that utilize decomposers in natural habitat or centralized wastewater treatment plants.

Most ornamental fishes are kept in small home aquaria, but public aquaria and research institutions also use larger systems for ornamental fishes. Many species of ornamental fishes are cultured to support basic and applied research at universities and other academic institutions. For example, a prominent model organism for developmental biology and biomedical research is the zebrafish (*Danio rerio*). As a result of high demand for research, highly efficient RAS have been developed for zebrafish, for example, the Aquatic Habitats Z-Hab system.

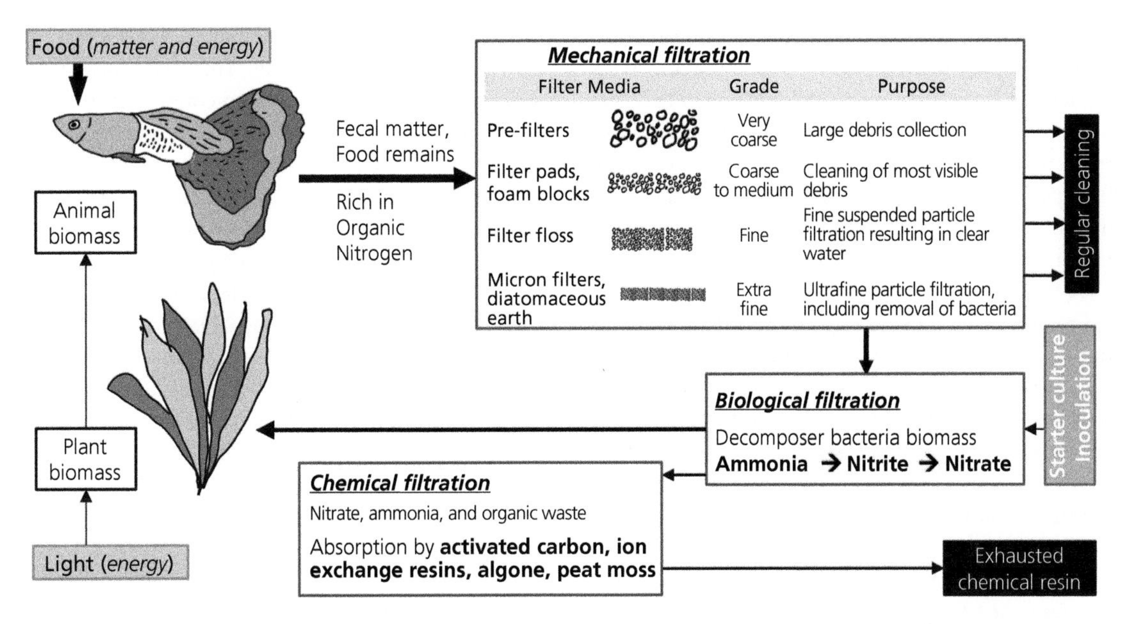

Figure 12.1 Three common steps for maintaining good water quality, nutrient recycling, and nitrogenous waste removal in aquarium microcosms are mechanical, biological, and chemical filtration. The capacity for filtration must match the amount of food and fish added to the system. Debris captured in mechanical filters and exhausted chemical resin must be regularly removed. If plants are present, then part of the food waste will be assimilated in plant biomass. Gas exchange and water losses and replacement due to evaporation, etc., are not considered in this figure. For biological filtration, only nitrogenous waste breakdown is sketched out, nutrient cycling of other elements is not considered. Grey boxes = inputs, black boxes = outputs.

Cichlids are another group of ornamental fishes that are used extensively in research. The explosive evolutionary radiation of cichlid species in the Great African Lakes led to many colourful variants that captured the interest of hobby aquarists and evolutionary biologists alike. Moreover, many ornamental species (e.g. gobies) are adapted to unique environmental conditions. The corresponding mechanisms of adaptation are studied by environmental physiologists to reveal how impacts of environmental change and stress on organisms, including humans, can be compensated.

12.3 FW ornamental fishes

In terms of trade volume, 90% of ornamental fishes are FW species and most of them are produced by aquaculture. Less than 10% of FW ornamentals are still provided by capture fisheries, which contrasts sharply with more than 90% of marine ornamental fishes still caught in their native habitat. Commercial capture fishery of ornamental fishes is primarily performed in subtropical and tropical areas characterized by a very high biodiversity. These areas include countries in the Amazon River drainage basin of South America, the East African Great Lakes region, the Congo River drainage basin in Central Africa, and drainage basins of major rivers in Southeast Asia. Southeast Asian countries are also major aquaculture producers of ornamental FW fishes, along with the Czech Republic and other industrialized countries.

Florida produces most ornamental FW fishes in the US. The production of almost 1000 different varieties of ornamental FW fishes represents the most valuable aquaculture commodity in Florida. It accounts for nearly half of all aquaculture sales in this state. Most ornamental aquaculture farms are in the southern part of Florida, south of Orlando, to take advantage of warm winters and the proximity to international airports at Orlando, Miami, Tampa, and Fort Myers. Proximity to major international airports facilitates the export of ornamental fishes from Florida by minimizing transportation time and stress. Shorter transport time and reduction in its associated stress maximizes the welfare of fish and profit to farmers as it reduces both mortality and health concerns. Ornamental aquaculture in Florida is supported by research and disease diagnostics at the University of Florida Tropical Aquaculture Laboratory (Florida Department of Agriculture and Consumer Services, 2016).

Ornamental FW fishes are raised in indoor (greenhouse) or outdoor tanks and ponds. When outdoor ponds are used then their size is much smaller than ponds used for aquaculture of seafood species and the depth usually does not exceed 2 m. Both earthen and concrete ponds are used. Sometimes, cage culture in larger lakes is also practised for producing ornamental fishes. Semi-intensive systems, and closed RAS are used for aquaculture of ornamental fishes. Outdoor ponds, tanks, and cage systems are semi-intensive systems that are managed similarly as for seafood species, including liming, fertilization, predator control, etc. However, since the ponds used for semi-intensive aquaculture of ornamentals are much smaller than those used for seafood species, they require less clean water as a resource. Furthermore, wastewater and sludge can be collected and properly disposed of (e.g. to wastewater treatment facilities) more effectively (Tlusty, 2002).

An encouraging trend in aquaculture of FW ornamental fishes indicates a shift from open, semi-intensive outdoor ponds to closed RAS facilities. The latter have become feasible because of improved husbandry approaches for many species and technological innovations. Closed RAS require energy to operate tools such as pumps for water recirculation and monitoring devices for abiotic parameters. They are most ecologically sustainable if this energy is generated by wind, solar, or geothermal sources. Closed RAS conserve clean water and minimize the dependence on natural ecosystem services. Additional advantages include better control of pathogens and prevention of diseases, minimizing escapes of farmed fish, and effective capture or recycling of waste.

FW aquaria are easier to maintain than marine aquaria, which is the main reason why nine out of ten households with ornamental fishes have FW rather than marine aquaria. The basic tools for a FW aquarium are an air pump connected to an air stone by plastic tubing, a heater connected to a thermosensor, a source of light, and a filter that includes mechanical, biological, and chemical stages. For FW aquaria that contain a small biomass of fish, the filter can be internal and does not require extra space, which makes FW aquaria much more space-efficient than marine aquaria (Figure 12.2).

The diversity of ornamental FW fishes is staggering and so are their requirements for space, food, and reproduction in captivity. Live-bearing cyprinodonts (toothcarp, e.g. guppies, *Poecilia reticulata*, Figure 12.3a) reproduce readily in small aquaria while captive reproduction of other species (e.g. discus cichlids, *Symphysodon spec.*, Figure 12.3b) depends on accurate tuning of key environmental parameters. Many ornamental fish species have relatively short generation times, that is, the time from hatching to reproduction. Short generation times accelerate domestication via artificial selection in captive breeding programmes. African annual killifishes of the genus *Nothobranchius* hold the record for the shortest known generation time of any vertebrate (17–18 days) (Blažek, Polačik, and Reichard, 2013).

Some ornamental FW fishes have undergone extreme artificial selection resulting in domesticated forms that differ greatly in colour and morphology (e.g. fin or body size, colour, and shape) from their wild ancestors. Examples include guppies (*P. reticulata*), discus cichlids (*Symphysodon spec.*), angelfish (*Pterophyllum scalare*), and betta fish (*Betta splendens*) (Figure 12.3). Hybridization has also been used to produce new varieties of ornamental FW fishes, in particular for cichlids. A prominent example is the flowerhorn cichlid, which is a hybrid cross of South American red devil (*Amphilophus labiatus*), trimac (*A. trimaculatus*), midas (*A. citrinellus*), and severum (*Heros severus*) cichlids (Figure 12.4). Breeding of new varieties of ornamental fishes represents an elaborate activity that can fetch very high prices for new varieties at breeding contests, which are professionally organized and widely attended for koi, guppies, bettas, and other ornamental fishes (Section 12.5).

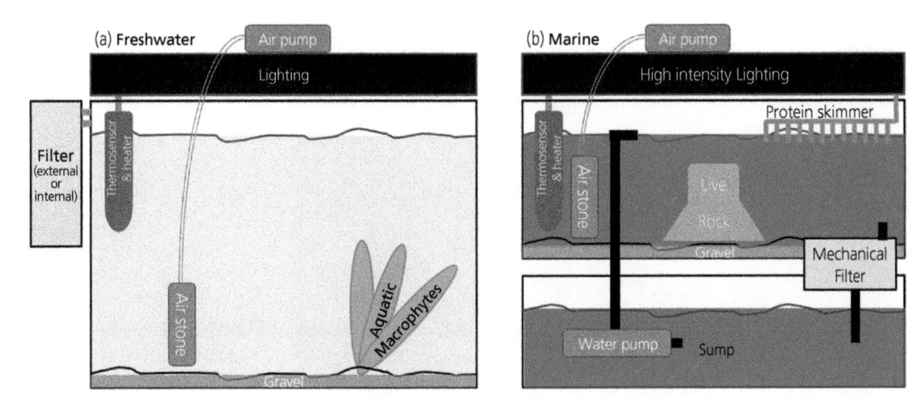

Figure 12.2 Basic setup of a freshwater aquarium (a) and a marine aquarium (b).

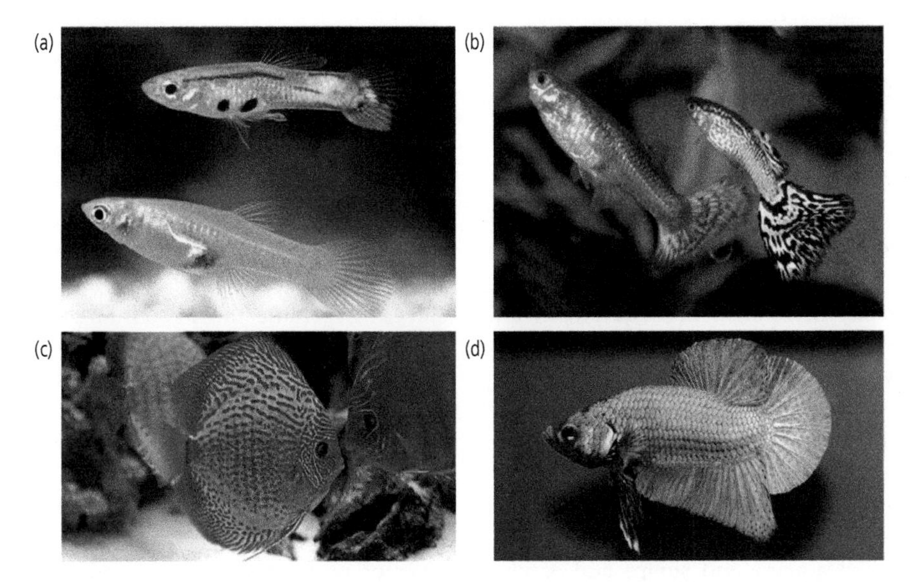

Figure 12.3 Freshwater ornamental fishes that have been rapidly domesticated into morphologically distinct variants by selective breeding. **A)** Wild Trinidad guppy (*Poecilia reticulata*) male (top) and female (bottom), photo credit: Wikimedia, CC-BY-4.0. **B)** Domesticated guppy (*P. reticulata*) female (left) and male (right), photo credit: Wikimedia, CC-BY-3.0. **C)** Domesticated Red turquoise discus cichlid (*S. aequifasciata*), variant 'Leopard Snakeskin', photo credit: Wikimedia, CC-BY-3.0. **D)** Domesticated betta fish (*Betta splendens*) variant, photo credit: Wikimedia, CC-BY-2.0.

12.4 Marine ornamental fishes

Marine ornamental fishes account for less than 10% of volume but more than 20% of the value of all traded ornamental fishes. These figures illustrate that marine ornamental fishes are, on average, significantly more expensive than FW ornamental fishes. The main reason for the higher price is the cost associated with capture and transport of marine fishes. Most marine ornamentals (>90%) are caught from their wild coral reef habitat in tropical and subtropical areas of the world, chiefly in the Coral Triangle area of Southeast Asia and Oceania. From there, shipping distances to major importers such as the US and EU are very long, and transport logistics is a major cost factor.

Only about 10% of marine ornamental species are currently amenable to production in aquaculture and even fewer can be produced for profit, which also reinforces the demand for unsustainable

Figure 12.4 Flowerhorn cichlid. This fish represents a hybrid of four South American cichlids characterized by wide variation in body colour and a greatly enlarged forehead. Photo by author.

marine fisheries capture methods. Florida and Hawaii are the main US aquaculture locations for marine ornamental fishes. In these states, many of the commercial and research hatcheries for marine ornamental fishes employ ecologically sustainable, intensive RAS.

Another reason for the higher price of marine ornamental fishes and invertebrates is the significantly greater investment into equipment for owners of marine aquaria when setting up their reef tank. Marine aquaria are technically more challenging to maintain, and they require more technological aids, energy, and attention from their owners than FW aquaria. Innovations that have made marine aquaria feasible for homeowners are non-corrosive plastic materials for plumbing connections between aquaria and sumps and acrylic materials for producing tanks that replace heavier and more break-prone glass.

Technical considerations specific to managing marine tanks include monitoring and maintaining salinity and ionic balance, minimizing evaporation, providing strong lighting, and creating water movement for detritus flushing if invertebrates containing photosynthetic symbionts and coralline algae are being cultured. Monitoring and treatment of animal waste is also more challenging for marine aquaria. Saline wastewater treatment requires additional space and technological aids compared to FW aquaria. A sump, which is often not much smaller than the actual aquarium, and a protein skimmer are essential for marine aquaria (Figure 12.2b). Sumps for decomposing bacteria and refugia with controlled lighting for photosynthesizing plants establish population densities of decomposers and primary producers that have sufficient capacity to recycle the animal waste generated in the marine aquarium.

Protein skimmers are necessary because the solvent capacity of seawater is significantly lower than that of FW. Seawater already contains 34 g of dissolved salt per litre (mostly sodium chloride but also trace amounts of most other chemical elements), which diminishes its capacity to dissolve proteinaceous waste excreted by fish and from deteriorating food remains. Therefore, proteins precipitate out more readily in seawater than in FW and precipitated protein aggregates become visible as a white foam floating on the water surface. This surface layer of proteinaceous foam limits gas exchange and impairs effective aeration while also blocking light from entering the water. Protein skimmers are designed to regularly remove this proteinaceous surface foam.

12.4.1 Anemonefish aquaculture

Various anemonefish (clownfish) species of the genus *Amphiprion* represent the most common and most popular marine ornamental fishes worldwide. Anemonefish are members of the family Pomacentridae, which also includes many other damselfish species. Aquaculture in Florida and other places now produces several anemonefish and other damselfish species.

Aquaculture methods of marine fishes are carefully customized for each species, but the general production cycle often follows that outlined here for anemonefish. Commonly, developmental stages are separated and reared in different tanks. Broodstock animals are held in breeding tanks such that spawning can be facilitated by maintaining high water quality, proper feeding, and appropriate temperature management. Under optimal conditions, repeated spawning of the same broodstock fish may occur as frequently as once every month.

After fertilization of eggs, the parents are generally removed from the tank. Alternatively, eggs are deposited by the female on a substrate (e.g. a plastic plate) that can be removed to transfer the fertilized eggs into a larval hatching tank. For anemonefish, hatching generally occurs eight to ten days after fertilization if temperature and other conditions are kept optimal (25°C). For other species, time to hatching can vary greatly. For most marine ornamental species, it is important to avoid salinity fluctuations and maintain constant seawater salinity of approximately 34 g/kg.

A major challenge for producing marine ornamental fishes by aquaculture is the small size and rapid yolk resorption of their larvae after hatching. Therefore, much research has focused on optimizing phytoplankton and zooplankton cultures that provide suitable live food for the larval stages of a given species. Because hatched larvae quickly use up their yolk, they must be fed with marine microalgae such as *Nannochloropsis spec.* For this purpose, separate algal cultures are often maintained in the same hatchery.

Circular tanks or small rectangular raceways with constant water current are used in conjunction with a significant flow rate to mimic water currents encountered by the larvae in natural reef habitat. Anemonefish larvae are grown to a size of 1 cm in length before they are transferred to larger circular tanks, which are used for the first phase of juvenile grow-out. During this phase, brine shrimp (*Artemia spec.*) are initially used as live food, but juveniles are gradually conditioned to feed on formula flake food.

Upon reaching 2–3 cm length, juveniles are transferred to larger raceways with stronger water current to promote rapid grow-out of juveniles to market size. For anemonefish, the whole production cycle takes between four and six months. Optimal water temperature for grow-out of juveniles to market size is 23 °C.

12.5 Advantages of ornamental fishes produced by aquaculture

Aquaculture can make a major contribution in reducing the negative impact of capture fisheries on ecosystems while also fostering research and development of technologies for wastewater recycling. It represents a more ecologically sustainable alternative to capture fisheries of both FW and marine ornamental fishes. While approximately only 10% of FW ornamental fishes are wild caught, the vast majority (>90%) of marine ornamental fishes are still caught from their native coral reef habitat. The most common environmentally destructive method of ornamental marine reef fish capture is cyanide poisoning. This method has been banned in many countries but is still widely used illegally.

Sodium cyanide solution is squirted on corals, sponges, anemones, and other reef invertebrates that fish use as shelter, which results in partial paralysis and enables capture of the fish. The cyanide is not only toxic to fish but also damages invertebrates and planktonic organisms that are exposed to it. This capture practice causes high mortality in fish and invertebrates and irreversibly damages affected reef ecosystems (Madeira *et al.*, 2012).

Because reef water movement rapidly disperses the toxic solution sprayed, high concentrations of sodium cyanide are used. An estimated amount of more than a thousand tons of sodium cyanide has been applied to coral reefs from the 1960s until the turn of the millennium (Mutia *et al.*, 2007). The half-life of cyanide in water is thought to be relatively short (days) before it escapes as a gas into the atmosphere. However, acute exposure of aquatic organisms to cyanide results in toxicity within seconds to minutes and the half-life of cyanide released from the water into the atmosphere is significantly longer (years).

Other environmentally destructive collection methods used to collect ornamental marine species include dynamite fishing and mining of corals and live rock with hammer and chisel (Livengood and Chapman, 2020). Environmentally destructive collection methods are used because of the high consumer demand and high market prices for ornamental marine reef fishes. Even though such methods are illegal in many countries, enforcement is hampered by the many intermediate steps in the commercial supply chain, which obscures the original collection source.

A global solution is needed to address this problem. Intensive, ecologically sustainable RAS at sites that are close to major markets are only part of the solution. Local fishers must be offered attractive alternative means to support their livelihoods and promote environmental conservation (Tlusty, 2002). Opportunities for their professional development and technical assistance should be provided and financed by multiple sources, including international development funds (Jayasankar, 1998; Mutia *et al.*, 2007).

Implementation of an effective global monitoring system for gathering accurate and timely information about the number and exact origin of all ornamental fishes is widely considered a key part of a global solution (King, 2019; Biondo and Burki, 2020). To that end, CITES established a working group on the trade of marine ornamental fishes, which fosters efforts of its nearly 200 member countries to scrutinize and better monitor the trade of ornamental species (Biondo and Burki, 2020).

Capture fishery of FW ornamental fishes is not as prevalent as for marine ornamentals since aquaculture meets 90% of the demand. Capture methods of wild FW ornamental fishes utilize various nets such as seines, trap nets, hand nets, and dip nets. In addition to nets, electrofishing and toxins such as rotenone are also used to capture ornamental FW fishes. By meeting most of the demand, aquaculture has helped minimize negative impacts of capture fishery on wild stocks of ornamental fishes and their habitat.

Transportation of wild-caught ornamental fishes to retail stores incurs large losses. Handling and crowding stress, prophylactic treatments with chemicals for pathogen control, and large fluctuations in abiotic parameters (temperature, pH, dissolved oxygen, etc.) are the main reasons of mortality during transport. Even if mortality can be avoided, stress associated with transport impacts animal welfare and health and may reduce the lifespan and performance of these fish in home aquaria.

Mortality during transport of ornamental fishes has been estimated as high as 80% for some species, for example, the Banggai cardinalfish (*Pterapogon kauderni*). Because transport-associated mortality is not considered in estimates, the actual number of wild-caught ornamental fishes may be much higher than outlined in Section 12.1. Therefore, overfishing leading to extinction of vulnerable species represents a real concern about capture fishery of ornamental fishes.

Ornamental aquaculture has successfully relieved pressure from capture fisheries on wild stocks of the golden dragon fish (Asian arowana, *Scleropagus formosas*), which is an endangered species on the IUCN red list. Other examples of ornamental aquaculture aiding significantly in species conservation efforts by relieving pressure on capture fisheries include Bala shark minnows (*Balantiocheilos melanopterus*), dwarf chain loach (*Ambastaia sidthimunki*), and tiger barb (*Puntius tetrazona*) (Tlusty, 2002).

Ornamental fish conservation efforts are coordinated by the Conservation, Awareness, Recognition and Responsibility, Encouragement & Education, and Support & Sharing (CARES) preservation programme to maintain stocks of endangered species by aquaculture. An example of a critically endangered cichlid is the Blackfin tilapia (*Sarotherodon linnellii*), which is endemic to Lake Barombi Mbo in Cameroon. Blackfin tilapia are propagated by hobby aquarists and researchers, including the author's laboratory, to prevent them from becoming extinct.

Conservation efforts supported by aquaculture also include restocking and transplantation activities aimed at reviving severely depleted natural populations of ornamental fish species by supplementation with fry produced in aquaculture hatcheries. As discussed in Chapter 6, the success of stock enhancement depends not only on the number of animals stocked but also, to an even greater extent, on restoring the natural habitat to a state that is conducive for re-establishment of viable populations. Restocking of ornamental barb species (*Barbus spec.*) in Sri Lanka represents an example for combining successful habitat restoration with restocking of ornamental fishes. However, restocking activities need to be planned carefully to minimize human-induced evolution, genetic dilution of wild stocks, and introduction of pathogens. As pointed out at the end of Chapter 6, stock enhancement should only be used as a last resort option.

Fish raised in aquaculture can be domesticated to emphasize preferred traits such as body and fin colour and shape. Domesticated bettas (*B. splendens*) are represented by dozens of strains with unique colours and fin types acknowledged by the International Betta Congress (IBC). Similar strains and varieties are registered for *Carassius auratus* by the various goldfish societies and clubs in many countries. Likewise, the different strains and varieties of domesticated guppies are recorded by the International Fancy Guppy Association (IFGA). Similar organized domestication efforts exist for many other species of ornamental fishes produced by aquaculture. Domesticated traits that confer an advantage to aquaculture fish over wild caught fish include lower susceptibility to handling stress, being accustomed to aquarium life and the presence of people, and easier rearing of offspring in captivity.

Other advantages of ornamental fishes produced by aquaculture over wild-caught ornamental fishes include their age, health, and pathogen-free condition. Fishes produced by aquaculture have a known age and are always sold as juveniles while fishes caught in the wild have an unknown age and remaining lifespan. The health of aquaculture fishes is constantly monitored and subject to periodic diagnostic screening for diseases. Moreover, aquaculture fishes are raised under pathogen-free conditions when closed indoor RAS are used. Therefore, fewer concerns exist regarding infectious diseases and parasites when purchasing ornamental fishes produced in aquaculture.

12.6 The impact of public media and perception on ornamental fishes

Films and public videos about coral reefs and the many species of marine fishes and invertebrates that inhabit them as well as documentaries and movies featuring aquatic life motivate people to pursue home aquaria as a hobby. Demand has increased steadily, which reveals that keeping aquatic animals as pets can represent a double-edged sword. On the one hand, pursuing this hobby educates us about the diversity of life in aquatic habitats and increases our resolve to support conservation to maintain this diversity. On the other hand, this hobby increases demand for ornamental aquatic species, which is particularly problematic for marine species that are mostly caught in their wild habitat. Since the supply of ornamental fishes is far from inexhaustible, increased demand has led to overfishing of some areas and endangerment of some ornamental species. Establishment of sanctuaries such as Marine Protected Areas (MPAs) helps sustain viable populations of susceptible species and can have spill-over effects to adjacent areas used for fishery.

Recent documentaries address this dilemma, but some also point out benefits of capture fisheries for FW ornamental fishes (Davenport, 2020). If practised in an ecologically sustainable manner using environmentally non-destructive capture methods and considering the health of wild stocks, then capture fisheries can make positive contributions to environmental health. For example, in the Amazon River basin the fishery of ornamentals offers indirect benefits to protecting rainforests by providing a source of income as an alternative to deforestation, agricultural development, and pollution by industrialization.

Public media clearly have a large impact on shaping consumer perceptions, decisions, and preferences. Information is key to decision making and, as discussed regarding seafood (Chapter 1 and Figure 1.2), information on whether ornamental fishes have been obtained by capture fishery or produced by aquaculture helps aquarists make better decisions. Outreach initiatives such as Tank Watch (https://forthefishes.org/) and Reef Count (https://www.reef.org/gafc) educate hobby aquarists about the source of ornamentals and help them make educated choices to reduce the demand for ornamental fishes captured using ecologically unsustainable methods. Exact metrics are scare on how the wealth of information currently available, whether accurate or not, influences demand for ornamental fishes. This difficulty is exemplified by the aftermath of the 2003 movie *Finding Nemo*.

Opinions on how *Finding Nemo* impacted wild anemonefish (*Amphiprion spec.*, Figure 12.5a) populations differ and are often portrayed as extremely polarized. Although the film's intended message appears to be conservation of fishes in their native habitat as opposed to captive confinement, it also portrays marine fishes as cute pets, which raised concerns about greatly increased demand for keeping anemonefish in home aquaria (Mutia *et al.*, 2007).

Increased demand for anemonefish correlates with local extinction of anemonefish populations in Southeast Asia. Conversely, *Finding Nemo* was suggested to have motivated the release of anemonefish from captivity in home aquaria into non-native marine habitat to which they don't belong, and in which they cannot survive. These paradoxical phenomena have been dubbed 'the Nemo effect'; however, a recent study concludes that fears about hobby aquarists being influenced by movies featuring animals are unsubstantiated (Veríssimo, Anderson, and Tlusty, 2020).

The perceived negative consequences of the 'Nemo effect' have motivated acceleration of research on anemonefish breeding programmes and commercial anemonefish aquaculture. Several species of anemonefish and other fishes in the family Pomacentridae are now successfully produced by commercial aquaculture. Hatchery-reared anemonefish have been produced since the 1990s—long before the film (Jayasankar, 1998), although it certainly benefitted their continued commercial production.

Unlike anemonefish, however, surgeonfishes (family Acanthuridae) cannot (at least not yet) be produced by aquaculture. They are also challenging to maintain in captivity. Palette surgeonfish (blue tangs, *Paracanthurus hepatus*), aka Dory in the 2016 movie *Finding Dory* (Figure 12.5b), have been bred in captivity at the Oceanic Institute at Hawaii Pacific University, which establishes proof of principle for aquaculture of this fish family. However, scaling up breeding programmes for *P. hepatus* to anywhere near commercial scale has not yet been possible.

Considerable efforts were made before and after the release of *Finding Dory* to educate existing and aspiring home aquarists about potential consequences of their actions and discourage them from keeping surgeonfish as aquatic pets. These efforts included recommending alternative species that are easier to keep in home aquaria. For example, domesticated betta fish strains having similarly striking blue colours as palette surgeonfish, require much less space, can be kept in easier to maintain FW aquaria, and are produced by aquaculture rather than caught in wild coral reef habitats. In addition, *Finding Dory* has triggered interest in supporting research initiatives for expanding captive breeding and aquaculture programmes for surgeonfish.

An important factor to consider in the ornamental fish trade is trendiness of certain species, for example, transient spikes in demand triggered by popular movies or other events. Transient spikes in demand for certain species can be accommodated by sufficient diversification of cultured species. This strategy improves the economic robustness of commercial ornamental fish farms. An example for a spike of interest followed by a period of reduced demand is the 'flowerhorn cichlid craze', which occurred in Malaysia in 2003 (Mutia *et al.*, 2007).

12.7 Invasiveness of ornamental fishes

Thousands of ornamental fish species have been translocated in large numbers to areas where they are not native. Often, ornamental fishes are exported from tropical and subtropical areas of the world to temperate climates. Tropical ornamental fishes cannot establish viable populations if escapes or releases occur in these non-indigenous habitats because of low winter temperatures that are outside their tolerance limit.

However, it is also common that ornamental fishes are translocated between areas with similar environmental conditions, for example, subtropical and tropical parts of the Americas, Southeast Asia, North Australia, and Africa. In Australia alone, forty-one alien (i.e. non-native) fish species representing more than 10% of indigenous FW fish species have established invasive populations. The majority of these species (thirty) are ornamental fishes introduced in recent decades (Corfield *et al.*, 2008), including nineteen cichlids, five live-bearing

(a)

(b)

Figure 12.5 Popular marine ornamental fishes. **A)** Anemonefish (clownfish) of the genus *Amphiprion* are among the marine ornamental reef fish that can be produced by aquaculture. This photo shows an *Amphiprion chrysopterus* specimen from Fiji, photo by author. **B)** Palette surgeonfish (blue tang, *Paracanthurus hepatus*), photo credit: Wikimedia, CC-BY 3.0.

cyprinodonts, four cyprinids, and Oriental weatherloach (*Misgurnus anguillicaudatus*).

The proportion of alien fish species established in the wild are even greater in other parts of Oceania. The increased number of ornamental fish species establishing invasive populations in Australia and Oceania parallels similar increases in Florida (including the Everglades National Park) and other southern states of the US. Because of the large number of ornamental fishes being translocated across the world, there is a high likelihood that some of them are invasive in certain areas or globally. Highly invasive ornamental fish species are documented in many parts of the world. The following paragraphs outline some examples to illustrate why they are regarded a major threat to global biodiversity and ecosystem stability.

Effects of invasive species on ecosystems include replacement of native species, disruption of trophic webs by heavy predation or competition, and resource depletion. Moreover, ornamental fishes, especially those collected by capture fisheries, may be vectors for invasive pathogen species that cause disease in wild animals. Invasive species also pose risks for the gene pool of related indigenous species if hybridization occurs.

The mosquitofish (*Gambusia holbrooki*) is a live-bearing ornamental cyprinodont fish that has become highly invasive in many parts of the world. This small fish (max. length of 6 cm) has been intentionally introduced to many areas across the globe to facilitate mosquito control. It reproduces rapidly and establishes large populations that greatly diminish the numbers of mosquito larvae, which represent their preferred food. Mosquitofish have altered the native FW fauna and ecosystems in most areas where they were introduced.

Other live-bearing ornamental cyprinodonts have also established invasive populations in many parts of the world, including swordtails (*Xiphophorus helleri*) and guppies (*P. reticulata*). Ornamental cichlids that are highly invasive and have colonized many parts of the world include tilapia species such as Mozambique tilapia (*Oreochromis mossambicus*) and Redbelly tilapia (*Tilapia zillii*). *O. mossambicus* is also an important species for seafood production (see Chapter 13). It is very euryhaline (has a very wide salinity tolerance) and has invaded both FW and marine environments. In southeast Asia, captive-bred flowerhorn cichlids (Figure 12.4) have become a prominent invasive species (Mutia *et al.*, 2007).

A FW ornamental catfish species that has had a large impact on non-native ecosystems as an invasive species is the janitor fish (*Pterygoplichthys spec.*). This genus of armoured catfishes is native to South America but used as an ornamental species throughout the world. It is very effective in cleaning algae growing on aquarium glass by utilizing them as a food source. Its sharp teeth scrape off algae from any substrate. Invasive janitor fish have established flourishing populations in lakes of Southeast Asia. In addition to changing the natural ecosystem, they have also interfered with net cage aquaculture and fisheries in those lakes. By feeding on algae growing on the nets their teeth damaged the nets, which resulted in massive escapes of cultured fish and significant economic losses (Mutia *et al.*, 2007).

Invasive ornamental fishes are not limited to FW but also include marine species. In Florida coastal waters alone, thirty-one non-native marine fish species have been recorded (Schofield, Morris, and Akins, 2009). They include the emperor angelfish (*Pomacanthus imperator*), raccoon butterflyfish (*Chaetodon lunula*), peacock grouper (*Cephalopholis argus*), devil firefish (*Pterois miles*), and red lionfish (*Pterois volitans*).

Lionfish are native to the Indo-Pacific region. They were introduced in the 1980s to the Atlantic Ocean near the Caribbean Islands and invaded coastal waters around Florida and most other parts of the Gulf of Mexico during the following decades. Recent records indicate that lionfish have spread throughout the Western Atlantic Ocean from New York to Rio de Janeiro. Lionfish are also an invasive species in the Eastern Mediterranean Sea and have recently been spotted in the Eastern Pacific Ocean, off the coast of California. In many cases, invasiveness is exacerbated by rapid environmental change and range expansion with invading species depicted as the 'winners' and native species as the 'losers' of climate change.

Management of invasive species is difficult and expensive and not many effective tools are available. Those that are available, for example, rotenone in small bodies of water, can only be applied in a restricted manner because they are toxic and/or have other harmful side effects. Significant ethical considerations, lack of information on side effects, and potential collateral environmental damage limit the application of such management tools. Therefore, prevention is the best management of invasive species and education of hobby aquarists about the consequences of releases or escapes of ornamental fishes into wild ecosystems is vital. Because of limited options for mitigation, current management is more focused on public education than control. However, there is a clear need to develop more effective monitoring and control methods for invasive species.

Key conclusions

- Aquaculture of ornamental fishes is rapidly expanding because of large demand and the much greater value of ornamental fishes relative to seafood fish (per kg fish produced).
- Aquaculture of ornamental fishes is well suited for utilizing ecologically sustainable RAS since the scale is much smaller than that of aquaculture for producing seafood.
- Technological innovations in materials (plastics, acrylics), lighting, protein skimming, and filtration have made marine aquaria readily accessible for hobby aquarists, but they are more challenging to maintain than FW aquaria.
- Thousands of different marine and FW ornamental fish species are being traded.
- Aquaculture accounts for more than 90% of all traded ornamental FW fishes but for less than 10% and fewer than fifty species of all traded marine ornamental fishes.
- Several species of ornamental FW fishes have been domesticated into many varieties that differ in fin and body size, colour, and shape. Examples include betta fish, FW angelfish, guppies, and discus cichlids.

Ornamental fish domestication is facilitated by short generation times.

- Research on husbandry and hatchery reproduction of many marine ornamental species is needed to address environmentally destructive methods of capture fisheries in coral reef habitats.
- Advantages of ornamental fishes produced by aquaculture include species and habitat conservation, reduction of transport-related stress and mortality, improved animal welfare, elimination of pathogens and better health, known young age, being accustomed to life in aquaria and the presence of people, and being amenable to domestication for artificial selection of desirable traits.
- Diversification of species is essential for ornamental aquaculture to accommodate trends triggered by public media or other events that can cause spikes in demand for certain species.
- Many ornamental species have become invasive. Their management and preventative measures for avoiding future invasions of additional ornamental species are very challenging but essential for supporting the ecological sustainability of ornamental fish aquaculture.

References

Adey, W. H. and Loveland, K. (2007). *Dynamic aquaria: building living ecosystems*, 3rd edn. Boston: Academic Press.

Biondo, M.V. and Burki, R. P. (2020). 'A systematic review of the ornamental fish trade with emphasis on coral reef fishes—an impossible task', *Animals*, 10, pp. 1–19.

Blažek, R., Polačik, M., and Reichard, M. (2013). 'Rapid growth, early maturation and short generation time in African annual fishes', *EvoDevo*, 4, pp. 24.

Clements, H., Valentin, S., Jenkins, N., Rankin, J., Baker, J. S., Gee, N., Snellgrove, D., and Sloman, K. (2019). 'The effects of interacting with fish in aquariums on human health and well-being: a systematic review', *Plos One*, 14, pp. e0220524.

Corfield, J., Diggles, B., Jubb, C., McDowell, R. M., Moore, A., Richards, A., and Rowe, D. K. (2008). *Review of the impacts of introduced ornamental fish species that have established wild populations in Australia*. Prepared for the Australian Government Department of the Environment, Water, Heritage and the Arts.

Davenport K. Wild caught ornamental fish: The trade, the benefits, the facts. *Ornamental Aquatic Trade Organization*. 2020:28.

Florida Department of Agriculture and Consumer Services. (2016). *Aquaculture best management practices manual*. Tallahassee: Florida Department of Agriculture and Consumer Services Division of Aquaculture.

Food and Agriculture Organization of the United Nations (FAO). (2020). *The State of World Fisheries and Aquaculture 2020*. Rome: Food and Agriculture Organization of the United Nations.

Jayasankar, P. (1998). 'Ornamental fish culture and trade: current status and prospects', *Fishing Chimes*, 17, pp. 9–13.

King, T. A. (2019). 'Wild caught ornamental fish: a perspective from the UK ornamental aquatic industry on the sustainability of aquatic organisms and livelihoods', *Journal of Fish Biology*, 94, pp. 925–936.

Livengood, F. A, and Chapman, E. J. (2007). The ornamental fish trade: an introduction with perspectives for responsible aquarium fish ownership. Gainesville: The Institute of Food and Agricultural Sciences.

Madeira, D., Narciso, L., Cabral, H. N., and Vinagre, C. (2012). 'Thermal tolerance and potential impacts of climate change on coastal and estuarine organisms', *Journal of Sea Research*, 70, pp. 32–41.

Mutia, T. M., Sunaryanto, A., Sujang, A. B., and Sulit, V. T. (2007). 'Review of the ASEAN ornamental fish industry: production, marketing trends, technological developments and risks', *Fish for the People*, 5, 13–27.

Schmale, M. C., Gibbs, P. D. L., and Campbell, C. E. (2002). 'A virus-like agent associated with neurofibromatosis in damselfish', *Diseases of Aquatic Organisms*, 49, pp. 107–115.

Schofield, P. J., Morris, J. Jr., and Akins, L. (2009). Field Guide to the Nonindigenous Marine Fishes of Florida. Washington, DC: National Oceanic and Atmospheric Administration.

Tlusty, M. (2002). 'The benefits and risks of aquacultural production for the aquarium trade', *Aquaculture*, 205, pp. 203–219.

Veríssimo, D., Anderson, S., and Tlusty, M. (2020). 'Did the movie *Finding Dory* increase demand for blue tang fish?', *Ambio*, 49, pp. 903–911.

CHAPTER 13

Freshwater fishes

'We see far fewer technical, economic and resource constraints for freshwater aquaculture than for ocean farming, and far greater potential for land-based fish farms to contribute to global food security.'

Belton, Little, and Zhang, in *The Conversation* (2021)

In addition to great technical and economic resource potential, freshwater (FW) aquaculture also has great promise for future ecological sustainability. Increasing support for mariculture (aquaculture expansion in the ocean) is based on claims that vast ocean resources are underutilized for producing seafood (and other purposes). Advocates in favour of open, semi-intensive mariculture systems argue that exploitation of wild ocean habitat for aquaculture should be greatly expanded to utilize 'free' marine ecosystem services at an unprecedented scale. This alarming mindset is bound to steer us into a direction of repeating the mistakes made during rapid expansion of large-scale, semi-intensive, open system FW and coastal aquaculture during the latter part of the twentieth century.

Consequences of such mistakes arise from improper consideration of the true long-term costs associated with assumingly 'free' ecosystem services. Mitigating these changes will require enormous resources over a long period; some changes are often irreversible. Ecologically sustainable expansion of aquaculture that places due emphasis on environmental stewardship to minimize reliance on ecosystem services and preserve natural resources for future generations is only possible by using available scientific and technological know-how globally, and in a timely manner. Rather than exploiting wild ecosystem services to an even greater extent, we should increase incentives for using new technologies for aquaculture intensification.

In the increased demand scenario pointed out in Chapter 4, the main ecologically sustainable way of greatly expanding global animal aquaculture production is by increased technology use for closed, intensive, recirculating aquaculture systems (RAS). Since aquaculture products are marketed globally, increased utilization of RAS should also be globally incentivized, which requires resources that offset the economic costs of technological infrastructure and professional development of human resources. Providing funds to offset these costs and distributing them equitably among developing and industrialized nations remains a challenge for global policy makers. Moreover, we need better implementation of incentivization approaches such as certificate programmes, subsidies that value shared community ecosystem services beyond national boundaries, professional development opportunities, and other innovative incentives.

RAS minimize the exploitation of natural ecosystems, including the use of clean water, and the potential for disease outbreaks while maximizing space efficiency and biosecurity. RAS are more economical (i.e. easier to implement and operate) for FW aquaculture than mariculture (Chapter 12). Moreover, FW fishes represent the largest segment of seafood production by aquaculture; culture methods for several key species have already been thoroughly optimized (Figure 13.1). Furthermore, most aquaculture species of FW fishes are positioned lower in the trophic web and, therefore, require fewer

A Primer of Ecological Aquaculture. Dietmar Kültz, Oxford University Press. © Dietmar Kültz (2022). DOI: 10.1093/oso/9780198850229.003.0013

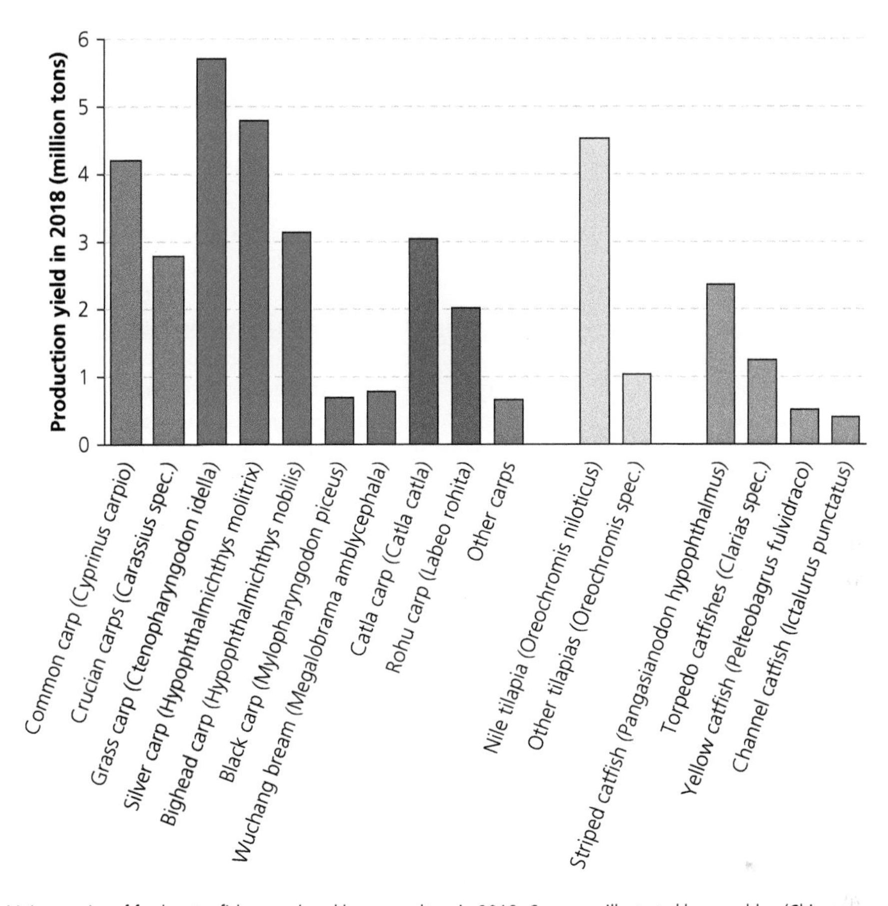

Figure 13.1 Major species of freshwater fishes produced by aquaculture in 2018. Carps are illustrated by grey, blue (Chinese carps), and red (Indian carps) bars, tilapia by tan bars, and catfishes by orange bars. Data source: (FAO, 2020).

resources than most catadromous, anadromous, and marine aquaculture fishes.

13.1 Carps

Carps are the dominant commodity of aquaculture animals. In 2018, 28 million tons of carp were produced by aquaculture (Figure 13.1). Carps belong to the order Cypriniformes, which comprises more than 2000 species. The greatest species diversity of Cypriniformes is in south-eastern Asia with approximately 600 species estimated in China alone. In contrast, Cypriniformes are not native to South America and Australia. Cypriniformes include species of a wide range of body sizes, including the smallest known fish (*Paedocypris spec.*, 7.9 mm max. length) and very large species such as the Siamese carp (*Catlocarpio siamensis*, 3.0 m length) and Mahseers (*Tor spec.*, 2.8 m length) (Mayden and Chen, 2010; De Silva and Wang, 2019).

Carp species used for aquaculture have been introduced to many non-native habitats across the world and become invasive pests in many cases. For example, common carp (*C. carpio*), which is native to Eurasia, has been introduced to all other continents except Antarctica (De Silva and Wang, 2019).

Currently, most carp aquaculture is performed in earthen ponds that are fertilized to increase yields. These ponds are open, semi-intensive aquaculture systems that rely heavily on ecosystem services. A less common alternative to pond culture is cage monoculture of carp in large FW reservoirs or lakes. Semi-intensive cage culture also strains the ecosystem service capacity of FW habitats and often causes eutrophication and pollution of common property resources. Therefore, currently the most common carp aquaculture systems have a considerable ecological footprint, and their long-term ecological sustainability must be significantly improved.

Since carps are the leading commodity produced by aquaculture, a shift to more ecologically sustainable production methods would have a large positive impact on global aquaculture sustainability. Intensive RAS units that utilize advanced technologies, environmentally friendly sources of energy (solar, wind, geothermal), and scientific know-how minimize the use of clean water and disposal of wastewater.

Common formula carp feeds contain significant amounts of fish meal and fish oil. In addition, aquaculture of several small carp species produces feed for more highly valued mandarin fish (*Siniperca chautsi*) that occupies a higher trophic level (Bunting, 2013). Using fish to feed fish is not an ecologically sustainable aquaculture practice and significant research currently focuses on developing plant-based diets that contain all ingredients, including polyunsaturated fatty acids, to support rapidly growing, healthy fish that occupy the lowest possible trophic level (Chapter 17).

13.1.1 Diverse feeding strategies of different carp species

Carp species used in commercial aquaculture have various feeding strategies and occupy different trophic levels within food webs. Understanding their position in food webs and their dietary requirements helps maximize productivity and minimize waste. The common carp (*C. carpio*, Figure 13.2a) is the oldest aquaculture species for which historical records exist (Chapter 2). This species has been domesticated longer than any other fish, which has resulted in multiple variants, including koi, the orange-coloured scale carp (*C. carpio var. flavipinnis*), partly scaled mirror carp (*C. carpio var. specularis*), and scaleless leather carp (*C. carpio var. nudus*).

The omnivorous common carp is native to Eastern Europe and Asia and consumes animal-, plant-, and detritus-derived food. Crucian carps (*Carassius spec.*) are also omnivores and are well-suited for aquaculture because their diet allows flexibility in food sources and positioning within food webs, and often it is easiest to wean omnivores to formulated dry pellet feeds.

Asian carp species with a central distribution range in China are commonly referred to as Chinese carps although they also occur in other parts of Asia. Commercially important aquaculture fishes in this group include the grass carp (*Ctenopharyngodon idella*, Figure 13.2b), silver carp (*Hypophthalmichthys molitrix*, Figure 13.2c), bighead carp (*Hypophthalmichthys nobilis*, Figure 13.2d), and black carp (*Mylopharyngodon piceus*, Figure 13.2e). An alternative common name for the grass carp is White Amur carp. This species is herbivorous; it feeds on plankton as fry and switches to macrophytes (larger aquatic plants) when juveniles reach about 15 cm in body length. To break down tough plant material from aquatic macrophytes grass carp have a well-developed pharyngeal grinding apparatus that resembles the molar teeth of mammals.

Adult grass carp feed continuously and can consume food in the amount of several times their body weight per day. As a result, they grow rapidly at warm temperature (>20 °C) but have a relatively short lifespan of ten years at the most. Grass carp have been introduced from Asia to many other parts of the world, including Europe and North America, where they have become invasive species. The main intent for their introduction into non-native habitats was to utilize their feeding behaviour and capacity for cleaning up eutrophic lakes. Harvesting them as seafood was a secondary objective promoted by supplemental feeding to accelerate growth. Meat quality of grass carp is very high and preferred over other carp species in many markets.

Like grass carp, silver carp are also herbivores, but they do not feed on macrophytes, instead preferring phytoplankton. Sometimes, silver carp can also be considered omnivores because they can feed on zooplankton. Plankton feeding is reflected by the silver carp feeding apparatus. Instead of a massive pharyngeal grinding apparatus, silver carp have long, fine, and closely spaced gill

Figure 13.2 Carp species used for aquaculture. **A)** Scaleless variant of common carp, photo from Wikimedia, CC0-1.0. **B)** Grass carp, photo by US FWS, Wikimedia CC0-1.0. **C)** Silver carp, photo by US FWS, Wikimedia, CC0-1.0. **D)** Bighead carp, photo by USGS, Wikimedia, CC0-*1.0*. **E)** Black carp, photo by USGS, Wikimedia, CC0-1.0. **F)** One of the Indian carp species used for aquaculture is the calbasu carp (*Labeo calbasu*), drawing by H. S. Thomas, Wikimedia, CC0-1.0.

rakers (cartilaginous extensions of the gill arch that are present at the opposite end of the gill filaments) that form a sieve to capture large amounts of tiny phytoplankton cells. Large amounts of water and phytoplankton are constantly passed across the gills of fish to support respiration, osmoregulation, acid-base regulation, and nitrogenous waste (ammonia) excretion of the fish. These physiological functions are performed by the epithelial cells lining the gill filament and secondary lamellae, while the gill rakers collect microscopic planktonic food that is packaged in mucus and passed down to the pharynx.

Bighead carp are omnivores with similar feeding habits to common carp. They feed opportunistically on phytoplankton, zooplankton, detritus, larger benthic invertebrates, aquatic plants, and readily consume formulated dry pellets. This species grows very rapidly and is highly invasive in the south-eastern part of the US and in other countries where average annual temperatures are high.

Black carp are also invasive in the US, where they were introduced in the 1970s, and in many other countries. This species is carnivorous, that is, it feeds on live animals, mostly small benthic invertebrates. Its diet includes snails, mussels, and crustaceans with hard shells or exoskeletons. To crush the shells and exoskeleton of its prey, black carp utilize a feeding apparatus that is different from that of other Chinese carps. Prey is crushed by gill rakers, which are much thicker, fused, sturdier, and spaced much farther apart from each other than the delicate and finely spaced gill rakers of silver carp.

Indian carps used for aquaculture include catla (*Catla catla*), rohu (*Labeo rohita*), mrigal (*Cirrhina mrigala*), and calbasu (*Labeo calbasu*, Figure 13.2f). Catla carp are omnivores that feed mostly on the surface and in the open water column. They consume phytoplankton, zooplankton, small invertebrates, and decaying macrophytes. Rohu carp are also omnivores with a very similar food preference as catla, but they tend to feed less on the surface and more in the open water column. In contrast to adult catla and rohu, adult mrigal and calbasu carp are bottom-feeding omnivores. These different feeding habits can be utilized in carp polyculture systems to maximize the utilization of resources.

13.1.2 Captive reproduction and culture of common carp

For carp monocultures, the specific conditions depend on the species being cultured and their feeding habits. All carp species grow best at temperatures of 20–30 °C (i.e. during summer season) but most can survive winter temperatures of 0–4 °C by entering a state of metabolic depression. Metabolic depression greatly reduces feeding, oxygen consumption, movement, and generation of waste.

Since general aspects of carp aquaculture are comparable, polyculture systems are easily feasible (Section 13.1.3). Specific parameters (optimal temperature for different developmental stages, feeds, etc.) vary somewhat between different carp aquaculture species but the overall culture practices are similar as those briefly outlined for common carp. One notable difference is that Indian carp species have a much narrower temperature tolerance with a minimum of 15 °C and an optimum just a bit higher (about 30 °C) than for other carps.

Carp reproduction utilizes broodstock selected for individuals having desired traits (growth rate, size, etc.), ideally from genetically heterogeneous stocks to avoid inbreeding depression. They are segregated by sex into different ponds several months before spawning takes place to prevent premature, unwanted reproduction. The two main methods of spawning are natural and artificial fertilization. Whether using natural or artificial spawning, ovulation can be induced by hormone treatment of broodstock (Box 13.1).

Box 13.1 Hormonal induction of ovulation and spawning

Carp ovulation and spawning can be induced and synchronized by injecting broodstock fish held in segregation ponds with crude pituitary extract derived as a by-product from processing *C. carpio* and other carps. Alternatively, injection of purified reproductive hormones induces and synchronizes spawning. Common hormones used for this purpose are recombinant human chorionic gonadotropin hormone (HCG) synthesized in vitro, and analogues of luteinizing hormone-releasing hormone (LHRH) and gonadotrophin-releasing hormone (GnRH). The latter is often accompanied by administration of dopamine antagonists. Induced spawning by hormone injection has enabled efficient captive breeding of Chinese and Indian carps and reduced the reliance of aquaculture on larvae collected from wild spawning habitat of these species. Furthermore, hormone-induced spawning doubles the number of possible spawning events per year without additional broodstock.

For artificial fertilization, males and females are anaesthetized, and eggs and milt are collected into a bucket by hand-stripping. The contents are then mixed in the bucket to maximize fertilization success, which is significantly higher than for natural spawning. The fecundity of all carps used for aquaculture is high, and a single female can produce more than 100,000 eggs.

Natural spawning uses spawning ponds that are populated with two to three females and four to five males when the water temperature warms to above 18 °C. This method works well for common carp but is less effective for Chinese and Indian carps. Carp spawning ponds are small (100–300 m^2). Keeping them dry and covered with grass prior to spawning eliminates aquatic pathogens and predators. Prior to stocking with broodstock animals the ponds are flooded with clean water to a depth of approximately 30 cm. Spawning usually occurs within one to two days and eggs attached to spawning substrate such as 0.5 m^2 mats made from plant or synthetic materials can then be removed and transferred to an indoor hatchery. Alternatively, broodstock fish can be removed and ponds used to grow larvae to the fry stage (about 2 cm in body length).

Fertilized carp eggs clump together because of a sticky proteinaceous coat that surrounds them. This sticky coat is removed before incubating fertilized eggs to maximize aeration and hatching success. Lowering egg adhesiveness is achieved by successive incubations with continued stirring in 0.3% solution of protein-denaturing urea for approximately one hour, 0.06% tannin solution for up to five times for twenty seconds, and clean fresh water for about five minutes.

During this incubation period the fertilized eggs detach from each other, swell, and undergo the first cell division. They are then transferred into 10-litre upwelling jars in indoor hatcheries (Figure 13.3), in which they are kept suspended in the water column by a constant flow of water from the bottom to the top of the jar. Upwelling ensures good aeration and development and enables collection of hatchlings via the water outlet at the top of the jar.

Yolk sac larvae hatch after incubation in upwelling jars for about three days at 24 °C. Decreasing the temperature will delay and increasing it will accelerate hatching. Hatched yolk sac larvae are reared in larger (200-litre) upwelling tanks and kept there for three to four days at a temperature of 20–24 °C. During this time, the yolk is absorbed, the swim bladder inflates, feeding starts, and swimming commences.

Free-swimming fry are transferred to small ponds or large indoor tanks for initial grow-out to fingerlings (about 5 cm in length), which takes approximately one month. They are then transferred to larger ponds for grow-out to market size (0.5–1.5 kg). Grow-out takes up to three years in more temperate climate with lower average annual temperature and less than one year in subtropical climate with high temperature throughout the year. Prior to stocking, both fry and final grow-out ponds are usually treated with quicklime, bleach, or insecticides that can be plant-based or synthetic to remove predators and competitors.

Extensive systems that rely on natural primary production in the grow-out ponds have the lowest stocking density and yield (about 200 kg/ha annually) but are the most ecologically sustainable open systems. If external fertilization using synthetic fertilizer or agricultural manure or both are used moderately, then the annual yield of carp increases to about 600 kg/ha. If supplemental feeding in the form of soaked low-quality grain or cereal/ rice

Figure 13.3 Upwelling jars are used in hatcheries for many aquaculture fishes, including salmonids, carps, and catfishes. Photo by US FWS, Wikimedia, CC0-1.0.

bran is provided, then the annual yield increases further to about 2000 kg/ha.

Ponds that are stocked at high density and supplied with complete, optimally formulated carp diets yield up to 10,000 kg/ha annually. However, under these conditions, periods of hypoxia (<2 mg/L dissolved oxygen, DO) or anoxia (0 mg/L DO) may occur in carp ponds. Therefore, they must be managed as semi-intensive systems by implementing aeration technologies. Aeration devices such as paddlewheel, fountain, or bubble aerators are necessary at high stocking densities and high rates of fertilization and feeding to prevent hypoxia and algal crashes due to eutrophication.

13.1.3 Carp polyculture systems

Carp polyculture has ancient origins in Asia out of necessity since many carp species, especially Chinese carps occur in the same habitat. Therefore, larvae collected from natural spawning grounds have almost always represented a mix of different carp species, which were then grown out together in aquaculture ponds as polycultures. Carp polyculture takes advantage of the different feeding niches of aquaculture carp species discussed above, which allows for more efficient resource utilization. While common carp are mostly produced in monoculture, Chinese and Indian carps are almost always grown out as polycultures of multiple carp species (Pillay and Kutty, 2005).

More recently, carp species have also been co-cultured with non-cyprinid taxa (e.g. tilapia) to diversify the range of fish species produced in a single pond even further. Polycultures of carp with crustaceans such as crayfish and crabs also exist. Despite the added value of better resource utilization by animals occupying different trophic levels in polyculture, all animals are still heterotroph consumers.

All animal aquaculture culture requires proportional amounts of primary producers and decomposers to balance trophic cycles either via ecosystem services or wastewater treatment systems. Polyculture of fish with primary producers contributes to balancing trophic cycles. A common example is rice–fish polyculture, which is similar to rice–crustacean polyculture (Chapter 11). Rice–carp polyculture has several advantages besides improved ecological sustainability. Rice production is enhanced by providing nutrients either directly (e.g. ammonia) or from decomposing fish waste (e.g. nitrates and phosphates). Moreover, growing fish in rice paddies produces seafood in addition to rice, which represents extra income that diversifies the source of revenue.

By removing both fish and plant (rice) biomass from the system during harvest, the burden for decomposers is reduced to the organic waste produced by the animals and food waste that cannot be assimilated directly by either rice or phytoplankton. Unfortunately, rice–carp polyculture is not compatible with high-yielding, short-stem rice varieties that require a water depth too shallow for aquaculture. Another disadvantage is that multiple rice cropping is not practical in polyculture systems. Furthermore, the use of insecticides and herbicides is limited to those benign to fish and their food organisms, and fish growth is more limited than in monoculture. Moreover, rice–fish polyculture systems need to provide refuges for fish during rice field drainage, which requires extra management. To encourage polyculture, these drawbacks must be offset by economic gains.

Polyculture of rice and carp has been expanded to rice–azolla–carp polyculture systems. Azolla (*Azolla spec.*) are aquatic fern (duckweed) species that float on the water surface. These floating plants are utilized to assimilate atmospheric nitrogen, and therefore, alleviate the need for the application of synthetic or organic nitrogen fertilizer to rice fields. Symbiotic cyanobacteria (blue green algae) that inhabit cavities within the azolla leaves are responsible for the fixation of atmospheric nitrogen. Azolla plants provide a natural food source for herbivorous and omnivorous carp species. Carp incorporate some of the atmospheric nitrogen fixed in the azolla plant into fish biomass and pass on the remainder by way of fish waste (chiefly ammonia) to the rice.

13.2 Tilapia

Tilapia are FW fishes belonging to the subfamily Tilapiinii, within the family Cichlidae and order Perciformes. Perciformes is the most speciose order of vertebrates consisting of almost 10,000 species. Many perciform species are marine. However, the family Cichlidae (cichlids) contains only FW and brackish water species, some of which (e.g. tilapia) are extremely euryhaline and tolerate salinity well above that of seawater. Their high salinity tolerance enables them to invade marine environments if they are translocated from their native FW habitat.

Almost 2000 species of cichlids have been described. Most cichlids have a native distribution range in Africa and most others are native to South and Central America. Many cichlids have attractive body and fin colours and interesting behaviours. These traits underly the high demand of cichlids as ornamental aquarium fish. Forty-six African cichlids have been classified as critically endangered and possibly extinct in their native habitat and some only survive as captive populations in aquaria (Kishe-Machumu *et al.*, 2018). Examples include *Hoplotilapia spec.* and the blackfin tilapia (*Sarotherodon linellii*). Efforts for preventing extinction are supported by conservation programmes actively involving many hobby aquarists, as well as research institutions.

African cichlids are divided into the two subfamilies Tilapinii and Haplochrominii. The Tilapinii include approximately 100 species while the Haplochrominii consist of over 1000 species that have rapidly evolved by adaptive radiation into diverse ecological niches. Due to rapid evolution, these fishes represent classical and powerful models for evolutionary biology research.

Tilapia species used in commercial aquaculture belong to the subfamily Tilapiini, which is native to Africa and consist of the three genera *Tilapia*, *Sarotherodon*, and *Oreochromis*. The distribution range of *Oreochromis* species is centred in Eastern Africa, while *Sarotherodon* species tend to be more abundant in Western Africa. These genera are distinguished by their feeding and breeding behaviours. Distinction based on breeding behaviour is most reliable since all Tilapinii species feed opportunistically on a wide variety of plant material, detritus, and (as fry) zooplankton.

The genus *Tilapia* consists of substrate spawners that provide parental care by guarding the eggs attached to an external substrate such as a plant leaf. Adults are preferentially macrophytophagous (i.e. they feed on large aquatic plants). Their gill rakers are robust and widely spaced with chisel-shaped teeth for chewing tough plant material.

The genus *Sarotherodon* (meaning brush toothed) are preferentially microphytophagous, feeding on phytoplankton and they provide parental care by paternal mouth breeding. They have thin and closely spaced gill rakers and teeth to rake algae from the water column.

The genus *Oreochromis* (meaning mountain coloured) comprises opportunistic omnivores that provide parental care by maternal mouth breeding (in some species both sexes perform mouth breeding). Tilapia feeding habits resemble those of carps and are a major reason for the superior suitability of tilapia for aquaculture.

Tilapia have many traits that make them an ideal aquaculture species and their production yield is second only to carps (Figure 13.1). Their high potential for economically efficient aquaculture and global marketing earned them the label 'aquatic chickens'. Three *Oreochromis* species are most used for aquaculture, although they are often hybridized with each other or other *Oreochromis* species: Nile tilapia (*O. niloticus*), Mozambique tilapia (*O. mossambicus*, Figure 13.4), and blue tilapia (*O. aureus*).

Like many aquaculture carps, tilapia species occupy relatively low levels in trophic webs, and they can be grown using ecologically sustainable plant-based diets. There is great potential for improving the ecological sustainability of culturing these species by replacing fish meal, fish oil, and other animal ingredients in their feeds with plant-based or synthetically derived nutrients (Chapter 17). In addition, fish meal can be produced sustainably as a by-product of tilapia aquaculture because only half of the fish is suitable for processing as seafood and fillets typically account for only a quarter of the fish weight.

13.2.1 Extreme environmental tolerance of tilapia species

Wide environmental tolerance (eurytopy) and rapid growth to a large size are major traits that are preferred for aquaculture species. Like carps, tilapia species excel in both traits. Tilapia have been used

Figure 13.4 Mozambique tilapia (Oreochromis mossambicus). Photo by Elizabeth Mojica.

extensively as models for basic research geared at understanding the molecular mechanisms of environmental stress tolerance. They are even more eurytopic than carps, although their maximum size is smaller compared to most carp species used for aquaculture.

High environmental stress tolerance of tilapia reduces the need for clean water relative to other species (e.g. salmonids), which is important for preserving clean FW resources. In addition, many tilapia species and hybrids tolerate brackish water and even seawater and high temperatures without a large reduction in growth rate or fitness. These key traits of tilapia make them highly resilient to coastal and arid habitat salinization and to global warming caused by anthropogenic acceleration of climate change.

Although not as speciose as the Haplochrominii, the Tilapinii subfamily includes species with the most extreme environmental tolerances of any fish. For instance, Lake Magadi tilapia (*Alcolapia grahami*) tolerate very alkaline water (pH >10) (Reite, Maloiy, and Aasehaug, 1974) while Blackfin tilapia (*Sarotherodon linellii*) tolerate very acidic water (pH <5). These species encounter these extreme pH values in their native habitats, Lake Magadi (Kenya) and Lake Barombi Mbo (Cameroon).

Mozambique tilapia (*Oreochromis mossambicus*) and Blackchin tilapia (*Sarotherodon melanotheron*) are extremely euryhaline. Even though they are FW fish, they tolerate a salinity of almost 4× seawater (120 g/kg), which they occasionally encounter in their native habitat, for example, Lake St Lucia (South Africa) and the Saloum estuary (Senegal).

Many tilapia species also survive hypoxia (very low DO levels) for at least several hours. In addition to high environmental stress tolerance, tilapia are the most disease-resilient fish produced in aquaculture. The incidence of infectious and parasitic diseases is lower than for other aquaculture fishes, although it has increased in recent years because of intensification and higher-density cultures in ponds and other open systems.

Treatment options for infectious diseases include acute alteration of salinity for several days, which is tolerated well by euryhaline tilapia but kills many pathogens. The main environmental parameter that limits the distribution of tilapia to tropical and subtropical climates is temperature. These fish do not survive temperatures below 10 °C and growth ceases at 16 °C. However, other than low temperature, tilapia tolerate stressful conditions very well without large impairment of growth and disease resilience. They even reproduce readily and without the need for artificial induction of spawning under conditions that deviate greatly from the optimum.

The same traits that are advantageous for aquaculture also represent the biggest problem with tilapia, as they confer a high propensity for invading non-native ecosystems. The flip side of high environmental stress tolerance and disease resilience of any species is a high potential for establishing feral populations in non-indigenous habitats. Tilapia have colonized tropical and subtropical habitats throughout the world. They are considered invasive pests in many areas, including Hawaii, California, and Florida, as well as parts of Asia, Oceania, and Australia.

Tilapia broodstock fish are kept in hatcheries or outdoor ponds and reproduce in almost monthly intervals if kept at optimal temperature at a male:female ratio of 1:3–10. Tilapia are the only group of aquaculture fishes that do not require hormonal injection for reliable reproduction in captivity, which increases their consumer acceptance. A large female weighing 1 kg produces up to 1500 eggs per clutch. The eggs and larvae are large and easily weaned to live and dry larval food after yolk absorption. Mouth-breeding of *Oreochromis* species is kept short and eggs are removed from the mouth within a few days after spawning to accelerate development of a new batch of eggs in the female.

Eggs are incubated in up- or down-welling jars or in containers on a shaking platform at 26 °C to accelerate embryo development. Hatching occurs in about a week and yolk is resorbed within another week. Free-swimming fry are then transferred to nursery ponds and grown out to fingerling size. Fingerlings are sexed and transferred to grow-out ponds to attain market size (200–700 g).

13.2.2 Tilapia aquaculture systems

Almost every possible aquaculture system can be adapted for tilapia as their culture conditions are very flexible. Consequently, there are many choices of tilapia aquaculture approaches, and many different systems, including monocultures and polycultures (Box 13.2), are in use throughout the world. Tilapia aquaculture systems span the entire spectrum from fully open, extensive systems to closed, intensive RAS. The following represents a brief overview of the main systems used for producing tilapia.

Box 13.2 Tilapia polyculture systems

Like for many other major seafood species, the most used aquaculture system for tilapia is open, semi-intensive culture in earthen ponds. In these pond systems, tilapia are often produced in polyculture with other species such as carps and shrimp. Polyculture of tilapia with shrimp reduces the incidence of shrimp diseases, presumably by better feed utilization and stimulation of beneficial microbial populations that compete with infectious pathogens. Polyculture of tilapia in rice paddies is also common. Moreover, tilapia are preferred species for aquaponics systems, in which the fish wastewater is used to fertilize hydroponically grown vegetable crops such as cabbages, basil, and tomatoes.

Extensive pond culture relies entirely on natural food and uses low-stocking densities. It is often performed in drainable earthen ponds. Because no pond fertilization and no supplemental feeding is performed, the burden on ecosystem service capacity is low and ecological sustainability is high. However, yields are also lowest of any aquaculture system, averaging approximately 0.5 tons per hectare pond surface area (tons/ha) annually. Extensive pond culture requires little maintenance and management and minimal upfront costs. This system of tilapia aquaculture is promoted to locally produce protein-rich seafood in developing countries, including in the native distribution range of tilapia in parts of Africa where sufficient water resources are available.

Semi-intensive pond culture employs similar drainable earthen ponds as extensive culture, but the ponds are often prepared artificially, and they are managed by application of chemicals for removing predators and competitors, by fertilization using organic and/or inorganic fertilizer, and by application of supplemental feeds. These culture systems sometimes utilize manure from adjacent livestock farms (chicken, duck, pig, etc.) as organic fertilizer in a polyculture system referred to as integrated animal agriculture (IAA). IAA systems were once heralded as a promising solution for increasing nutrient recycling, but difficulties of balancing manure inputs with aquatic ecosystem service capacity and concerns about food safety have recently reduced enthusiasm for this form of polyculture.

Semi-intensive systems span a wide range of intensification with production costs and yields depending on how much fertilizer and supplemental feed is being used. Yields of 2–5 tons/ha are typical for systems that rely exclusively on fertilizer, while up to 10 tons/ha are possible by using both fertilizer and supplemental feeding with agricultural by-products (cereal/rice bran, etc.). Although semi-intensive systems are most common for tilapia aquaculture, they rely heavily on ecosystem services and are the least ecologically sustainable form of aquaculture. Solutions for technology intensification need to be implemented in such systems to parallel the intensification of fertilization, feeding, and stocking and alter their mode of operation from open to semi-closed or closed systems.

Technological intensification does not always support a shift from open to closed systems. For example, the most common technology used for semi-intensive aquaculture ponds is aeration technology, which enables further intensification by increasing the amount of fertilizer, stocking density, and supplemental feeding with nutritionally optimized diets (formulated dry pellets). Water exchange pumps that supply ponds with clean water from natural sources also support intensification of open systems. When aeration and water exchange pumps are used then the yield of semi-intensive tilapia production can be further increased up to 100 tons/ha at the expense of an added cost to ecosystem service capacity.

Cage culture is another form of semi-intensive tilapia aquaculture. Like semi-intensive pond systems, it produces high yields but also relies heavily on ecosystem services. Cage culture is performed in large, deep reservoirs, lakes, or rivers. Therefore, compared to pond culture, it is much more difficult to implement technologies that promote a shift from open to closed culture systems.

Cages used for tilapia culture may have an inner compartment enclosed by a coarse-meshed net and an outer compartment enclosed by a fine-meshed net to separate offspring due to uncontrolled reproduction from grow-out fish. The fine mesh is prone to biofouling, which impedes water exchange and access to natural food. Consequently, regular cleaning is required, which adds to management costs. These costs are offset by very high yields of up to 3000 tons/ha by using cages that extend deep into the water column.

Raceway production systems are another form of semi-intensive tilapia aquaculture that are commonly operated as open systems. Raceway production yield of tilapia can reach similarly high yields as cage culture. Even though open, flow-through raceways are usually more technology-dependent methods than ponds or cage systems, they are still relatively easy to manage if no water recirculating technology is used. Open raceways rely on a large supply of clean water from the natural environment, while a large volume of wastewater is discharged untreated or minimally treated on a continuous basis because raceways are operated at very high flow rates. Nevertheless, raceways are well suited for transitioning from open to closed systems by integrating water recirculating, filtration, wastewater treatment, and disinfection technologies.

Although closed systems have a much smaller ecological footprint than open systems, the necessary technology is expensive to implement and maintain. The high cost currently disincentivizes their use since the corresponding ecosystem services are not nearly valued as highly as the technologies able to substitute for those services. Proper valuation of ecosystem services by estimating the true costs of our current way of seafood production for future generations remains a challenge that transcends national boundaries and needs urgent and more tangible attention from policy makers.

Recirculating aquaculture systems (RAS) are closed, intensive water reuse systems that typically have higher yields than semi-intensive pond culture but lower yields than open cage and raceway cultures (500–1000 tons/ha). Their ecological footprint is much lower than that of pond, cage, or raceway cultures. RAS are intensive systems because intensification is not just limited to increasing stocking rates, fertilization, feeding, and aeration, which are also characteristic for semi-intensive systems, but it also includes implementation of water reuse and wastewater treatment technologies that greatly reduce the ecological footprint. The ecological sustainability of RAS is highest when wastewater discharge and the amount of clean water needed are minimal, the waste materials (particulate waste, exhausted filter materials, etc.) are disposed of or recycled with minimal impact on natural ecosystems, and environmentally friendly sources of energy are used.

RAS have many advantages in addition to clean water conservation and high ecological sustainability. Since they are closed systems with maximal control over the production conditions, optimal parameters can be maintained even if they are different from the outside environment. Indoor RAS such as greenhouses enable accurate temperature control and efficient oxygenation of recycled water. Removal of odours resulting from the consumption of detritus and plankton by starvation for three days in clean water is easiest in closed RAS and increases market value. RAS also facilitate the maintenance of pure genetic lines of broodstock, which is critical for production of male-only hybrids (Section 13.2.3).

Additional significant advantages of RAS include greater biosecurity via disease control and integration of disinfection and water sterilization procedures, increased food safety, isolation of pure lines of broodstock, and containment of genetically engineered strains and non-native species. Because RAS are technology intensive, utility failure has a potentially devastating impact on their operation. Therefore, semi-intensive operation in an open mode may be implemented as a temporary backup solution in case of utility failure if a suitable water source for temporary switching to a flow-through mode of operation is available nearby as a backup.

13.2.3 Minimizing unwanted reproduction of tilapia

One of the main constraints of tilapia aquaculture is their tendency towards stunting (reduced growth) and early maturation when held at high density. Stunting results from crowding and premature reallocation of metabolic energy from muscle growth to gonadal development. Reproduction starts as early as three months of age when fish have attained just 10% of their market size. This problem is compounded by frequent spawning in monthly intervals under diverse environmental conditions. Offspring resulting from uncontrolled reproduction represents different size classes that overpopulate grow-out systems and compete for resources with the fish initially stocked for grow-out to market size.

To address unwanted reproduction of tilapia, genetic manipulation is most effective, which is why containment of aquaculture stocks in closed RAS is increasingly critical. I first briefly outline several alternative strategies for minimizing unwanted reproduction before discussing genetic manipulation of tilapia. Manual sexing of fingerlings before segregation into separate grow-out ponds for males and females exists but is a very labour-intensive and time-consuming method. Moreover, accurate determination of males and females is very difficult at this age and success rates are not 100%, which only reduces the problem in the initial stages of grow-out but does not eliminate it. Often, only males are grown-out to avoid mixing of males and females unless ponds at different sites are used. Males grow faster than females but 50% of the fingerlings are wasted when only males are kept after manual sexing.

Another solution to unwanted reproduction and overpopulation of grow-out ponds is polyculture with predator fish that feed on tilapia fry. This practice requires accurate estimation of the optimal ratio of predators versus tilapia stocked, often predator fish contribute to the overall fish population 1–10%. Other cichlids (e.g. *Hemichromis fasciatus*, *Cichla ocellaris*), catfish (*Clarias spec.*), snakehead fish (*Channa spec.*), and Nile perch (*Lates niloticus*) have been used for this purpose. In most cases, the market value of predator fish is significantly higher than that of tilapia, which incentivizes this form of polyculture.

Sex reversal by administration of steroid hormones during a critical time of tilapia development has also been used to produce nearly all male populations. It appears that sex can be determined hormonally between one and three weeks post hatching. Androgens (testosterone derivates) are administered for about a month during this critical window of time in the form of feed additives or as a bath application to achieve masculinization. Oestrogens can be used to achieve feminization, but their use is not common because male tilapia grow faster and larger than females. As with manual sexing, the success rate is between 90 and almost

100% but some fish with the opposite sex usually remain, which only reduces but does not eliminate the problem of unwanted reproduction. In addition, the use of steroid hormones is subject to strict regulation in many countries and may negatively impact consumer acceptance.

13.2.4 Tilapia hybrids and transgenics

Hybridization is a method of genetic manipulation that can yield all male hybrids between species with different sex chromosome systems if pure genetic lines are used and non-chromosomal mechanisms of sex determination are suppressed. This phenomenon was discovered unintentionally but has had a transformative tangible impact, illustrating the critical importance of unbiased basic research (Figure 13.5). Hybridization of tilapia species occurs naturally at range margins of populations where different species overlap. However, in captivity it is achieved by artificial fertilization after hand-stripping and mixing eggs and milt. Some tilapia species (e.g. *O. niloticus*, *O. mossambicus*, *O. spilurus*) have XY sex chromosomes with XX homogametic animals being females, just as in humans. Other tilapia species (e.g. *O. urolepsis hornorum*, *O. aureus*, *O. macrochir*) have WZ sex chromosomes with ZZ homogametic fish being males. Hybridization of tilapia species with XY and WZ sex determination systems often

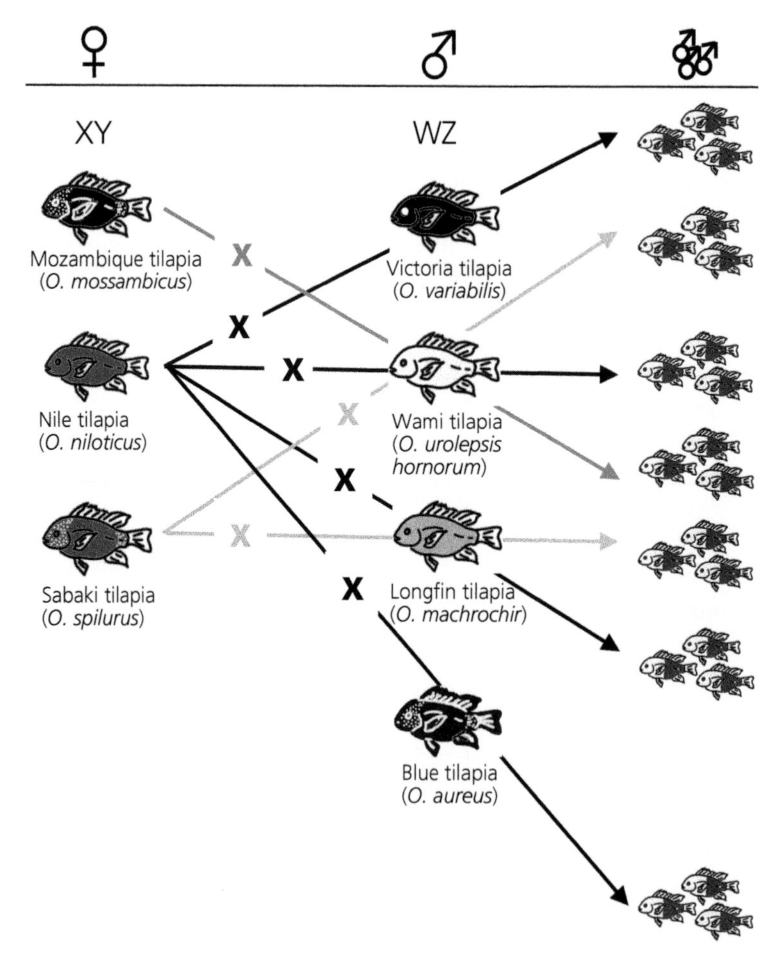

Figure 13.5 Select tilapia hybrid crosses of different *Oreochromis* species that result in all or nearly all male offspring. Data sources: summarized in (Pillay and Kutty, 2005; Suresh and Bhujel, 2019).

produces all male offspring. However, there are exceptions since sex determination in tilapia also depends on factors other than sex chromosomes and only certain genetic lines produce 100% male hybrids.

Tilapia hybrids are also produced for improving other production traits such as growth and colour, for example, red tilapia, which grows faster, has a better feed efficiency, and is more marketable than other tilapia. Red tilapia is obtained by extensive hybridization of occasionally occurring *O. mossambicus* mutants having reddish-golden body colour with *O. niloticus*. Additional crosses with other *Oreochromis* species may be performed to fix and propagate the desirable production traits mentioned. As a result, many different red tilapia strains are now in use for aquaculture. Red tilapia strains are sometimes marketed under different names that increase their consumer appeal, including FW snapper, because they resemble highly valued marine species such as red snapper.

Genetic engineering of tilapia has been extensively used in basic research to establish causal links between phenotypes (traits) of interest and specific genomic loci. In addition, genetic engineering approaches for tilapia that were developed by basic research have been translated to improve tilapia aquaculture. For example, a transgenic *Oreochromis spec.* hybrid has been generated in Cuba by insertion of an additional copy of a tilapia growth hormone gene, which increases growth rate by 1.5-fold. Likewise, a transgenic Nile tilapia (*O. niloticus*) strain has been generated in the UK by inserting a salmon growth hormone gene into its genome. The resulting increase in growth rate is 3.5-fold.

Containment of transgenic strains is one of the challenges associated with genetic engineering of aquaculture organisms, which can be addressed most effectively by using RAS technology. Chromosome manipulation approaches such as production of triploid, androgenic, or gynogenic tilapia and generation of YY supermales have also been explored but results so far have been too variable for their adoption on a commercial scale.

13.3 Catfishes

Catfishes (order Siluriformes) are represented by approximately 2500 species and are distributed on all continents except Antarctica. Most catfish are omnivorous, bottom-feeding, FW species, but some also inhabit brackish water and two families (Ariidae and Plotosidae) include marine species. Catfishes have several characteristic morphological traits. Their skin is either naked or covered by bony plates rather than scales. Their head contains four pairs of barbels, some of which can be very long and whisker-like, hence the common name catfish. The barbels are used to locate and inspect food.

Some catfish are poisonous and avoided by aquaculturists, especially when they are aggressive and attack humans, or their poison may be lethal, for example, the Asian stinging catfish (*Hetoropneustes fossilis*) and the striped eel catfish (*Plotosus lineatus*). Many catfish are small (maximum length of <12 cm) but some species such as Asian shark catfish (*Pangasionodon spec.*) can be very large. The largest catfish is the European catfish or Wels (*Silurus glanis*), which can attain a length of 3 m and is farmed in Southern Europe (Figure 13.6). The four most common catfish species used for aquaculture are the low-eyed shark catfish (*Pangasionodon hypophthalmus*), torpedo catfish (*Clarias spec.*), yellow dragon catfish (*Peltobagrus fulvidraco*), and channel catfish (*Ictalurus punctatus*) (Figure 13.1).

13.3.1 Channel catfish

The channel catfish (*Ictalurus punctatus*) is an important aquaculture species in the US where most of this species is produced. F1 (first-generation offspring) hybrids resulting from crosses of female *I. punctatus* and male blue catfish (*I. furcatus*) are also used for aquaculture. These hybrids show improved growth, higher survival in captivity, and increased yield at harvest.

Aquaculture of channel catfish and their hybrids is carried out in the south-eastern states of Mississippi, Louisiana, Arkansas, and Alabama. Channel catfish have been exported from North America to several other continents, but other catfish species are produced more commonly in Asia or elsewhere.

Besides open, semi-intensive pond culture, open raceways, cage culture, and tank culture are also performed but to a much lesser extent than pond culture. Polyculture with other fish species is less common than for carps and tilapia. However, in the

Figure 13.6 Catfish are aptly named as their long barbels resemble the whiskers of a cat, as displayed by this European catfish (Wels catfish, *Silurus glanis*). Photo by Milos Prelevic (Unsplash).

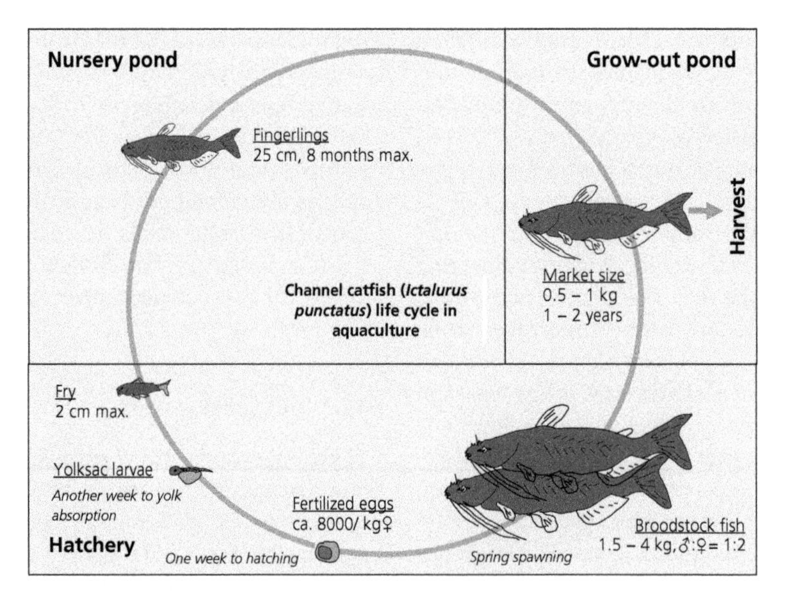

Figure 13.7 Life cycle of the channel catfish (*Ictalurus punctatus*) in aquaculture. Development proceeds in a clockwise direction starting in a hatchery, then in a nursery pond, and completes in a grow-out pond.

south-eastern US polyculture with rainbow trout and polyculture with rice and soybean are sometimes practised. Channel catfish aquaculture can be divided into three phases: hatchery, nursery ponds, and grow-out ponds (Figure 13.7).

Channel catfish hatcheries are operated at 26 °C to accelerate spawning of broodstock and embryonic development after egg fertilization. Broodstock fish are fed formulated, nutritionally optimized diets to promote gonadal development and gamete maturation. They are spawned naturally in spawning ponds, pens, or large aquaria. Nesting sites are provided to facilitate spawning. Females are usually injected with pituitary hormones to induce ovulation and spawning (Box 13.1). Artificial spawning is common for hybrids with females being stripped of eggs and males sacrificed to remove milt for fertilization. A large female weighing several kg can produce 10,000 eggs, which are usually incubated in hatchery jars or troughs to minimize losses.

Under these conditions larvae hatch in about a week and yolk is absorbed within a week after hatching. As soon as the yolk sac is absorbed fully, fry start to feed and swim freely. They are transferred to nursery ponds or tanks within a week. Sometimes, fry are reared in intermediate rearing troughs or tanks to a size of about 3 cm before transfer to nursery ponds. Nursery ponds are smaller than grow-out ponds and used to produce juveniles (fingerlings). On average, shallow nursery ponds cover an area of about 150 × 200 m.

Fry are fed protein-rich (45%), finely ground, formulated diets that are more taxing on water quality than the floating, larger, formulated pellets used for feeding fingerlings and adults (30% protein content). Using floating, pelleted feeds maximizes feed efficiency, increases food availability for a longer time of feeding, and minimizes loss of feeds as water quality-compromising waste.

Fingerlings are transferred from nursery to grow-out ponds when they have reached a size of 10–15 cm length. At this point the fingerlings are about half a year in age and they are grown for another year or longer before harvest. Grow-out ponds are larger and slightly deeper than nursery ponds, typically covering an area of 200 × 300 m. The size of these ponds represents a compromise between ease of management (small ponds are preferable) and construction cost (large ponds are cheaper). From hatching to harvest, catfish are grown for 15–24 months to attain a market size of at least 0.5 kg.

As with other semi-intensive pond systems, earthen nursery and grow-out ponds are intermittently drained and treated to remove predators, pests, and competitors using chemicals (e.g. rotenone) before stocking with catfish. Ponds are also fertilized, and supplementary feeding is performed multiple times daily at rates of up to 5% of fish biomass per day. Overfeeding has been identified as a common problem of catfish aquaculture (Pillay and Kutty, 2005).

Chemical pre-treatment, fertilization, and overfeeding of catfish ponds all represent practices that strain FW ecosystem service capacity. Tank culture is more amenable than pond culture to implementing water and nutrient recycling and utilizing RAS technology. Moreover, commercial grow-out of channel catfish in large 6-m-diameter tanks produced the same yield as a comparable pond culture.

Catfish harvest from grow-out ponds takes place using seines similarly to the harvesting of other FW fishes. Fish are removed from the seines using a mechanical basket and cranes and transferred into containers filled with clean, aerated water for live shipment by truck to processing plants. An alternative method of harvest is draining ponds, raceways, cages, or tanks before removing the fish.

13.3.2 Torpedo catfishes

Two species of the genus *Clarias* (*C. batrachus* and *C. macrocephalus*) that are native to Southeast Asia are important aquaculture catfish (Figure 13.1). They are mostly farmed in their native distribution range in extensive or semi-intensive ponds. The Nile catfish (African sharp-tooth catfish, *C. gariepinus*) is used for aquaculture in Africa and the Middle East, which represent the native distribution range for this species.

Reproduction and grow-out of *Clarias* species are similar as described for Channel catfish. Ovulation-inducing pituitary hormones and artificial fertilization are used to increase fertilization and hatching rates. Males must often be sacrificed to obtain enough milt. Of note for ecological sustainability prospects, experimental-scale RAS at very high stocking density appears both feasible and profitable for torpedo catfishes (Pillay and Kutty, 2005).

Torpedo catfishes tolerate brackish salinity and hypoxia without significant impairment of growth. The ability to tolerate hypoxia is due to an air-breathing organ associated with the swim bladder that allows these catfish to gulp atmospheric air and use it for respiration. They use their air-breeding organ and the scaleless leathery skin in lieu of the gills for gas exchange when exposed to air. This physiological capacity enables torpedo catfishes to tolerate extensive periods of emersion on land and migrate long distances to colonize new FW ponds and riverine systems. The hardiness of these catfish has encouraged overfeeding in aquaculture, which diminishes the ecological sustainability of their production and can trigger disease outbreaks that cause significant mortality.

The high environmental stress tolerance of torpedo catfishes combined with their introduction to non-native habitat has also made them invasive species in several parts of the world. The walking catfish (*C. batrachus*) is an invasive species in Florida and the African sharp-tooth catfish (*C. gariepinus*) is an invasive species in Brazil. *C. gariepinus* was also introduced to several Southeast Asian countries even though the native *C. batrachus* and *C. macrocephalus* are well suited for aquaculture. Introduction of non-native species despite the presence of similar local species is questionable from the perspective of ecological sustainability. The import of *Clarias* species as ornamental fish and their sale in pet stores or for seafood aquaculture has now been banned or requires special permits in several countries.

13.3.3 Shark catfishes

Shark catfishes represented by the family Pangasiidae have become the dominant aquaculture catfish in recent years. Species in this catfish family can grow very large. The Mekong giant catfish (*Pangasianodon gigas*) is one of the largest FW fish and may reach a length of 3 m. *P. gigas* is native to the Mekong River delta and endangered because of overfishing. Another species native to Southeast Asia is the low-eyed catfish (iridescent shark catfish, *P. hypophthalmus*), which now dominates catfish aquaculture. Like *Clarias* species, it uses the swim bladder and skin as accessory surfaces for gas exchange when exposed to air (Figure 13.1).

Reproduction of *Pangasianodon* broodstock uses pituitary hormone induction and is similar to other catfish. It is often performed in hatcheries although sometimes broodstock is kept in floating cages in a large river. Grow-out of shark catfish by open cage culture in large rivers and deltas was the preferred form of aquaculture during the twentieth century.

The turn of the millennium witnessed an almost complete shift to semi-intensive pond culture that resembles systems used for other catfish. The nursery and grow-out ponds used for riverine shark catfish are subject to much higher water exchange rates than ponds used for other catfish species. The high water exchange rates of these open pond systems are a major limiting factor for the ecological sustainability of shark catfish aquaculture. In addition, mortality in nursery ponds can be as high as 90%, which adds to the burden placed on ecosystem services provided by decomposing bacteria.

Key conclusions

- FW aquaculture has greater potential than mariculture for intensifying production in ecologically sustainable systems that utilize available know-how and technologies.
- FW aquaculture fishes are positioned lower in trophic webs and, thus, require fewer resources than most catadromous, anadromous, and marine aquaculture fishes.
- Carps, tilapia, and catfishes are the most important taxa of FW fishes used for aquaculture. Semi-intensive, open pond culture that relies heavily on ecosystem services is still most common.
- Carp, tilapia, and catfish aquaculture is often divided into hatchery, nursery, and grow-out phases.
- Reproduction of carp, tilapia, and catfish broodstock is performed by natural spawning or artificial fertilization. In carp and catfish, injection of pituitary hormones induces ovulation and spawning.
- Carp and tilapia species dominate animal aquaculture and are often raised in polycultures to better utilize resources by taking advantage of different feeding habits and food preferences.
- Polyculture of carps and tilapia with rice and azolla improve resource utilization by integrating primary producers in addition to heterotroph consumers into aquaculture nutrient cycles.
- Tilapia, carps, and catfishes have high environmental tolerance and are ideal for a wide variety of aquaculture systems. On the other hand, their high environmental stress tolerance promotes invasiveness.
- Premature reproduction of tilapia is problematic for aquaculture and promotes invasiveness. It has been addressed by producing single-sex (mostly all male) populations using multiple approaches.
- Catfish aquaculture has shifted from an initial focus on channel catfish in the second half of the twentieth century to Asian torpedo and shark catfishes, which have superseded channel catfish in value and yield.

References

Belton, B., Little, D. and Zhang, W. (2021). 'Farming fish in fresh water is more affordable and sustainable than in the ocean', The Conversation [online]. Available at http://theconversation.com/farming-fish-in-fresh-water-is-more-affordable-and-sustainable-than-in-the-ocean-151904 (accessed 23 June 2021).

Bunting, S. W. (2013) Principles of *sustainable aquaculture: promoting social, economic and environmental resilience*. Abingdon: Routledge.

De Silva, S. and Wang, Q. (2019). 'Carps', in Lucas, J. S., Southgate, P. C., and Tucker, C. S. (eds.) *Aquaculture*, 3rd edn. Oxford: John Wiley & Sons, Ltd, pp. 339–362. doi:10.1002/9781118687932.ch23.

Food and Agriculture Organization of the United Nations (FAO). (2020). *The state of world fisheries and aquaculture 2020*. Rome: Food and Agriculture Organization of the United Nations. Available at: http://www.fao.org/documents/card/en/c/ca9229en.

Kishe-Machumu, M.A. Natugonza, V., Nyingi, D. W., Snoeks, J., Carr, J. A., Seehausen, O., and Sayer, C. A. (2018). 'The status and distribution of freshwater fishes in the Lake Victoria basin', in Sayer, C. A., Máiz-Tomé, L., and Darwall, W. R. T. (eds.) Freshwater biodiversity in the Lake Victoria Basin: guidance for species conservation, site protection, climate resilience and sustainable livelihoods. Cambridge: IUCN, pp. 41–64.

Mayden, R. L. and Chen, W.-J. (2010). 'The world's smallest vertebrate species of the genus Paedocypris: a new family of freshwater fishes and the sister group to the world's most diverse clade of freshwater fishes (Teleostei: Cypriniformes)', *Molecular Phylogenetics and Evolution*, 57(1), pp. 152–175. doi:10.1016/j.ympev.2010.04.008.

Pillay, T. V. R. and Kutty, M. N. (2005) *Aquaculture: principles and practices*, 2nd edn. Oxford: Wiley-Blackwell.

Reite, O., Maloiy, G. and Aasehaug, B. (1974) 'pH, salinity and temperature tolerance of Lake Magadi tilapia', *Nature*, 247(5439), pp. 315–315. doi:10.1038/247315a0.

Suresh, V. and Bhujel, R. C. (2019) 'Tilapias', in Lucas, J. S., Southgate, P. C., and Tucker, C. S. (eds.) *Aquaculture*, 3rd edn. Oxford: John Wiley & Sons, Ltd, pp. 391–414. doi:10.1002/9781118687932.ch23.

CHAPTER 14

Anadromous fishes

'Identification of certain fish species as undesirable often shows considerable cultural bias: common carp and tilapias, for example, are considered as invasive pests in some countries whereas salmonids introduced outside their native range are almost never categorized in that way.'

De Silva and Wang, in *Aquaculture: farming aquatic animals and plants* (2019)

Migratory fish that spend part of their lifecycle in freshwater (FW) and part in the ocean are referred to as diadromous fish. Diadromy is a life-history strategy that expands the ecological niche repertoire and geographic distribution range of a species. Two different forms of diadromy are anadromy and catadromy. Catadromous fish reproduce in the ocean. Their offspring migrate to brackish and FW during the early stages of development before returning to the ocean as juveniles or adults. Anadromous fish spawn in FW and complete the first part of their lifecycle in lakes or rivers. They migrate as juveniles to the ocean, where they grow to maturity and spend most of their adult life before returning to FW for reproduction.

Salmonids (salmon, trout, and char species) are classical examples of anadromous fishes, although some salmonid species have resident FW populations that do not migrate to seawater (SW). Other anadromous species that are of commercial interest for aquaculture are some sturgeon species (family Acipenseridae). Several species of sturgeon are produced in aquaculture for caviar production. Sturgeon meat is also consumed as seafood. Other fish families that include anadromous species are sticklebacks (Gasterosteidae), herrings (Clupeidae), basses (Moronidae), smelts (Osmeridae), and lampreys (Petromyzontidae).

Anadromous fish are highly valued in many countries. Their consumption as seafood represents a century-old tradition in large parts of the world. A main reason for this long-standing preference for anadromous species is the relative ease by which they could be obtained by simple capture fisheries that take advantage of their migratory behaviour. Salmon, for example, can be caught in large numbers at convenient river locations on their upstream migration to spawning grounds. The ease of their capture and long-standing fisheries in some parts of the world, as well as barriers in migratory corridors and loss of access to spawning and fry/fingerling habitat due to dam construction and habitat degradation, led to severe declines of salmon populations during the twentieth century.

Salmon overfishing was a consequence of the steep increase in human population growth, an increase in per capita consumption of seafood leading to increased demand, and more efficient capture and processing technologies. Many government- and state-funded salmon and trout hatcheries have been built to produce juveniles to augment wild populations of salmonids. In many cases, salmonid hatcheries have produced trout to stock lakes and rivers for sport fishing purposes even if the species was not native to the corresponding habitat. In some areas this practice has continued to the present day and is often less subject to scrutiny by policy makers and conservationists than introduction of other non-indigenous species.

For salmonid aquaculture aimed at seafood production, both phases of the anadromous lifecycle must be accommodated unless the species of

A Primer of Ecological Aquaculture. Dietmar Kültz, Oxford University Press. © Dietmar Kültz (2022). DOI: 10.1093/oso/9780198850229.003.0014

interest can form resident FW populations. Anadromous salmonids undergo a physiological transformation during development, which is called smoltification or parr-smolt transformation. Parrs are juvenile salmon limited to FW while smolts have gained the physiological capacity to enter SW (Section 14.1.2). The hatchery phase takes place in FW while the grow-out phase of salmon and anadromous trout takes place in SW. Mariculture grow-out coupled with FW hatchery production of smolts adds logistical, financial, and infrastructure costs to salmonid aquaculture. These costs are offset by the high prices fetched for salmonid seafood products compared to other fish.

In addition, costs can be minimized by utilizing the migratory homing behaviour of anadromous salmonids for sea ranching in lieu of cage-based offshore grow-out. Since sea ranching does not confine domesticated salmon, this form of aquaculture is only suitable for native species with genetic population structures that closely reflect those of corresponding wild populations. Moreover, straying complicates sea ranching because some sea ranched fish migrate to watersheds other than the one in which they were released as juveniles and compete for mates and spawning habitat with wild fish. Therefore, sea ranching is not amenable to selective breeding, genetic engineering, and disease resistance improvements of strains used for this form of aquaculture.

14.1 Salmon

Despite increasing awareness of the benefits of ecologically more sustainable aquaculture of native species, the production of a single salmonid species, Atlantic salmon (*Salmo salar*), has greatly outpaced that of all other salmonid species. Aquaculture of *S. salar* accounts for more than all the other species taken together. Moreover, from 2010 to 2018, *S. salar* aquaculture yield has doubled while production of other species has remained stable or only moderately increased (Figure 14.1).

Most Atlantic salmon is produced in its native distribution range, with Norway accounting for half of the production in 2018. But salmon are non-native to the second-leading salmon-producing country, Chile, which accounted for a quarter of the production in 2018. The remaining quarter of Atlantic salmon is produced mostly in the UK, Canada, Australia, Ireland, the USA, and Iceland. Among these countries, Australia and the Pacific coasts of Canada and the USA lie outside the native distribution range of this species.

The other two salmon species that contribute significantly to commercial seafood production are Coho salmon *(Oncorhynchus kisutch)* and Chinook salmon (*Oncorhynchus tshawytscha,* Figure 14.1). Both species are Pacific salmon that have a native distribution range along the Pacific coasts of North America and Northern Asia. Several other species of Pacific salmon are also used for aquaculture, especially in Japan, but to a lesser overall extent than Coho and Chinook.

Oddly, in 2018, Atlantic salmon dominated seafood production in the Pacific Northwest of Canada and the USA, which is the native distribution range of Coho and Chinook salmon, while countries leading the production of Chinook salmon (New Zealand) and Coho salmon (Chile) lie outside the native distribution range of these Pacific salmon species. It is not clear whether salmon have irreversibly colonized the Southern hemisphere, to what extent they have altered the corresponding ecosystems, and whether it is both feasible and beneficial to manage them as invasive species. More research is needed to answer these questions and identify native species that are most suitable for aquaculture in the Southern hemisphere.

Environmental concerns about salmon aquaculture are prominent, receiving much attention by the public media, NGOs, conservationists, and policy makers. A significant portion of these concerns could be alleviated by two strategies of aquaculture development:

1. The exclusive use of native species for aquaculture along with proper management of genetic diversity.
2. Increased containment of domesticated salmon that have been altered by selective breeding, genetic engineering, hormonal, or chemical treatments, or by other human interventions for improving aquaculture production traits.

Increased containment, greater biosecurity, and reduced ecological and spatial footprints can be

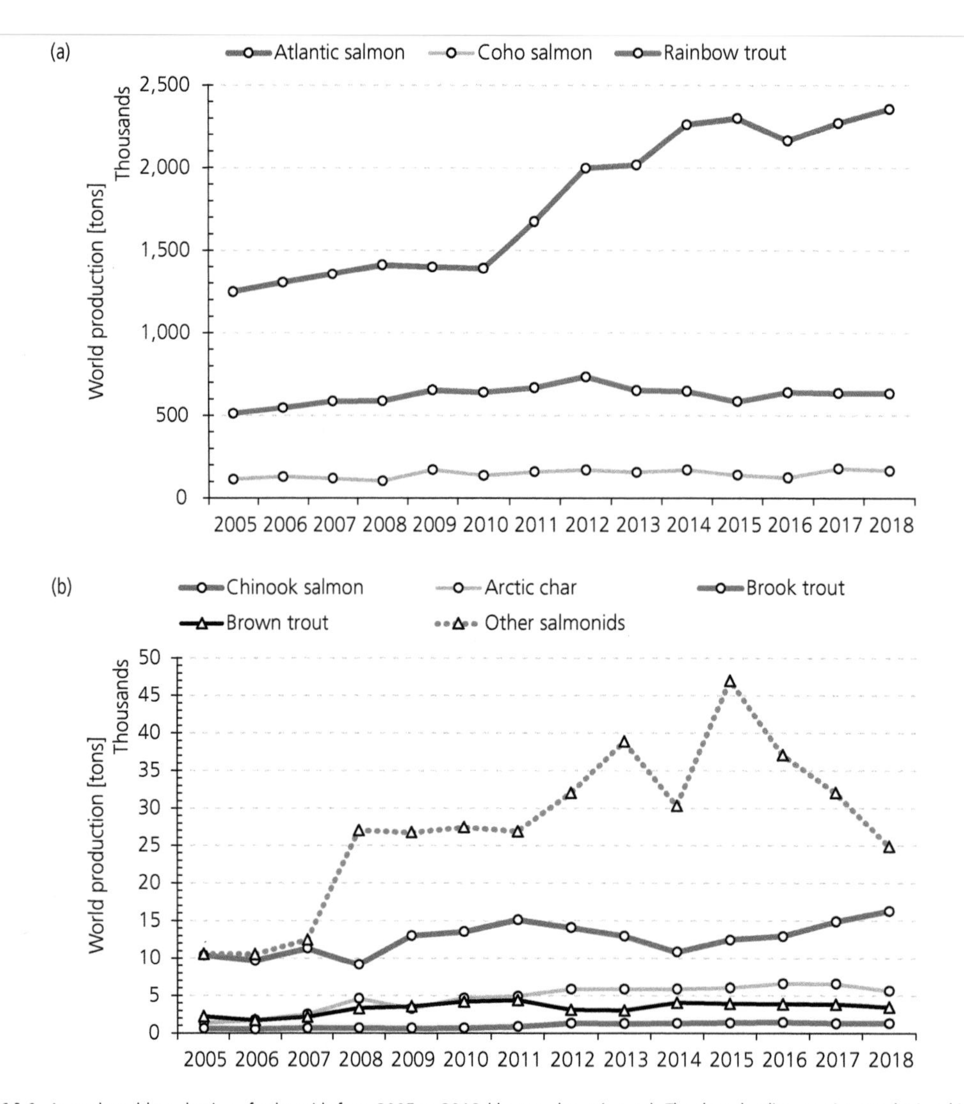

Figure 14.1 Annual world production of salmonids from 2005 to 2018 (thousand metric tons). The three leading species are depicted in Panel A. The next four most produced species and the cumulative remainder (other salmonids) are shown in Panel B. Note the almost hundredfold difference in *y* axis scale. Data source: OECD.Stat.

achieved by using closed, intensive, recirculating aquaculture systems (RAS). Well-designed and managed RAS address concerns about escapes, pollution, and visual pollution caused by semi-intensive net-pen cages in coastal marine waters. Since genetic engineering has produced GMO salmon (and other fish) that outperform conventional strains in aquaculture (see Chapter 6), aquaculture systems that are able to fully contain them are needed in addition to unbiased, well-designed studies that evaluate food safety and animal welfare aspects associated with GMOs.

The most effective and space-efficient containment systems such as RAS are also the most expensive, especially if they are operated with SW, which is much more corrosive than FW and requires more elaborate technologies for wastewater recycling than FW (see Chapter 12). Temporary flow-through backup solutions that maintain 100% containment in case of technology failure are also more difficult to implement for marine than for FW RAS. Moreover, salmonids are carnivores that occupy high levels in trophic webs, which renders their production more resource intensive than that of herbivorous fishes. For these reasons, anadromous species of salmonids may not be the best focus for future development of aquaculture.

If the human world population continues to rise, then a shift in consumer preference to herbivorous fish seems inevitable. It remains to be seen how the century-old, deeply engrained preference for salmonids can best be transformed towards species that can be produced more sustainably. Developments in this regard are encouraging as more sustainable FW species already comprise three-quarters of all fish aquaculture (Chapter 13).

Along with a historical origin of a strong preference for salmonids as seafood, the millennia-long capture fisheries of salmonids in the Northern hemisphere gave rise to cultural traditions and shaped the socio-economic structures of indigenous civilizations in Asia, Europe, and North America. Ecologically responsible stewardship of wild salmon populations has developed as an integral part of their capture fisheries and led to the emergence of long-standing cultural traditions. This cultural aspect is exemplified by native peoples of the Pacific Northwest in Canada and the US. Annual salmon runs are an integral part of the spiritual and cultural identity of these peoples. They have been the basis of their economic livelihood and trade for millennia.

Displacement of wild salmon in the marketplace by farmed salmonids impacts an important economic pillar of these traditional societies. Furthermore, the practice of farming salmon, restricting their migrations and free movement, and animal welfare aspects associated with stress and disease in captivity may be in opposition to traditional customs and beliefs. Nevertheless, the demand for salmon as seafood is increasing, which increases the pressure on wild populations. Therefore, tribal facilities contribute to producing smolts for salmon restocking programmes aimed at conservation and restoration of native salmon populations and support of capture fisheries and sea ranching (Young, 2011).

Similar aquaculture approaches are used for all anadromous salmonid species. For semi-intensive and intensive seafood production, a FW hatchery phase is followed by a marine grow-out phase. For restocking and sea ranching programmes, only the FW hatchery phase is performed before smolts are released into the wild. Since Atlantic salmon are the main salmonid species used for aquaculture, the following paragraphs briefly illustrate the two production phases for *S. salar*.

One notable difference between Atlantic salmon and Pacific salmon is that the former produce kelts (multi-year spawners) while all Pacific salmon die after spawning. Atlantic salmon kelts are mostly females that return to the ocean after annual reproduction and contribute to spawning during multiple years. The percentage of females that survive the annual spawning and return as kelts in wild populations can exceed 10% for females but is generally lower for males and depends on the river system. For aquaculture, *S. salar* kelts are used in selective breeding programmes because they have a higher fitness and fecundity than the rest of the population and they can be reconditioned for future spawning after stripping of the gametes.

14.1.1 Hatchery phase in FW

In the wild, the first phase of salmon development takes place in shallow tributaries of FW rivers. Migration of wild salmon to their FW spawning grounds have been impaired by many weirs, which can be by-passed if fish ladders are installed (Figure 14.2a). In aquaculture, the initial FW phase takes place in hatcheries that operate either as open, flow-through systems or RAS.

Broodstock fish are selected based on desirable performance traits (growth, fitness, disease resistance, etc.) and genetic diversity. They are conditioned, i.e. acclimated from SW to FW, for at least four weeks prior to spawning in large tanks or raceways with sufficient flow of high-quality water sourced from a spring, aquifer, or RAS. Conditioning also includes feeding of a protein-rich diet that contains enough carotenoids to enhance the survival and orange colour of eggs.

At many salmon farms ovulation and spawning are rarely induced by injection of pituitary hormones but rather by manipulation of photoperiod or by monitoring natural maturation. Males and

Figure 14.2 Salmon migrating upstream a river to reach their freshwater spawning grounds. Photo by Brandon@Greener_30 (Unsplash).

females are anesthetized using tricaine (MS-222) followed by hand-stripping of eggs and milt into a bucket. The method of dry mixing gametes at a ratio of approximately 1 mL milt per 5000 eggs without adding water is used to maximize fertilization success. A notable difference for Pacific salmon species (and trout) is that broodstock animals are killed without using MS-222 or other chemicals before stripping the gametes since they cannot be reconditioned for multiple spawning. This practice allows for processing of meat and carcass mass from broodstock animals after stripping.

Fertilization takes place within a few minutes of mixing the gametes and during this period the gametes are left undisturbed. Afterwards, water is added to the fertilized eggs, which induces uptake of water leading to egg swelling and hardening. Upwelling jars (Figure 13.3), trays, or trough systems are used to incubate the fertilized, hardened eggs at 10 °C at a density of approximately 7000 eggs/L until hatching. An advantage of using upwelling jars is that they require less water and space. The main advantage of tray and trough systems is that they permit regular removal of dead and infected eggs when chemical treatments are prohibited and fungal infections are prevalent. Exposure to light is minimized during egg incubation to avoid DNA damage and the associated increased mortality and incidence of developmental abnormalities. Under these conditions, salmon embryos develop to the eyed stage after about three weeks and larvae hatch after another two weeks.

Hatched yolk sac larvae of salmonids are called alevins (Figure 14.5), which are transferred to small incubators. They develop a functional gastrointestinal (GI) tract (including gut) within a week after hatching and can begin to feed even before the yolk is completely absorbed. Before feeding commences the alevins swim up to the surface of the incubator tray to inhale air for inflating the swim bladder. This stage, known as the swim-up stage, is the most critical period in salmon aquaculture when mortality is highest.

Larval mortality can be reduced by providing proper substrate for keeping larvae immobile in a water current, which maximizes yolk utilization and development. Feeding *ad libitum* with starter feed (crumble formulated feeds) multiple times daily is critical for development of proper feeding habits while the remaining yolk is absorbed. Complete absorption of yolk may take up to a month after feeding has started since yolk will be used up more slowly when external feeding occurs.

At a water temperature of 12–16 °C it takes approximately two months until salmon fry (parr that are smaller than fingerling size) have fully absorbed the yolk. At that time, they are transferred to indoor tanks or small raceways. Sloping earthen ponds that permit unidirectional water movement at a constant rate to encourage development of swimming muscles are also sometimes used for growing fry. However, temperature and other parameters can be controlled more efficiently in indoor tank or raceway systems, in particular if RAS are used.

The use of RAS for FW hatchery aquaculture of salmon has increased in recent years because of the capacity to control water quality parameters, possibility to place them in closer vicinity to major markets, higher stocking densities, and greater biosecurity. RAS stocking densities of 100 kg fry per m^3 (10% of water volume!) can be achieved if oxygen infusion or ozonation systems are employed (Purser, 2017).

Size-grading of fry is performed on a regular basis to transfer large fish to separate grow-out tanks or raceways, adjust stocking density, adjust feed pellet size and ration, minimize aggression, and minimize formation of social hierarchies. The frequency of grading represents a trade-off between the benefits stated here and additional management costs as well as negative impacts on growth resulting from grading stresses (handling and pre-grading starvation) that are imposed on the fish. Some salmon farms discard low-performing (bottom 20%) fish during grading because they are unlikely to perform well during grow-out. If fertilization takes place in spring, then maturation of fry into salmon parrs occurs within six months at a temperature of about 16 °C. The juvenile parrs are then ready in the fall to transform into smolts by undergoing the physiological processes of smoltification (Figure 14.3).

14.1.2 Smoltification

Smoltification is the physiological process that prepares salmon juveniles for entering SW. It occurs naturally before and during downstream migration of smolts from rivers towards the ocean. If sea-ranching or restocking programmes are pursued, then release of fry in FW rivers is often practised one or two months prior to smoltification to facilitate homing behaviour. However, for sale or transfer to marine grow-out cage farms smolts are produced in the hatchery. Smoltification alters the structure and function of tissues that are responsible for osmoregulation, i.e. the maintenance of body water and salt homeostasis, in fish.

The main tissues involved in fish osmoregulation are the gill epithelium, kidney, and GI tract. Smoltification is induced by an intrinsic developmental programme in combination with extrinsic clues such as the seasonal change in photoperiod and water temperature (McCormick, 2012). These extrinsic clues are utilized in aquaculture, for example, in Norwegian salmon farms, to induce and synchronize smolting by controlled alteration of photoperiod. During smoltification the osmoregulatory tissues that control transepithelial water and salt movements are restructured to support a reversal of active ion and water transport when the fish enters SW. Table 14.1 summarizes the main physiological changes that take place during this reversal.

Figure 14.3 Atlantic salmon (*Salmo salar*) parr (top) before, and smolt (bottom) after smoltification. Photo reproduced with kind permission from Steve McCormick (McCormick, 2012).

Table 14.1 Environmental and physiological changes that occur during smoltification. The unit for salinity given in the table is grams of salt per kg of water (g/kg), which roughly corresponds to parts per thousand (ppt)

Parameter	Parr(→ signifies direction)	Smolt(→ signifies direction)
Environmental salinity	Freshwater, FW (~0 g/kg)	Seawater, SW (~34 g/kg)
Passive salinity gradient (passive loss/gain of ions)	Fish body fluid → FW	SW → Fish body fluid
Passive osmotic gradient (passive loss/gain of water)	FW → Fish body fluid	Fish body fluid → SW
Blood and interstitial fluid osmolality (~salinity)	310 mosmol/kg (~9g/kg)	320 mosmol/kg (~9g/kg)
Active ion (salt) transport	FW → Fish body fluid (active salt absorption to counteract passive diffusional loss)	Fish body fluid → SW (active salt secretion to counteract passive diffusional gain)
Secondary active water transport	Fish body fluid → FW (counteract passive hydration)	SW → Fish body fluid (counteract passive dehydration)

Smoltification ensures that salmon can maintain their internal body water and ion concentrations (osmolality) constant at approximately 315 mosmol/kg (equivalent to ~9 g/kg salinity). This constancy of the ionic and osmotic milieu inside the fish is called osmotic homeostasis and maintained by the physiological process of transepithelial osmoregulation. The ability to maintain osmotic homeostasis distinguishes osmoregulators (most vertebrates) from osmoconformers (most invertebrates).

Because FW has a lower (~0.1 g/kg) and SW a higher (~34 g/kg) salinity than salmon blood (9 g/kg), migration from FW to SW reverses the ionic and osmotic gradients across the skin, gill, and other epithelia. This reversal is counteracted by corresponding reversals of active (metabolic energy-expensive) transport processes of water and salt in the opposite direction (Kültz, 2015). Because the osmotic and ionic gradient is much steeper in SW (25 g/kg) than FW (9 g/kg), metabolic energy requirements for active transepithelial transport are higher. Even though energy spend for osmoregulation represents only a fraction (2–5%) of resting metabolic rate, greater allocation of energy towards osmoregulation in SW affects other energy-dependent processes including respiration, growth, and behaviour of the fish.

Smoltification is completed just before juvenile salmon enter SW (i.e. it represents a developmental preadaptation programme). This physiological process not only alters molecular transport processes in osmoregulatory tissues but also easily recognizable traits. Parrs are yellowish in body colour with characteristic broad black stripes on their sides. The appearance of juvenile salmon changes to a silver body colour in smolts and an increase in black fin margins can be observed (Figure 14.3). For this reason, smoltification is also referred to as 'silvering'.

Most fish in a salmon hatchery smolt within the first year after hatching but smaller parrs (~10%) are delayed and may require an additional year to transform into smolts. First-year smolts weigh 25–100 g while second-year smolts can be significantly larger. Some salmon farms in Northern Europe keep fish longer in FW RAS and only transfer them to marine cages when they have reached a weight of 500 g or more. This practice shortens the time of marine cage grow-out and reduces exposure to sea lice and infectious pathogens. Triploid single sex smolts are sometimes produced to prevent reproduction of escapees from mariculture cage farms (see Chapter 6 for a discussion of triploids).

14.1.3 Grow-out phase in mariculture

The second phase of salmon aquaculture is semi-intensive grow-out of smolts by mariculture. Floating, anchored sea cages (or net pens) in coastal areas such as fjords, sounds, bays, and more exposed open ocean sites farther offshore are the

Figure 14.4 Mariculture cages for grow-out of Atlantic salmon (*Salmo salar*) in coastal marine waters have large diameters (10–50 m) and are several meters deep. Cages are often covered with a net to provide protection from large seabirds (not shown).

commonly used aquaculture systems for this phase (Figure 14.4). Cages are open systems that rely on ecosystem services (e.g. tidal flushing) to balance the effect of salmon feed and waste on marine nutrient cycles. Cages for Atlantic salmon and some species of Pacific salmon are used for mariculture in temperate regions throughout the world where water temperature is optimal (12–15 °C).

Since the start of commercial salmon aquaculture in the 1970s the impacts of salmon cage farms on marine ecosystems have steadily increased, including local eutrophication, proliferation of sea lice and virus populations that infect wild salmon, alteration of trophic webs and competition by escapees, and deterioration of sediment quality under cages. Consequently, ecosystem services have reached a limit in many grow-out areas and alternatives for cage grow-out are now being researched, including RAS. Moving cages to different sites and employing fallowing strategies to allow the areas under cages to recover from the effects of faecal accumulation are increasingly used to facilitate ecosystem recovery at mariculture sites.

Recently, concerns about chemical and visual pollution of near-shore coastal habitat and interference with marine mammals and other wildlife have inspired the development of salmon cage culture systems that are suitable at sites farther offshore or as fully submersed systems. However, management of cages for salmon mariculture require considerable resources and the corresponding costs and logistical complexity increase greatly for open ocean and fully submersed salmon grow-out systems. Motorized boats are needed to regularly access cages for stocking, feeding, cleaning, routine monitoring, and harvesting.

Smolts are transported in oxygenated containers or plastic bags in Styrofoam boxes by truck and then by boat (or sometimes by airplane) from hatcheries to the marine cages. Transport and handling stress should be minimized as much as possible and smolts need to be monitored frequently after stocking into marine cages when they are particularly susceptible to diseases. Stress-induced mortality is a significant concern during transport and stocking.

Routine cleaning of the netting on cages is essential to maintain sufficient water exchange for these open systems as biofouling organisms that grow on the cage mesh or nets can obstruct water flow significantly over time and need to be removed regularly. Furthermore, biofouling increases the weight of cages, which complicates their management. Regular cleaning prevents hypoxic conditions, facilitates efficient dispersal of waste into the marine environment, and maintains the weight of cages within a manageable range. Monitoring at appropriate intervals is essential for disease and sea lice control and for removing sick and dead fish.

Regular monitoring is also necessary for recognizing and repairing damage to cages caused by attacking predators (e.g. sharks, morays, marine mammals, birds) and poaching. Cage damage can be minimized by using sturdy cage mesh materials such as strong plastics to reduce economic losses and environmental concerns due to escapes of farmed fish. Consequences of escapes of farmed aquaculture salmon have been extensively debated, especially if non-native species are used for aquaculture, for example, Atlantic salmon on the Pacific coasts of the US and Canada and Pacific salmon in the Southern hemisphere (Hindar and Fleming, 2007; Young, 2011).

At optimal temperature and feeding conditions, the SW grow-out phase takes generally between eight and eighteen months to produce salmon with a harvest weight of 6–8 kg before sexual maturity is reached. Even though the weight of Atlantic salmon is much greater at maturity (up to 30 kg) it takes several more years to attain this weight. Moreover, sexual maturation is undesirable because it diverts energy from muscle growth to gonadal development. In addition to reduced growth, meat quality is diminished as indicated by a change in texture and loss of flavour and colour. Harvesting of farmed salmon is done by lifting the cage with a crane and concentrating the salmon in a small area at the bottom from which they are removed by pumping over a dewatering screen and then into super chilled water in totes.

14.1.4 Salmonid feeds

Salmonid feeds have been extensively researched to optimize the amounts of essential nutrients, maximize feed conversion ratio and growth, and minimize waste excretion. This research has benefitted other sectors of aquaculture, which either use salmonid feeds or base the development of diets for other fish species on approaches pioneered for salmon and trout diets. Salmon diets are tailored to different stages of the life cycle but generally contain between 38–50% protein, 15–35% lipid, and 8–13% carbohydrate (Purser, 2017).

Optimization of nutrient composition of salmon feeds has been accompanied by improvements in the delivery of feeds. This was possible by detailed monitoring of feeding behaviour and adjustment of feeding regimes to best accommodate salmon feeding behaviour in mariculture cages. These improvements have resulted in significant increases of feed use efficiency over the past two decades. Formulated salmon diets have been manufactured in the form of moist (20–50% water content) and dry (12% water content) pelleted feeds (Pillay and Kutty, 2005).

Dry feed pellets are more environmentally friendly than moist feeds. Dry pellet buoyancy can be adjusted during pellet manufacturing so that pellets float or slowly sink in the water column, making them available for consumption longer than rapidly sinking pellets. Dry pellets are more water-stable than moist pellets—an advantage for water quality. An important ingredient of salmon feeds is crushed shrimp waste or purified astaxanthin or canthaxanthin to provide carotenoid pigments that yield the preferred orange meat colour. Salmonids cannot synthesize carotenoid pigments and therefore must obtain them in their diet.

Although soybean meal and other plant-based ingredients are increasingly used for manufacturing dry pelleted fish feeds, fish meal and fish oil still account for a significant portion of formulated dry pelleted salmon feeds. About half of this fish meal and fish oil is now generated by processing waste products that are left over after filleting of seafood fish. This significant improvement in recycling of leftover fish carcasses has greatly reduced the amount of fish meal and fish oil derived from capture fisheries' by-catch.

14.1.5 Sea ranching of salmon

The ecologically most sustainable form of salmon aquaculture is FW hatchery RAS combined with extensive marine grow-out by sea ranching. This approach is used for salmon restoration but not cost-efficient for most commercial seafood production since losses at sea are large. This form of salmon aquaculture is used in Alaska for pink and chum salmon and in California for Chinook salmon.

A major concern of this approach is a potential genetic dilution of wild stocks due to hatchery bias towards a small number of parents giving rise to an unproportionally large number of offspring. One solution to address this concern is mass capture of broodstock animals before they arrive at their spawning grounds followed by mass spawning using artificial fertilization in a hatchery. Obviously, this practice should be restricted to a small percentage of tributaries that represent spawning grounds for natural salmon runs in any given area. A mass spawning approach generates genetically highly diverse offspring, which could then be released downstream of the spawning habitat of their parents to maintain local population structure.

Captured broodstock animals may be used to fertilize spawning habitat and mimic their natural death after reproduction in the wild. Salmon carry marine nutrients into FW ecosystems to fertilize forests and provide nutrients for phytoplankton that supports early larval development of FW fish species. Alternatively, after stripping the eggs and milt, broodstock fish could be processed into fish meal, fish oil, fertilizer, and, if meat quality is sufficient, seafood. A significant drawback that cannot be alleviated by mass broodstock capture are the artificial selection conditions during larval development in the hatchery, which differ from those in natural spawning grounds. This difference may affect the fitness of larvae in wild habitat. More research is needed to develop an optimal strategy addressing these concerns.

14.2 Trout

Trout species that are important for aquaculture are classified in the same two genera as salmon and the additional genus *Salvelinus* (chars). Pacific trout and salmon are grouped within the genus *Oncorhynchus*, while their Atlantic counterparts are consolidated in the genus *Salmo*. Chars (*Salvelinus spec.*) have a circumpolar distribution in Arctic and temperate parts of the Northern hemisphere. The most important trout species for aquaculture is the rainbow trout (*Oncorhynchus mykiss*) (Figure 14.1). The other three trout species of note for commercial aquaculture are Arctic char (*Salvelinus alpinus*), brook trout (*Salvelinus fontinalis*), and brown trout (*Salmo trutta*).

Many trout species form non-anadromous populations that complete their entire life cycle in FW. In salmon species, such land-locked populations are the exception while in trout they are much more common. Nevertheless, some trout species are represented by populations with facultative anadromous life histories, for example, steelhead trout is the anadromous form of rainbow trout (*O. mykiss*), and sea trout is the anadromous form of brown trout (*S. trutta*).

In 2018, the leading producers of rainbow trout were Turkey and Iran followed by Chile, Norway, and Peru. In the US, most trout aquaculture is performed in Idaho, where the resident FW form of rainbow trout is produced. Idaho has abundant geothermal springs that have a year-round water temperature of 15 °C and a flow rate of four million litres per minute. The topography of Idaho provides convenient natural slopes that support natural water movement through raceways without the need for pumps. These conditions are ideal for trout aquaculture using open, flow-through raceway systems. The productivity of these open, flow-through raceways depends on the ecosystem service capacity for trout wastewater treatment.

14.2.1 Rainbow trout hatcheries

Trout hatcheries operate similarly as salmon hatcheries and use exclusively FW. Rainbow trout is the preferred species because it grows faster, has a higher tolerance of reduced water quality and infectious diseases, and utilizes artificial dry pelleted food better than brook trout and brown trout. The optimal temperature for development of trout embryos and fry and for grow-out to market size is 15 °C, which is slightly higher than the temperature optimum for Atlantic salmon. At this temperature growth is fastest while metabolism of trout, which are cold water fish, is not negatively affected by heat stress.

Trout development proceeds through the same stages as those experienced by salmon. Fertilized eggs develop into early embryos, late embryos (eyed eggs), yolk sac larvae (alevins), advanced fry (yolk sac fully absorbed), fingerlings (parrs), juveniles, to adults (Figure 14.5). Trout broodstock are kept segregated by sex in ponds, raceways, or tanks before stripping and artificial fertilization of eggs as described for salmon.

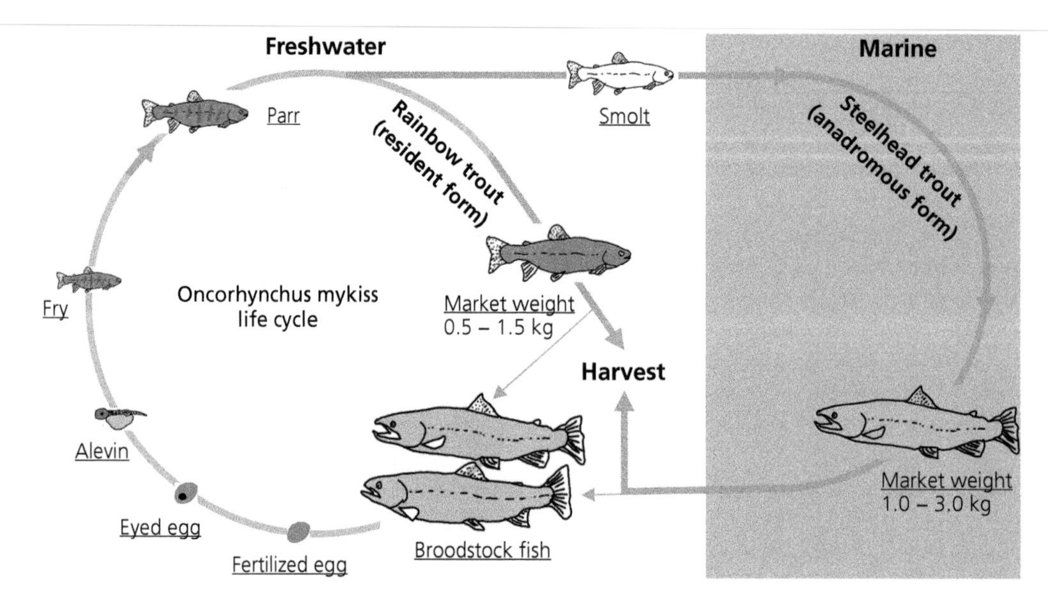

Figure 14.5 Life cycle of resident freshwater rainbow trout and anadromous steelhead trout (both *Oncorhynchus mykiss*).

All female offspring can be obtained by fertilizing eggs from genotypic and phenotypic XX females with masculinized genotypic XX females that produce milt. Masculinization is induced by injection or oral administration (with feed) of male steroid hormones (androgens). Commonly, 3 ppm of 17-alpha-methyltestosterone (MT) or 17-alpha-methyldihydrotestosterone (MDHT) are administered. Milt from masculinized females can be cryo-preserved for convenient distribution to hatcheries. Production of all female offspring is also possible for other salmonid species used for aquaculture. The advantage of this method is that unwanted reproduction can be minimized, and gonadal development delayed, both of which prolong the grow-out period and produce larger fish. Moreover, offspring produced as seafood will not be subject to hormonal treatment, which is limited to parental broodstock (Judycka *et al.*, 2021).

Incubation of *O. mykiss* eggs is usually performed at a higher density than for salmon (about 9000 eggs/L) because trout eggs are smaller than salmon eggs. Upwelling jars (Figure 13.3) are used if large numbers of eggs are incubated and space is at a premium. Incubation pans or troughs are used if chemical treatments with antifungal compounds are impractical or prohibited and regular monitoring and handling to rapidly remove dead eggs is necessary.

Trout alevins hatched in up-welling jars are transferred to fry-rearing tanks immediately after hatching, while alevins hatched in pans or troughs are usually kept there until they reach the swim-up stage. In the latter case the empty eggshells are removed after hatching. Alevins are supplied with starter feed before the yolk sac is fully absorbed and transferred to small indoor raceways or tanks with higher water flow rate after the yolk has been fully absorbed. They are size-graded while developing into fingerlings (parrs). After two months most fingerlings have reached a size of about 5–10 cm. At that time their immune system has fully developed, they are being vaccinated against common diseases, and they can be moved to larger outdoor raceways for grow-out.

14.2.2 Rainbow trout grow-out

Most rainbow trout are grown-out in FW raceways, but some trout production involves grow-out to large sizes in SW. If anadromous broodstock of rainbow trout (steelhead trout) has been used in the hatchery, then smoltification occurs in FW when parr become smolts and prepare for SW entry. In that case, smolts are gradually acclimated to full-strength SW followed by translocation to marine cages (Figure 14.5). This process is similar as described for salmon. If the supply of suitable FW is limited, for example, in some areas of Australia, SW grow-out of rainbow trout in flow-through pond or raceway systems is preferred in coastal areas with abundant supply of SW or brackish water that has the proper temperature.

The anadromous form of rainbow trout (steelhead trout) can reach a weight of 20 kg in the wild, which is significantly larger than resident FW animals. Similar size differences are also observed for anadromous vs. resident forms of other trout species. Maximum sizes are attained when anadromous trout spent four years in SW. Steelhead trout and sea trout (the anadromous form of brown trout) are produced in the same way as salmon and compete with salmon on many markets. However, because FW aquaculture is logistically easier and more cost-efficient than mariculture, most trout aquaculture is done in FW.

Grow-out in outdoor raceways occurs over a period of five to ten months. The harvest weight of rainbow trout is usually 0.5–2 kg. During grow-out fish are graded and sorted into size classes that are kept in separate raceways. Many grow-out facilities use automated equipment for sorting (Figure 14.6).

Trout aquaculture is more suitable for ecologically sustainable RAS than salmon farming since it can be done entirely in FW. Although research on salmon diets indicates that it may be possible to adapt these carnivores to a plant-based diet,

Figure 14.6 Trout raceway aquaculture. **A)** Raceways used for fry are often located indoors. **B)** Raceways for grow-out are larger, have higher flow rates, and are usually located outside. **A)** photo credit: Stephanie Raine, Wikimedia, CC0 1.0. **B)** Photo credit: USFWS, Wikimedia, CC0 1.0.

this process is more difficult than in herbivorous FW fishes that feed naturally on plant-based diets. Salmonid aquaculture also requires more resources, higher flow rates, better water quality, and higher-quality feeds than aquaculture of herbivorous FW fishes discussed in Chapter 13. Therefore, the most cost-efficient, ecologically sustainable increases in seafood production beyond current levels can best be achieved using herbivorous or omnivorous FW fishes.

To accommodate consumer preference for carnivorous fish species, which in part is due to their high omega-3 PUFA levels, significant research efforts have focused on developing plant-based diets for salmonids and marine carnivores (Naylor *et al.*, 2021). Future advances in developing domesticated salmonid strains that perform well on plant-based diets in closed RAS will support ecologically sustainable salmonid aquaculture intensification. In addition, extensive or semi-extensive mariculture of salmonids represents an ecologically sustainable aquaculture solution if it does not strain marine ecosystem service capacity. However, the scope of such ecologically sustainable extensive aquaculture is limited to local markets and insufficient to address a global rise in seafood demand.

Key conclusions

- Anadromous fish hatch and develop in FW and migrate to SW where they spend most of their life before returning to FW for spawning.
- Most salmon are strongly anadromous while most trout are non-anadromous and reside in FW throughout their entire life. However, exceptions are numerous in both cases.
- Atlantic salmon (*Salmo salar*) dominates aquaculture of anadromous fish with the majority being produced in its native distribution range.
- Most Pacific salmon are produced in the Southern hemisphere, outside their native distribution range.
- In contrast to salmon capture fisheries, which have millennia-long cultural roots in many parts of the Northern hemisphere, salmon aquaculture has originated relatively recently (in the 1970s).
- Salmon aquaculture consists of a FW hatchery phase and a SW grow-out phase; thus, aquaculture of anadromous species requires two very different culture environments and systems.
- The most critical stage of salmon development when mortality is highest is the alevin stage, at which yolk sac larvae are trained to consume starter feeds.
- Smoltification (silvering) is the physiological process of salmonid parr transforming to smolts that are ready to enter SW.
- Smoltification is essential for the anadromous life history of salmonids and precedes translocation of aquaculture salmon from FW hatcheries to open marine cages for grow-out.
- Rainbow trout (*Oncorhynchus mykiss*) is the second most produced aquaculture species of salmonids; most rainbow trout are produced in open, flow-through FW hatcheries and FW grow-out raceways although anadromous trout (e.g. steelhead trout) are grown-out in mariculture.
- Salmonids are carnivores that occupy high levels in trophic food webs, require high water quality, strong flow, and high-quality feeds.
- From an ecological sustainability perspective, salmonids are more challenging and more resource-intensive than herbivorous FW fishes.
- Consumer preference and demand for salmonids is high, which has fuelled research to develop more sustainable feeds, reduce water usage and waste, and minimize reliance on ecosystem services.

References

De Silva, S. and Wang, Q. (2019). 'Carps', in Lucas, J. S., Southgate, P. C., and Tucker, C. S. (eds.) *Aquaculture: farming aquatic animals and plants*, 3rd edn. Hoboken: Wiley-Blackwell, pp. 339–362. doi:10.1002/9781118687932.ch23

Hindar, K. and Fleming, I. A. (2007). 'Behavioral and genetic interactions between escaped farm salmon and wild Atlantic salmon', in Bert, T. M. (ed.) *Ecological and genetic implications of aquaculture activities*. Dordrecht: Springer Netherlands, pp. 115–122. doi:10.1007/978-1-4020-6148-6_7

Judycka, S., Nynca, J., Hliwa, P., and Ciereszko, A. (2021). 'Characteristics and cryopreservation of semen of sex-reversed females of salmonid fish', *International Journal of Molecular Sciences*, 22(2), p. 964. doi:10.3390/ijms22020964

Kültz, D. (2015). 'Physiological mechanisms used by fish to cope with salinity stress', *The Journal of Experimental Biology*, 218(Pt 12), pp. 1907–1914. doi:10.1242/jeb.118695

McCormick, S. D. (2012). '5 - Smolt physiology and endocrinology', in McCormick, S. D., Farrell, A. P., and Brauner, C. J. (eds.) *Fish physiology*, vol. 32, *Euryhaline fishes*. Oxford: Academic Press, pp. 199–251. doi:10.1016/B978-0-12-396951-4.00005-0

Naylor, R. L., Hardy, R. W., Buschmann, A. H., Bush, S. R., Cao, L., Klinger, D. H., Little, D. C., Lubchenco, J., *et al.* (2021). 'A 20-year retrospective review of global aquaculture', *Nature*, 591(7851), pp. 551–563. doi:10.1038/s41586-021-03308-6

Pillay, T. V. R. and Kutty, M. N. (2005). *Aquaculture: principles and practices*, 2nd edn. Oxford: Wiley-Blackwell.

Purser, J. (2017). 'Salmonids', in Lucas, J. S., Southgate, P. C., and Tucker, C. S., *Aquaculture: farming aquatic animals and plants*, 3rd edn. Hoboken: Wiley-Blackwell, pp. 363–389.

Young, N. (2011). *The aquaculture controversy in Canada: activism, policy, and contested science.* Vancouver: UBC Press.

CHAPTER 15

Catadromous and marine fishes

'The effects on milkfish population genetics of the current "fry" fishery, and the soon-to-be possible sea-ranching of hatchery-produced larvae will need to be monitored.'

Bagarinao, in *Environmental Biology of Fishes, vol. 39* (1994)

Catadromous fishes are born in full-strength seawater (SW) in the ocean but migrate to brackish and freshwater (FW) as fry or juveniles to develop into adults before returning to the ocean. The best examples are river eels. However, it is important to realize that catadromy and anadromy are idealized concepts. Intermediate forms of migratory behaviour are exhibited by many fish species and considerable variation in diadromy exists even within a single species. Aquaculture species of fishes that are sometimes considered marine often have catadromous (e.g. milkfish, mullets) or semi-catadromous life cycles.

The chapter epigraph illustrates that collection of fry in brackish migration corridors and hatchery production of large quantities of fry must be managed carefully to support resilient genetic population structures of catadromous species. Collection of fry from wild habitats is still the only commercially feasible method for river eel aquaculture. Conversely, hatchery production and release into coastal ponds now contributes the majority of milkfish fry for aquaculture. Both approaches require careful management to avoid genetic alteration of wild populations as pointed out for milkfish nearly three decades ago (Bagarinao, 1994).

Movements of semi-catadromous species between FW, brackish, and SW biomes are determined largely by trophic conditions rather than developmental programs or reproductive behaviours. Such species are also referred to as amphidromous (Zydlewski and Wilkie, 2012). They include many aquaculture fishes that are frequently considered marine species. Numerous species of flatfish, perciform fishes (e.g. seabasses and seabreams), and pufferfish used for aquaculture are either amphidromous or catadromous (Figure 15.1).

Mariculture hatcheries for amphidromous and strictly marine species are used to condition broodstock and produce fry while coastal cage culture is commonly used for grow-out (Figure 15.2). Only six species of marine or amphidromous fishes have yielded aquaculture harvests of more than 100,000 tons in 2018: European seabass (*Dicentrarchus labrax*), Gilthead seabream (*Sparus aurata*), Japanese seabass (*Lateolabrax japonicus*), large yellow croaker (*Larimichthys crocea*), pompano (*Trachinotus ovatus*), and Japanese amberjack (*Seriola quinqueradiata*). Various species of flatfish, other seabreams, and groupers are also commonly produced by aquaculture (Figure 15.1). Except for the flatfish (order Pleuronectiformes), these species belong to the order Perciformes and are high trophic level carnivores.

Important catadromous aquaculture species outside the orders Perciformes and Pleuronectiformes are river eels (*Anguilla spec.*), milkfish (*Chanos chanos*), and mullets (Mugilidae, Figure 15.3). These fishes reproduce in the ocean and are commonly grown out by aquaculture in FW or brackish water. Their larvae drift with currents into marine or brackish inshore habitat (bays, lagoons, estuaries, mangrove swamps, salt marshes) where they metamorphose into juveniles. After metamorphosis, they are ready to enter FW.

A Primer of Ecological Aquaculture. Dietmar Kültz, Oxford University Press. © Dietmar Kültz (2022). DOI: 10.1093/oso/9780198850229.003.0015

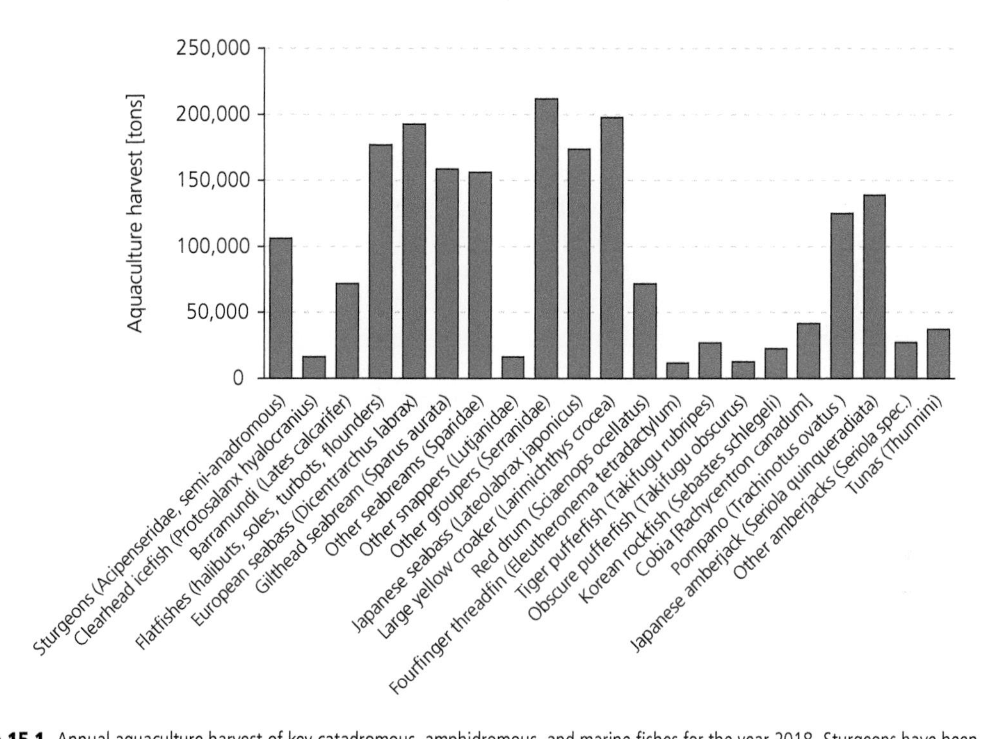

Figure 15.1 Annual aquaculture harvest of key catadromous, amphidromous, and marine fishes for the year 2018. Sturgeons have been included for comparison since they are the only significant (semi-)anadromous aquaculture fishes other than salmonids. Species summarized in this figure are not considered in detail in this book. Data source: OECD.stat.

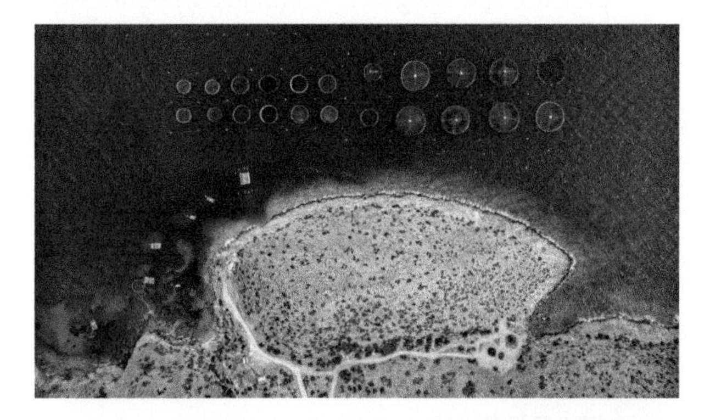

Figure 15.2 Aerial view of a coastal mariculture farm featuring different cage sizes. Photo by Alex Antoniadis (Unsplash).

Figure 15.3 Bluespot mullet (*Crenimugil seheli*). Photo by author.

Milkfish (*C. chanos*) and grey mullet (*Mugil cephalus*) can be spawned in captivity with and without induction by pituitary hormones. Nevertheless, some aquaculture farms still collect milkfish and mullet fry from their natural habitat in areas where they are highly abundant. River eels and milkfish represent the most important catadromous aquaculture species, with annual harvests that are significantly higher than those of other catadromous, amphidromous, or strictly marine fishes. Therefore, the remainder of this chapter focuses on river eels and milkfish.

15.1 River eels

Two species of river eels commonly used for aquaculture are the Japanese eel (*Anguilla japonica*) and the European eel (*A. anguilla*). Other species, including the American eel (*A. rostrata*), short-finned eel (*A. australis*), and tropical river eels (e.g. *A. borneensis, A. bicolor, A. marmorata*) are also used. River eels are a highly priced delicacy in Japan and parts of Europe. Eels have been commercially cultured in Japan and Western Europe for over a century. However, the extent of eel aquaculture is limited because their culture depends on the collection of juveniles (glass eels or elvers) from the wild.

Aquaculture of river eels (*Anguilla spec.*) currently relies entirely on their reproduction and early development in wild habitat. Significant research efforts are aimed at closing the life cycle of eels in captivity (Section 15.1.3). However, current commercial eel production still represents a combination of capture fishery of glass eels or elvers and aquaculture grow-out to harvest size.

15.1.1 Eel life cycle

The natural life cycle of river eels is complex and very difficult to mimic in captivity. European and American eels reproduce in the Sargasso Sea near the island of Bermuda. The larvae are carried with the Gulf stream to the coasts of North America and Western Europe (Figure 15.4). Japanese eels reproduce in the Western Pacific Ocean at the Suruga Seamount near the islands of Guam. Its larvae are carried with the North equatorial and Kuroshio currents towards the coasts of continental East Asia, Taiwan, and Japan (Figure 15.5).

Fertilized eel eggs are buoyant and float freely in the water column. Their movement with currents ensures optimal conditions for development, including stable saturated oxygen, temperature, pH, and salinity. After hatching, eel yolk sac larvae are immersed in phytoplankton upon which they feed. Once the yolk sac has been fully absorbed the larvae develop into preleptocephalus larvae, which grow by extreme dorsoventral expansion into leptocephalus larvae. The name of these larvae refers to the small head relative to the dorsoventrally highly compressed body.

The dorsoventral body compression of leptocephalus larvae confers a selective advantage to larval migration. Leptocephalus larvae are translucent, which minimizes predation pressure. Their large surface area facilitates movement with water currents while minimizing their volume and weight, which ensures that the larvae remain in the upper parts of the water column, which is rich in phytoplankton. This larval morphology maximizes migration velocity and access to food during development. Depending on the destination, migration of leptocephalus larvae can take a year or even longer. During their migration they develop into glass eels.

Glass eels are also translucent to retain the selective advantage of minimizing their susceptibility to predation. Although many glass eels enter estuaries and eventually FW, recent research has shown that FW entry of *Anguilla* species is not mandatory and full development in brackish water or SW may also occur (Zydlewski and Wilkie, 2012). In that regard, eels are similar to catadromous milkfish and mullets and the boundary to amphidromous (semi-catadromous) species is blurred.

Glass eels metamorphose into elvers, which have a yellow-brownish colour and are no longer translucent. Elvers and subsequent developmental stages are preferential carnivores but also feed on detritus if necessary. The metamorphosis into elvers happens when glass eels enter estuaries and other brackish migration corridors. It prepares them for life in FW.

Elvers grow into yellow eels that migrate upstream to colonize FW creeks, ponds, and lakes. Yellow eels are greenish brown on the dorsal and

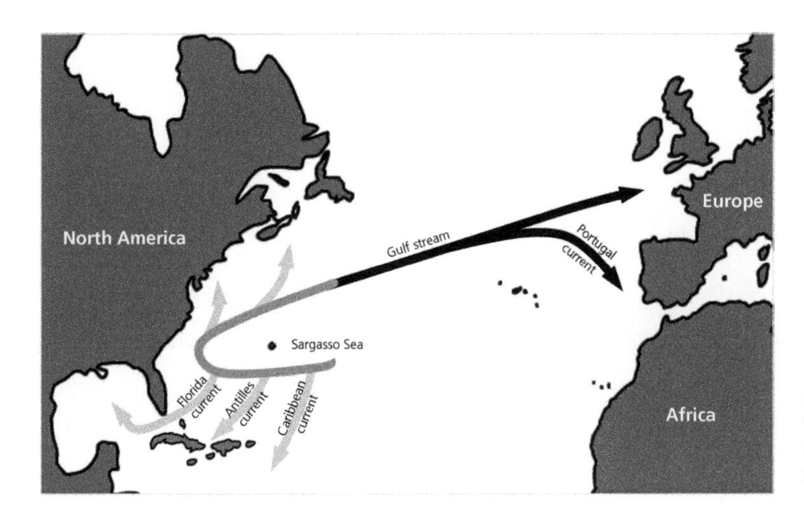

Figure 15.4 Migration routes of European eel (*Anguilla anguilla*, black arrows) and American eel (*A. rostrata*, grey arrows) along major currents of the Atlantic Ocean.

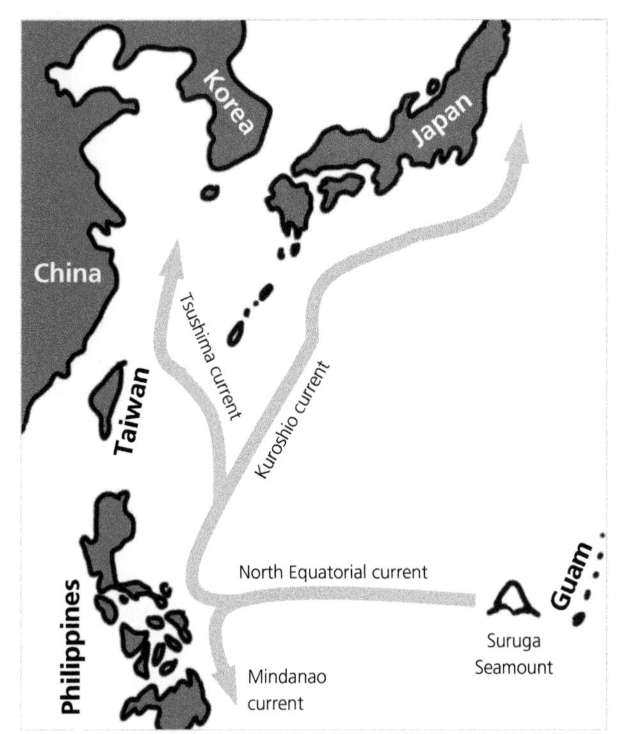

Figure 15.5 Migration routes of Japanese eel (*Anguilla japonica*) along major currents of the Western Pacific Ocean.

yellowish on the ventral part. After growing out in FW for many (commonly 12–18) years, yellow eels transform into silver eels that are ready to migrate downstream and start the journey to their marine spawning habitat (Figure 15.6). Besides their silvery body colour, silver eels can be recognized by their bulkier head and body, and an increased fat content that represents the energy storage reservoir for their ocean migration and gonadal maturation.

Yellow eels and silver eels can perform respiratory gas exchange through their skin, in which small, fine scales are embedded. Moreover, their snake-like body shape facilitates rapid movements on land. Consequently, eels can migrate large distances over land at night if humidity is high and temperature is low. The ability to migrate over land greatly facilitates the colonization of FW habitats by yellow eels and the return to major rivers for

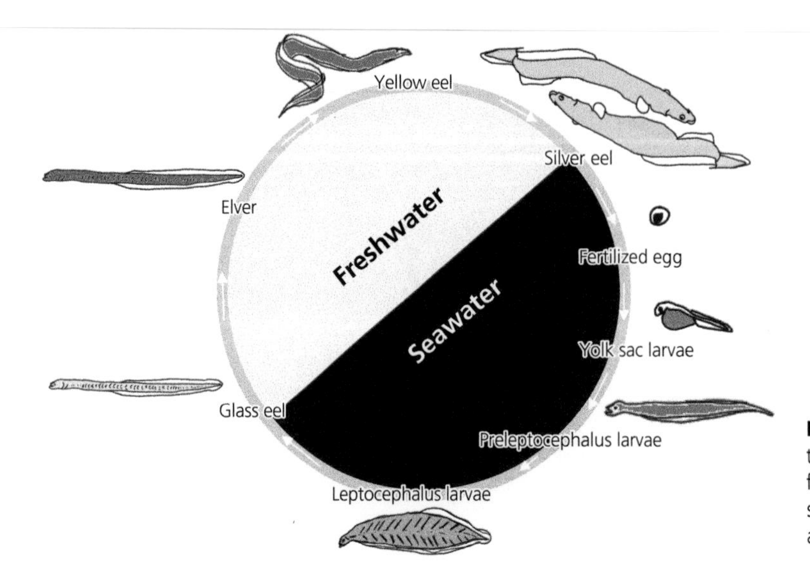

Figure 15.6 Life cycle stages of eels. The transitions between seawater (SW) and freshwater (FW) takes place at the glass eel stage, while the FW to SW transition occurs at the silver eel stage.

downstream migration of silver eels towards the ocean.

15.1.2. Aquaculture grow-out of eels

Commercial eel aquaculture is currently limited to grow-out of glass eels or elvers into yellow eels. This period of eel development corresponds to the FW phase of their life cycle (Figure 15.6). Eel aquaculture is mostly performed for providing seafood. Nevertheless, production of juvenile yellow eels in nurseries is also performed for restocking FW habitats to support eel conservation and sport fishing. Even when eels are farmed with the primary objective of producing seafood, farmers may use some of the animals for restocking FW rivers and lakes to replenish natural populations and support future harvest of glass eels.

After arrival of glass eels at the farm site, they are first quarantined in small 3000–4000-litre tanks, examined for diseases, and treated by bath incubation in antibiotics or sterilizing agents if necessary. They are grown from about 0.5 g to reach 5 g in these nursery tanks and weaned from moist paste to small dry formulated pellet feeds before being distributed into juvenile tanks that are twice as large as nursery tanks.

Eel grow-out occurs in ponds (traditional) or RAS (now the most common form in Europe). Ponds are made of concrete or earthen and are small, initially having a surface area of 100–300 m^2. Larger eels are thinned out periodically into larger ponds of approximately 1000 m^2 (one-seventh the size of a soccer field). Elvers weighing about 25 g can also be exported to be used for valliculture, which is a traditional form of aquaculture in Mediterranean coastal, brackish lagoons (e.g. in Italy).

Eel ponds are about 1 m deep and are stocked at high density (100 elvers per m^3) to maximize the use of space. Water supply may be static or flow-through. Greenhouses are sometimes used to prevent escapes from ponds. If outdoor ponds are used then they are often enclosed by barriers that prevent eels from crawling over land to explore novel, less-crowded FW habitat.

The best temperature range in eel ponds is 20–28 °C, depending on the species grown. When temperature is high (e.g. in Southern Japan), then grow-out to market size (150–600 g) occurs rapidly over a period of six months. Lower temperatures prolong the grow-out period. The optimal grow-out temperature is significantly higher for *A. japonica* (25–28 °C) than for *A. rostrata* and *A. anguilla* (20–23 °C).

In Northern and Central Europe eel farms often use heaters to increase water temperature during the colder months, which accelerates grow-out. For these heated eel aquaculture systems, it is economical to employ RAS technology, which is commonly used for eel aquaculture in Europe. In addition to saving energy for heating, RAS also increase biosecurity by implementing water sterilization via UV irradiation or ozone treatment steps. RAS are often comprised of circular concrete or fibreglass tanks holding up to 10,000 litres of water when eels of approximately 50 g are stocked for grow-out to final market size. Smaller tanks are used for smaller eels.

Besides training for surface feeding on formulated pellets at daytime instead of being nocturnal bottom feeders, glass eels and elvers are also trained to feed at the same location to make food delivery most convenient and facilitate size grading. Elvers and larger developmental stages are fed formulated dry pellets automatically several times daily. When eels congregate at the feeding spot at specific times, capture for sorting into size classes (grading) is convenient. Size grading is necessary about every one to two months as eels grow unevenly when stocked at high density and to prevent cannibalism.

Size grading also supports a more continuous eel harvest that is spread out over much of the year. It is performed manually or by using automatic grading equipment. Eels enter a pipe connected to a conveyer belt from the feeding spot and are sorted into different ponds based on their size. This process is also used for harvest except eels that have reached market size will be collected and kept in clean water tanks without feeding for several days to purge their stomach and improve flavour. Slow-growing animals may require eighteen months to reach the market size of 150–600 g, even when temperature conditions are optimal.

Yields depend on water exchange rates and can be as high as 2 kg per m^2 pond/tank surface area with 40% daily water exchange. The main species of eel produced are *A. japonica* and *A. anguilla* (Figure 15.7), although illegal trade with glass eels of *A. rostrata* could mask the contribution of this species to overall production. China is the leading eel producer with over 200,000 tons of annual *A. japonica* production. *A. anguilla* is mainly produced in Europe but with lower annual yields of approximately 5000 to 6000 tons. Japan and Western Europe are the biggest seafood markets for eels.

Indonesia contributes significantly to the production of river eels, mainly by extensive or semi-intensive aquaculture of native species, which include *A. bicolor*, *A. nebulosa*, *A. interioris*, *A. borneensis*, *A. celebesensis*, *A. marmorata*, *A. obscura*, and *A. megastoma* (Sugeha *et al.*, 2008). Indonesia's eel production equaled that of the EU in 2009 but dropped to fewer than 1000 tons annually in recent years. A major reason for this decline is the easier and more profitable culture of other species, in particular milkfish (Figure 15.7).

15.1.3 Eel aquaculture challenges

As mentioned, eel aquaculture relies heavily on capture fishery of glass eels and the costs of glass eels account for a major portion of the overall costs. Glass eels or elvers are caught in coastal areas near estuaries in East Asia, Northeast America, and Western Europe. Glass eel supply can vary greatly from year to year, which causes large fluctuations in the costs of this essential resource for eel farmers.

Lack of glass eel supply or prohibitively high costs can lead to economic losses that threaten the viability of commercial eel farms. Unfortunately, the attractive financial aspect has caused an increase in poaching, illegal capture, and trade of glass eels in many parts of the world. The uncontrolled capture of glass eels has further exacerbated the problem and increased pressure on natural eel populations. Another possible reason for the decline of wild riverine eel populations are flood defence systems that are installed on rivers and impede eel migration.

A fully closed eel life cycle is currently still not commercially feasible. The bottlenecks are efficient maturation of broodstock in captivity and raising the unique leptocephalus larvae. However, there is significant progress in developing hatchery reproduction methods for Japanese and European eels (Tanaka *et al.*, 2003; Kagawa *et al.*, 2005; Masuda *et al.*, 2012; Sørensen *et al.*, 2016). These ongoing research efforts are inching closer towards commercial hatchery production of glass eels, which would help alleviate their illegal capture, reduce costs for eel farmers, and relieve pressure on wild eel populations.

Diseases represent a significant problem for eel aquaculture because of a high prevalence of pathogens in wild-caught glass eels, high levels of organic compounds (e.g. fish meal) in feeds, and the species' intrinsic disease susceptibility.

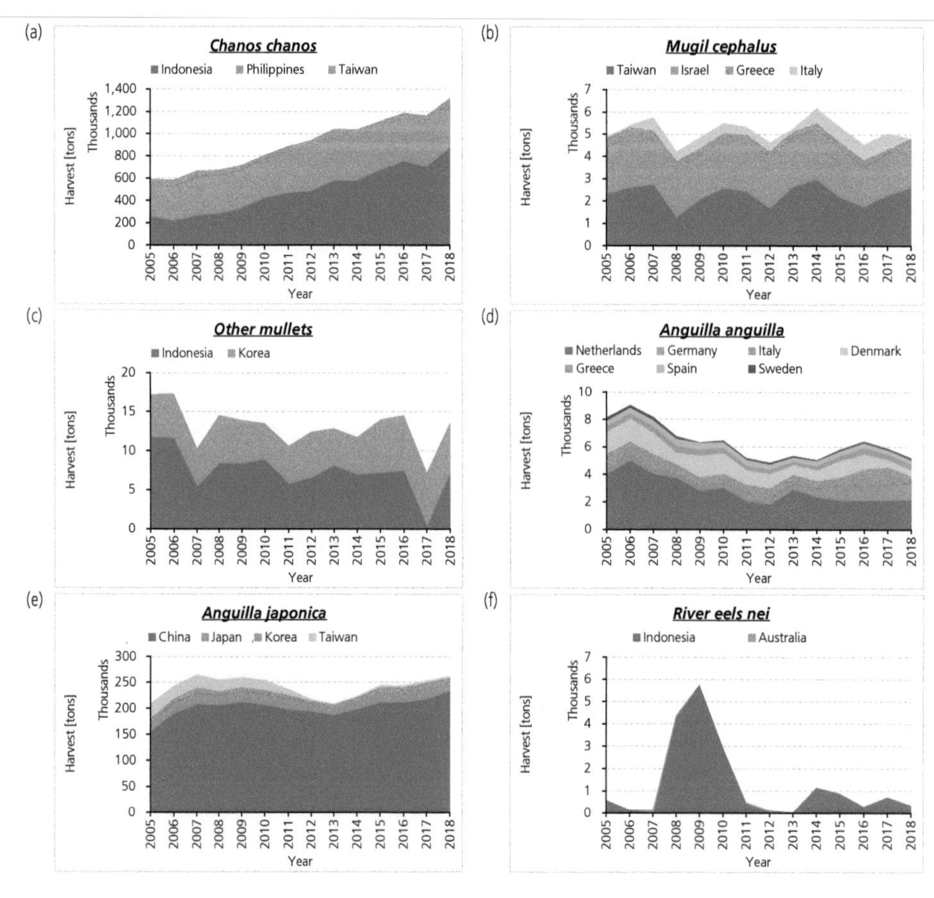

Figure 15.7 Yields of catadromous aquaculture species 2005–2018 in leading producer countries. **A)** milkfish, **B)** flathead grey mullet, **C)** other mullets, **D)** European eel, **E)** Japanese eel, **F)** other eels.

Diseases include infectious diseases caused by various pathogens (see Chapter 18), cancer-like proliferative diseases (e.g. cauliflower disease), and nutritional diseases. Treatments are available for some infectious diseases, but they are often based on application of drugs, stains, or sterilizing chemicals that are classified as human carcinogens or teratogens and illegal for aquaculture use in many countries. Examples include malachite green, methylene blue, and formalin.

15.2 Milkfish

The milkfish (*C. chanos*, Figure 15.8) has a long history of aquaculture in Southeast Asia, Oceania, and Hawai'i. Because of its catadromous lifestyle and high abundance in coastal ponds of the tropical Pacific and Indian Oceans this species was easy to notice, confine, and manage, and sustained the livelihood of coastal societies in this region for hundreds of years (Liao and Leano, 2010). Milkfish is one of the few aquaculture species for which pure capture fishery never played as significant a role as aquaculture.

Milkfish form schools that can be very large, but they also occur solitary or in small groups in nearshore coastal waters. This species grows rapidly and matures to reproductive age in five years, which makes it attractive for aquaculture. Large specimens can reach a maximal size of 1.5 m and 20 kg.

Unlike many other migratory and marine aquaculture fish, milkfish are not carnivores but occupy lower levels of the trophic chain as opportunistic omnivores. They feed on microbial mats growing

Figure 15.8 Milkfish (*Chanos chanos*). Photo by Bryan Harry, USNPS, Wikimedia CC0-1.0.

on detritus and other surfaces (periphyton), phytoplankton, and zooplankton. From this perspective, milkfish aquaculture has greater potential for ecologically sustainable development than aquaculture of most other species of migratory or strictly marine fish that are currently farmed.

Milkfish are highly tolerant of many forms of environmental stress and grow well over a wide range of environmental conditions. In that regard they resemble common FW aquaculture fishes such as tilapia and carps. One of the few environmental limitations for milkfish is their lower temperature limit of 20 °C, which restricts their distribution to tropical and subtropical latitudes. An additional advantage of milkfish is their low intrinsic disease susceptibility, which facilitates aquaculture biosecurity.

Since the 1970s, large investments have been made in the Philippines, Taiwan, Indonesia, and Hawai'i (USA) in terms of infrastructure, research, credit, and training in support to the milkfish industry. Significant research and development on farming systems, breeding, and fry production technologies in these countries resulted in increased productivity of milkfish aquaculture.

15.3.1 Milkfish life cycle

Milkfish only spawn in the ocean in full-strength SW near the outer boundary of tropical coral reefs. Often, reproduction is synchronized with the new or full moon phases and takes place at night. Spawning can take place year round as water temperatures in natural milkfish habitat are high throughout the year and not subject to large seasonal fluctuations (Bagarinao, 1994). Milkfish fecundity is very high as a single female produces millions of eggs. The eggs have a very small diameter of 1.1–1.2 mm and are buoyant. While drifting with currents towards the shore and coastal lagoons the eggs develop into yolk sac larvae that are still very small (about 3.5 mm in length) at hatching. Hatching of yolk sac larvae takes place 1–1.5 days after fertilization.

Milkfish yolk sac larvae are pelagic and feed on phytoplankton during their migration to inshore areas. They typically arrive in coastal lagoons, wetlands, mangrove swamps, and estuaries within three weeks after fertilization (Figure 15.9). At that time, they have grown to a length of 1–2 cm and have fully resorbed the yolk sac while feeding on zooplankton in addition to larger phytoplankton. After another one to two weeks the fry undergo metamorphosis, which prepares them for entering FW and changes feeding behaviour from pelagic phytoplankton feeders to bottom feeders.

This is the stage during which large numbers of milkfish fry are caught for stocking milkfish aquaculture ponds. Their capture is done using finely meshed dip nets or drag seines. During their natural life cycle, milkfish fry can spend several months in these coastal wetland nurseries, which are often brackish and characterized by tidal salinity fluctuations (e.g. mangroves). Therefore, milkfish are highly euryhaline and tolerate salinities ranging from FW to at least 45 g/kg. Milkfish fry and juveniles can swim upstream in estuaries to colonize lakes and grow out in FW. They return to the ocean as juveniles or adults when they have reached a size that is too large for shallow brackish and FW nursery habitats (Figure 15.10).

Most milkfish aquaculture is performed in coastal earthen ponds that closely mimic the natural nurseries for juvenile development. In fact, many of these aquaculture systems have been diverted from mangroves and other natural coastal wetland

Figure 15.9 Anchialine fresh- and brackish water ponds near Anaehoomalu Bay on the Big Island of Hawai'i are utilized during juvenile development by catadromous milkfish and mullets. Photo by author.

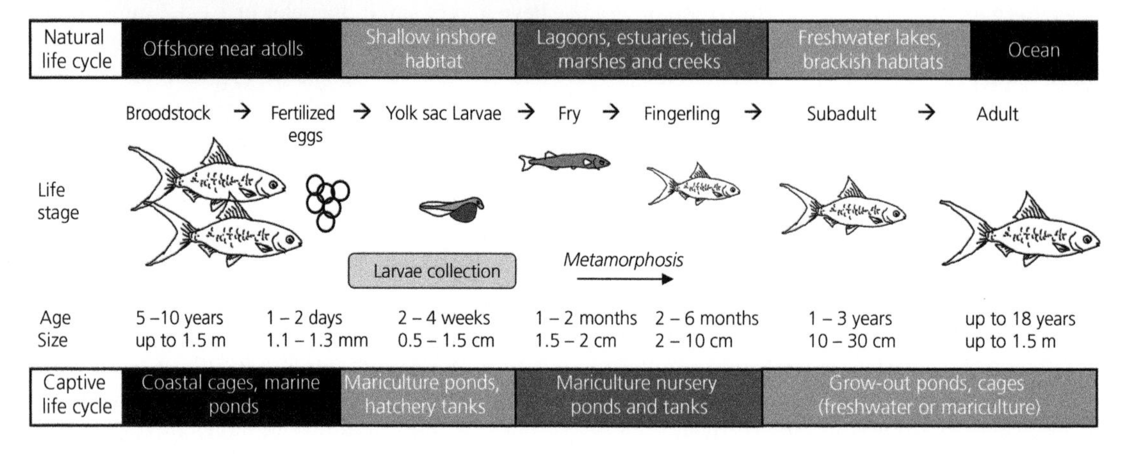

Figure 15.10 Catadromous life cycle of milkfish (*Chanos chanos*). Note that life cycle stages are not drawn to scale. Habitats occupied by wild populations are indicated at the top while the corresponding captive rearing systems are indicated at the bottom.

areas to semi-intensive aquaculture ponds that are managed by controlled water flow, fertilization, supplemental feeding if necessary, and periodic dry-out and quicklime application. Often, milkfish polyculture with shrimp, mullets, and other euryhaline species is performed in these ponds. Even if milkfish monoculture is the initial intent, it is difficult to keep out other species when pond water is regularly renewed by incoming tidal flows that sweep in larvae and juveniles from other species.

15.3.2 Milkfish nursery ponds and hatcheries

As mentioned, fry can be collected from natural habitat to stock nursery ponds. As demand for milkfish fry has increased, natural supply has decreased and become unpredictable. Therefore, methods for milkfish reproduction in captivity have been developed. Production of fry from captive broodstock has alleviated the reliance on wild fry and significantly increased milkfish aquaculture yields. Broodstock fishes that are five to seven years of age are maintained and spawned naturally in SW cages. Older broodstock can also be spawned in ponds, tanks, and raceways but spawning under these conditions is induced by administration of pituitary hormones and fertilization success is reduced. Fertilized eggs are collected from these breeding systems and transferred to nursery ponds or hatchery trays and tanks where they develop into fry.

Larvae are kept in small, shallow (50–100 cm), semi-intensive outdoor nursery ponds or in hatchery trays or small tanks in intensive or semi-intensive indoor systems. If outdoor ponds are used, care must be taken to avoid heat stress due to the shallow depth of the ponds, which promotes light penetration and high-density natural phytoplankton production. Larvae are weaned to small phytoplankton (microalgae) food starting two days post hatching (dph) to develop proper feeding behaviour by the time the yolk sac is absorbed

(5 dph). Between two to three weeks post hatching fry are weaned gradually to formulated dry feeds that supplement live zooplankton and phytoplankton food.

Since milkfish larvae are tiny and start feeding before the yolk sac is absorbed fully, a critical aspect of milkfish hatchery operation is the production of live food. Culture systems for food organisms such as microalgae (*Nannochloropsis spec.*, *Chlorella spec.*, etc.), rotifers, and brine shrimp are integrated into commercial milkfish hatcheries to provide a stable food supply.

After one month rearing at the hatchery fry attain a length of 2–3 cm. They are then transferred to larger circular tanks, raceways, or earthen nursery ponds. Milkfish hatcheries often operate as independent facilities that only produce fry, which is then sold to separate grow-out facilities to provide a stable supply of milkfish fry. During transport from hatcheries to nurseries handling stress due to hypoxia, and fluctuations in temperature, salinity, and pH must be avoided to maintain high survival rates. Long transport or transit times require feeding with proper live or dry pelleted food.

15.3.3 Milkfish grow-out

Milkfish nurseries are often part of, or situated near, grow-out facilities that consist of ponds categorized by their size and purpose. Nursery ponds are smallest, transitional ponds are intermediate in size, and grow-out ponds are largest. Fry are initially stocked into nursery ponds to grow them to fingerling size. Fingerlings are transferred to transitional ponds and, as they grow larger, to grow-out ponds, net pens, or cages.

Fry grow to fingerling size (5–8 cm) in nursery ponds over a period of 1–1.5 months. At this size they are transferred to transitional ponds or directly to grow-out ponds where they are grown to final market size. Transitional ponds are used to temporarily stunt juveniles and prolong the grow-out period to enable more continuous harvest. Milkfish ponds are fertilized to encourage growth of masses that consist of benthic algae and periphyton forming at the pond bottom. They represent the main food source for milkfish.

Ponds for semi-intensive grow-out are usually deeper (1–3 m) than extensive pond systems (<0.5 m) and require supplemental feeding and aeration. Besides providing a larger volume of water, deeper ponds are less susceptible to fluctuations in temperature and salinity. Semi-intensive grow-out of milkfish is also performed in net pens in shallow lakes or bays and in cages that are positioned in deeper brackish or marine coastal habitats.

Supplemental feeds for semi-intensive grow-out include rice and corn bran and formulated dry feeds. Formulated milkfish feeds have been extensively optimized over the past decades and include floating dry pellets as well as semi-floating and sinking pellets of various sizes. Commercial formulated milkfish feeds are tailored towards the specific nutrient requirements of different developmental stages.

Milkfish reach market size in two to four months under semi-intensive grow-out conditions. Their harvest size is 20–40 cm corresponding to a weight range of 0.25–0.5 kg. Harvesting can be done in complete batches or partially. For complete harvests, ponds are drained, or cages are emptied and reconditioned. Pond reconditioning includes repeated dry-out, raking, ploughing, and application of pesticides (e.g. quicklime, tea seed cake) and fertilizer. Cage reconditioning includes dry-out and removal of biofouling organisms and epiphytic growth.

Partial milkfish harvesting is performed using size-selective seines or gillnets to capture larger fishes and retain smaller fish. This method of harvest yields more uniform and, on average, larger fishes and prolongs the production period by spreading out the harvest over a longer time. Depending on whether extensive or semi-intensive systems are used the annual yield of milkfish aquaculture varies between 500 and more than 10,000 kg/ha. Milkfish are marketed fresh as a whole or processed by chilling, deboning, freezing, descaling, smoking, drying, fermentation, or a combination of these processing methods.

15.2.3 Sustainable development of milkfish aquaculture

Countries that led aquaculture production of milkfish in 2018 are also the countries with the richest tradition of milkfish aquaculture in areas where milkfish is native (Figure 15.11). These countries are most active in research aimed at improving milkfish aquaculture. Spawning and larval rearing of milkfish in captivity were first developed in the late

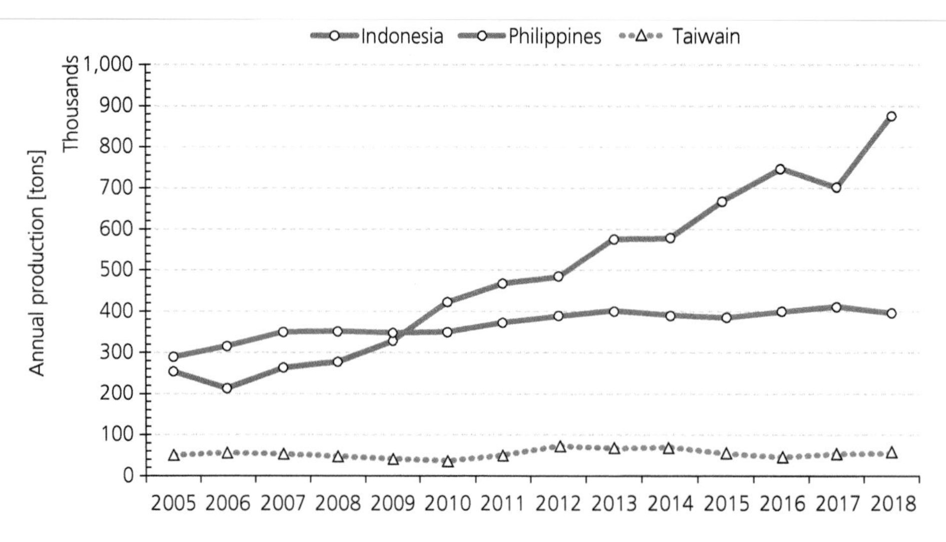

Figure 15.11 Milkfish aquaculture production 2005–2018 by country. Data source: OECD.stat.

1970s/early 1980s. Formulated, dry milkfish feeds were optimized and improved over the past few decades to maximize feed efficiency and minimize food waste.

Genetic strains of milkfish have been developed by selective breeding, including colour variants such as a golden colour strain, which seemingly commands higher market prices. But genetic improvement of milkfish by selective breeding or other approaches are not extensively utilized. Opportunities in this regard do exist as significant phenotypic and genetic heterogeneity has been documented for several geographic milkfish variants.

Along with these advances, production systems for milkfish have been intensified, mostly by converting shallow, extensive pond systems into deeper semi-intensively managed pond systems. In addition, cage and net pen culture systems are increasingly utilized for milkfish grow-out. These semi-intensive systems have increased the burden on ecosystem services as higher fertilization and feeding rates and greater stocking densities generate significantly more waste. In very rare instances, RAS are used to recycle water used for milkfish aquaculture, and wastewater treatment solutions are not commonly deployed downstream of milkfish ponds.

Economic investments that factor in the value of ecosystem services can address these technical challenges. However, milkfish aquaculture faces several additional challenges. Although herbivorous/omnivorous milkfish provide better opportunities for ecologically sustainable aquaculture of marine fish species than marine carnivores, consumer preference and demand currently incentivize production of carnivorous species with higher commercial values (Figure 15.1).

Processing methods that increase consumer acceptance also represent a challenge for milkfish aquaculture. Milkfish have tiny, intramuscular pin bones that reduce consumer interest. Therefore, refined deboning methods for milkfish would increase market acceptance. In addition, phenomena caused by anthropogenic acceleration of climate change such as rapid sea level rise and global warming have negative impacts on coastal milkfish aquaculture systems that depend on tidal flows, especially shallow ponds that are most susceptible to temperature fluctuations, droughts, and tidal flow.

Investment in technology-intensive solutions such as RAS or outdoor wastewater treatment recirculating systems would overcome these challenges associated with semi-intensive, open systems. However, these solutions require economic incentives that equal the value of common good ecosystem services, which are currently provided free of charge or at very low cost. Eco-labelling and health certificates are now used by major milkfish-producing countries to trace the use of antibiotics and drugs for pathogen control of aquaculture fish.

Key conclusions

- Catadromous fishes reproduce in the ocean and migrate as larvae or juveniles to coastal brackish and inland FW habitats with high nutrient productivity before returning to the ocean as adults.
- Catadromous migratory behaviour varies between and within many coastal marine species. Some species display true developmentally encoded catadromy (river eels, milkfish, mullets) while others' migrations are amphidromous (i.e. opportunistically driven by trophic conditions).
- Aquaculture of marine amphidromous species focuses on carnivores in the orders Perciformes and Pleuronectiformes and has lower global yields while requiring more resources and having a greater ecological footprint than the leading aquaculture fish species.
- Two species of catadromous river eels are mainly used for aquaculture and have long histories as seafood in Europe and Asia: the European eel (*A. anguilla*) and the Japanese eel (*A. japonica*).
- *A. anguilla* reproduces in the Sargasso Sea of the Atlantic Ocean, while *A. japonica* spawns at Suruga Seamount near Guam in the Pacific Ocean. The larvae develop over a period of many months on their migration to the coasts of Western Europe and Eastern Asia, respectively.
- River eel aquaculture depends on capture of glass eels or elvers when they arrive on the coasts and prepare to enter FW. Significant research advances have been made in developing methods for captive eel reproduction but their adoption on a commercial scale is not yet possible.
- High market prices for eel have incentivized technology-intensive RAS approaches in Europe but most eel aquaculture is performed in China using semi-intensive approaches.
- Catadromous milkfish (*C. chanos*) is a primarily herbivorous species that has many advantages for aquaculture, including high stress tolerance, disease resistance, and rapid growth.
- Milkfish aquaculture has a century-long history in Southeast Asia, driving research on captive reproduction and feed optimization. Specialized hatcheries now produce most of the milkfish fry for aquaculture, which relieves pressure on natural populations but requires careful genetic management.
- Milkfish grow-out is mostly performed in semi-intensive ponds or cages but is amenable to ecologically sustainable development if water recycling or treatment solutions are implemented.
- Consumer acceptance for milkfish is lower than for marine carnivores. The relationship of consumer preference versus ecological footprint of carnivore/herbivore aquaculture requires efforts in consumer education and improved seafood processing.

References

Bagarinao, T. (1994). 'Systematics, distribution, genetics and life history of milkfish, *Chanos chanos*', *Environmental Biology of Fishes*, 39(1), pp. 23–41. doi: 10.1007/BF00004752

Kagawa, H., Tanaka, H., Ohta, H., Unuma, T., and Nomura, K. (2005). 'The first success of glass eel production in the world: basic biology on fishes reproduction advances new applied technology in aquaculture', *Fish Physiology and Biochemistry*, 31(2–3), pp. 193–199. doi: 10.1007/s10695-006-0024-3

Liao, I. C. and Leano, E. M. (eds.) (2010). *Milkfish aquaculture in Asia*. Sorrento: World Aquaculture Society.

Masuda, Y., Imaizumi, H., Oda, K., Hashimoto, H., Usuki, H., and Teruya, K. (2012) 'Artificial completion of the Japanese eel, *Anguilla japonica*, life cycle: challenge to mass production', *Bulletin of Fisheries Research Agency*, 35, pp. 111–117.

Sørensen, S. R., Tomkiewicz, J., Munk, P., Butts, I. A. E., Nielsen, A., Lauesen, P., and Graver, C. (2016). 'Ontogeny and growth of early life stages of captive-bred European eel', *Aquaculture*, 456, pp. 50–61. doi: 10.1016/j.aquaculture.2016.01.015

Sugeha, H., Suharti, S. R., Wouthuyzen, S., and Sumadhiharga, K. (2008). 'Biodiversity, distribution and abundance of the tropical Anguillid eels in the Indonesian waters', *Marine Research in Indonesia*, 33, pp. 129–137.

Tanaka, H., Kagawa, H., Ohta, H., Unuma, T., and Nomura, K. (2003). 'The first production of glass eel in captivity: fishes reproductive physiology facilitates great progress in aquaculture', *Fish Physiology and Biochemistry*, 28(1), pp. 493–497. doi: 10.1023/B:FISH.0000030638.56031.ed

Zydlewski, J. and Wilkie, M. P. (2012). 'Freshwater to seawater transitions in migratory fishes', in McCormick, S. D., Farrell, A. P., and Brauner, C. J. (eds.) *Fish physiology: euryhaline fishes*. Oxford: Academic Press, pp. 253–326. doi: 10.1016/B978-0-12-396951-4.00006-2

PART 3

Water Quality Parameters

CHAPTER 16

Abiotic parameters

'I know the ways, I know the plays, and also know where the players gather; I know they were secretly drinking wine but publicly preaching water.'

Heine, Deutschland ein Wintermärchen (1844) *transl. from German*

Nineteenth-century poet Heinrich Heine may have meant more than just stressing the ubiquitous importance of water, yet the chapter epigraph illustrates that clean water was plentiful and undervalued before the beginning of the twentieth century. The value of clean water is much more appreciated today as demand for it has soared dramatically. A high quality of clean water critically depends on a range of abiotic and biotic parameters that are closely monitored and managed in semi-intensive and intensive aquaculture. These water quality parameters are discussed in Chapters 16–18.

Abiotic parameters are non-living components that define the environment inhabited by an organism. Aquaculture organisms (like any others) perform best when they are held in optimal environments (optimal water quality) and not subjected to environmental stress. Therefore, it is critical to know the optima for key environmental factors for the species to be cultured and manage water quality accordingly.

16.1 Temperature

Temperature is a measure of heat intensity (thermal energy) that is proportional to the kinetic energy of random molecular movement (Brownian motion). In theory, molecules stop moving and thermal energy is zero at the lowest possible temperature of 0 Kelvin (K), which can be approached very closely experimentally but cannot be attained in praxis. This theoretical temperature of absolute zero (0 K) corresponds to –273 °C and –460 °F.

The three units of temperature use different scales and reference points. The International System of Units (SI) unit for temperature is Kelvin. It is named after William Thompson (1824–1907), who became Lord Kelvin in 1892. He was an Irish-Scottish mathematician who calculated the absolute zero temperature as –273.15 °C or –459.67 °F. Although Kelvin is the SI unit for temperature, the Celsius scale is most used in scientific publications, which is why it was adopted throughout this book.

The other two temperature units were defined earlier than the Kelvin scale and are based on different reference points. The reference points for the Celsius temperature scale are the freezing and boiling points of water, which are 0 °C (273 K) and 100 °C (373 K), respectively. This scale was introduced by the Swedish physicist Anders Celsius in 1741 as the centigrade temperature scale, the unit of which now bears his name. The relationship to the Kelvin scale is simple (°C = K – 273.15).

Three reference points were used by the German physicist Daniel G. Fahrenheit to define the Fahrenheit temperature scale in 1724. The first reference marks 0 °F (255 K) as the temperature of a mixture of ice, water, and ammonium chloride upon reaching equilibrium. The second reference point is the same as for the Celsius scale, that is, the freezing point of water corresponding to 32 °F (273 K). The third reference point is human body temperature defined as 98.6 °F (309 K, 36.6 °C). The relationship to the

A Primer of Ecological Aquaculture. Dietmar Kültz, Oxford University Press. © Dietmar Kültz (2022). DOI: 10.1093/oso/9780198850229.003.0016

Kelvin and Celsius scales is more complex (°F = 9/5 (K – 273) + 32; °F = C × 9/5 + 32).

16.1.1 Temperature range experienced by aquatic organisms

The range of temperature on earth is narrow compared to that of the universe. Stars are very hot and range in temperature from 3000 to 20,000 °C while the mean temperature of the universe is very low at –235 °C, just 38 K above absolute zero. This low average temperature of the universe indicates that most of the vast space between stars is extremely cold and has a very low density of molecules.

In contrast, the temperature range of the earth crust and atmosphere (the biosphere) is very small, spanning generally between –100 °C and +100 °C. The cold pole on earth is Antarctica with a record low air temperature of –89 °C, which was recorded on 21 July 1983 at the Russian Vostok station. The heat pole on earth is in North Africa with a record high air temperature of 58 °C, which was measured on 13 September 1922 at El Azizia, Libya. This record high air temperature was nearly broken at Death Valley in California, US, in summer of 2021 and may have fallen by the time you read this book.

Because water acts as a heat sink, the range of temperatures in aquatic habitats is even smaller than that in terrestrial habitats and temperature fluctuations or extremes are generally less severe in aquatic ecosystems. Most liquid water on earth has a temperature between –2 °C (Arctic and Antarctic saline waters) and +40 °C (shallow tropical and subtropical lakes).

Nevertheless, there are aquatic habitats on earth that have much higher temperatures. For example, hot springs and geysers such as those in Yellowstone National Park have water temperatures close to the boiling point. At hydrothermal vent ecosystems the water temperature far exceeds the boiling point (at 1 atm pressure), but water remains liquid because of the high hydrostatic pressure at several hundred meters depth (Figure 16.1).

Global temperatures have increased in recent decades due to rapid anthropogenic acceleration of global warming. This increase affects the physiologies of almost all organisms, including fish, invertebrates, plants, and microbes, all of which are poikilothermic (ectothermic) organisms, that is, organisms that cannot physiologically regulate their internal body temperature. Only mammals and birds can control the temperature of their bodies; they are homeothermic endotherms, that is, species that can generate enough metabolic heat to be able to maintain a stable internal body temperature that can be much higher than that in their environment. Nevertheless, a few large fishes such as tunas, some sharks, and opah are regional homeotherms that can elevate temperature in parts of their bodies to temperatures above ambient.

16.1.2 Temperature tolerance of aquaculture organisms

The distribution range and activity of poikilothermic organisms is directly dependent on the temperature of their environment. Thus, aquaculture

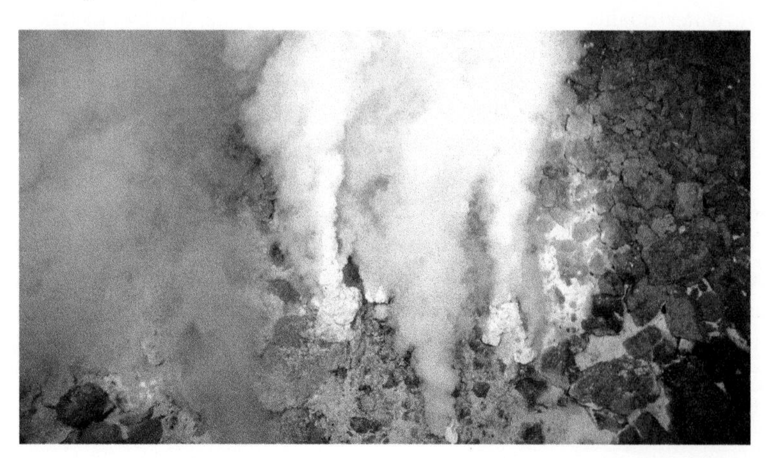

Figure 16.1 Marine hydrothermal vent (Champagne vent) of the white smoker type emitting hot gases that increase the temperature of water in the close vicinity to greater than 100 °C. The water is kept liquid as the boiling point is increased because of the hydrostatic pressure. Photo by NOAA, Wikimedia CC0-1.0.

is affected by climate change in that farms initially located to capitalize on an ideal temperature for a given species may not be ideally located in the future. Another potential effect of climate change is that groups of people with traditional access to wild fish may turn to aquaculture if wild populations decline or their distribution range is shifted because of climate change.

Because the health, growth, and well-being of poikilothermic organisms is greatly dependent on temperature, frequent monitoring of temperature is necessary in any aquaculture system. The relationship between temperature and growth (and other aspects of organismal metabolism) is often expressed as the Q_{10} effect. The Q_{10} effect defines the relative increase in metabolic rate with every 10 °C increase of temperature. As a rule of thumb, metabolic rate tends to double with every 10 °C increase in temperature within the permissible temperature range of a particular organism. However, actual Q_{10} values vary widely depending on species and which aspect of metabolism, that is, which enzymatic reaction or physiological process is considered (Somero, Lockwood, and Tomanek, 2017).

Some aquaculture organisms have a wide temperature tolerance range, e.g., common carp (*Cyprinus carpio*) tolerates temperatures of 4–40 °C (Golovanov and Alexey, 2007). These species are referred to as eurytherms. Other species tolerate narrow temperature ranges, such as brown trout (*Salmo trutta*), which only tolerates temperatures of 4–10 °C during early development. These species are referred to as stenotherms (Réalis-Doyelle *et al.*, 2016). For aquaculture, it is critical to know the thermal optima and tolerance ranges of each species and at different developmental stages, all of which may differ substantially.

The temperature tolerance range depends on the performance parameter that is being considered. It is widest for survival, intermediate for growth, and narrowest for reproduction (Figure 16.2). Aquaculture systems are designed to maintain a temperature that is optimal or near optimal for growth. Stricter temperature control may be necessary for maintaining broodstock to support reproduction.

16.1.3 Temperature monitoring and adjustment in aquaculture

Temperature monitoring is straightforward and can be done with small, inexpensive equipment. Electronic temperature sensors are preferable and less hazardous than conventional mercury-based thermometers. They can also be conveniently connected to alarm systems and wireless networks to notify aquaculture farm personnel if temperature deviates from permissible conditions and corrective measures are needed to prevent losses.

Aquaculture systems that operate at ambient temperature are most ecologically sustainable because little to no additional energy is needed for temperature control. Such systems are usually located in subtropical and tropical areas with small seasonal temperature variation. Temperature control is only practical in recirculating aquaculture systems (RAS) that are operated indoors (e.g. in greenhouses). Such systems are primarily used in temperate regions that are subject to large seasonal temperature fluctuations.

Heating is more commonly used and the choice for locating a facility is often favoured by heating rather than cooling being required. Heaters are much less expensive, technologically less complex, less prone to failure, and require less maintenance than chillers. Any temperature control device used for aquaculture should be made of inert materials to avoid introducing heavy metals or other contamination into the water. For that reason, water heaters are commonly coated with inert glass.

Cooling units often use copper coils for efficient heat exchange, which represent a potential source of contamination and should be avoided. For heating or cooling units and any other technologies such as pumps and lighting used in RAS, a source of clean, renewable energy is preferred to maximize the ecological sustainability of these systems.

16.2 Photoperiod

Photoperiod is the portion of time during a twenty-four-hour day that an organism is exposed to light. It is often expressed as the ratio of light (L) to dark (D) hours. For example, the photoperiod of a day

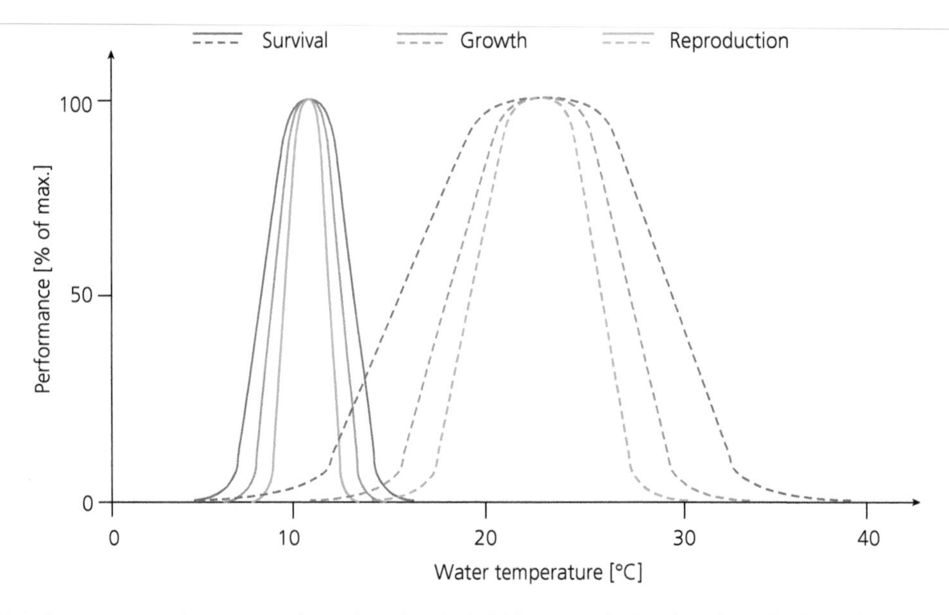

Figure 16.2 The temperature tolerance range of stenothermal species (solid lines, e.g. Atlantic cod, *Gadus morhua*) is much narrower than that of eurythermal species (dashed lines, e.g. common carp, *C. carpio*). Moreover, temperature tolerance varies depending on season and what physiological performance parameter is assessed (Shelford's law of tolerance, black versus grey lines).

with twelve hours of daylight would have a photoperiod of 12L:12D. In RAS where fish are reared at high density a 24L:0D photoperiod is sometimes used to offset the high cost of RAS production. The advantages of continuous lighting include continuous feeding activity and avoidance of hypoxia associated with lack of photosynthesis of algae at night, which reduces costs for aeration. However, such unnatural life cycle may impair animal welfare, long-term performance and growth, and meat quality.

The relationship between temperature and photoperiod is readily evident when comparing temperature in an area exposed to sunlight to that in an adjacent shady area. Some of the energy contained in sunlight is converted to kinetic energy of molecules present in air, water, and solid matter, which increases their Brownian motion and, hence, temperature. Therefore, a relatively inexpensive means of crude temperature control for aquaculture systems is manipulating the extent of sunlight and shade that reaches the water surface. In most cases, however, outdoor aquaculture systems aim to maximize the amount of light received to encourage primary productivity (photosynthesis). Moreover, providing shade is often impractical for large outdoor systems. Thus, photoperiod can only be manipulated efficiently in indoor RAS.

Photoperiod is particularly important for temperate species that inhabit areas subject to large seasonal changes in day length. Such changes are tightly correlated with seasonal temperature changes. They represent important cues for synchronizing reproduction in a way that ensures optimal development. Development proceeds more rapidly at high temperature, which is why most temperate species reproduce in Spring. Reproduction of aquaculture broodstock in hatcheries can be facilitated by increasing temperature and adjusting photoperiod to mimic cues that trigger natural reproduction.

Photoperiod is critical for managing semi-intensive pond aquaculture systems that receive large amounts of fertilizer and/or feeds. Input of these nutrient sources stimulates primary production and causes eutrophication and, in extreme cases, blooms of microalgae and cyanobacteria. Apart from the potential to produce toxic metabolites, such blooms can cause a reduction of oxygen in the water (Section 16.5.1) leading to fish kills.

Eutrophication and microbial blooms are promoted by a long photoperiod and high temperature. They are particularly prevalent during the summer months in temperate climates. Such blooms reduce light penetration and limit photosynthesis to shallower depths, which reduces the oxygen production in the pond and may lead to the formation of a near bottom hypoxic zone in deeper ponds. Unless well-mixed, shallow systems are used, water samples should be taken from different depths to account for possible differences in oxygen concentration, temperature, and salinity due to stratification.

Eutrophication causes increased production of zooplankton and communities of aerobic decomposing bacteria that consume oxygen. Production of these organisms is beneficial from the perspective of providing live food for fish and invertebrates that feed on zooplankton or detritus covered with aerobic bacteria. However, it also increases overall oxygen consumption rates in the pond, which may outweigh photosynthetic capacity on cloudy days with low light and cause hypoxia.

Hypoxia is most prevalent during the early morning hours at dawn because photosynthesis is minimal at night, and essentially zero if there is no moonlight. In contrast to photosynthesis, respiration continues at high rates throughout the night if large amounts of plants, phytoplankton, zooplankton, aerobic bacteria, and animals are present in the pond. Therefore, oxygen depletion during night-time peaks at dawn. For this reason, aerators are often operated during the early morning hours to prevent hypoxia (Figure 16.3).

16.3 Turbidity

Turbidity refers to the extent of cloudiness or opaqueness of water. It is caused by tiny, undissolved particles that are suspended in the water column but not visible to the naked eye. Turbidity is directly related to eutrophication of aquaculture ponds and other water bodies used for aquaculture. It is not entirely an abiotic parameter but rather a collective measure of abiotic and biotic particles that are suspended in the water column.

Eutrophication increases the nutrient (trophic) content of water, which is reflected in an increase in phytoplankton concentration. Phytoplankton consists of many species of tiny microalgae that are suspended in the water column and impair light transmission (i.e. increase turbidity). Small zooplankton, suspended detritus particles, and bacteria also increase turbidity.

The Trophic State Index (TSI) is used to evaluate water bodies by biological productivity and nutrient content on a scale from 0 to 100 (Carlson, 1977). Three main TSI categories are distinguished. Oligotrophic water bodies have a low TSI (0–40) and are referred to as clean, good-quality water. Mesotrophic waters have fair water quality with an intermediate TSI of 40–60, and eutrophic waters have poor water quality with a high TSI of 60–100.

The TSI can be derived from measuring nutrient (phosphorous, nitrogen) and chlorophyll concentrations, but the easiest way to determine TSI is by measuring turbidity. Turbidity is measured using a Secchi disc—a white plastic disc connected to a line that is lowered into the water until it is no longer visible. The depth at which visibility of the disc disappears is the Secchi depth. Oligotrophic waters have a Secchi depth of at least four metres. This depth is decreased for mesotrophic (1.5–4 m) and eutrophic (<1.5 m) waters.

The turbidity of aquaculture ponds can be reduced by using coagulants that bind to small, suspended particles and promote their aggregation to larger particles that sediment more readily and/ or can be removed more easily by mechanical filtration. Coagulants can also be used to promote the formation of bioflocs (see Chapter 5.5). Coagulants that are sometimes used in aquaculture are calcium and aluminium sulphates, but their use is not widespread.

Turbidity can also be reduced by sedimentation of suspended particles or by mechanical filtration. Sedimentation rate is high in turbid water supplied by highly productive estuaries and rivers. When flow rates are high, then sediment particles are suspended in the water column even when they are heavier than water. However, they will sink and accumulate at the bottom when large sedimentation or settling ponds with low flow rates are used. If brackish water is used for mariculture, then sediment and sludge disposal can be problematic because of the high salt load of such sediment and

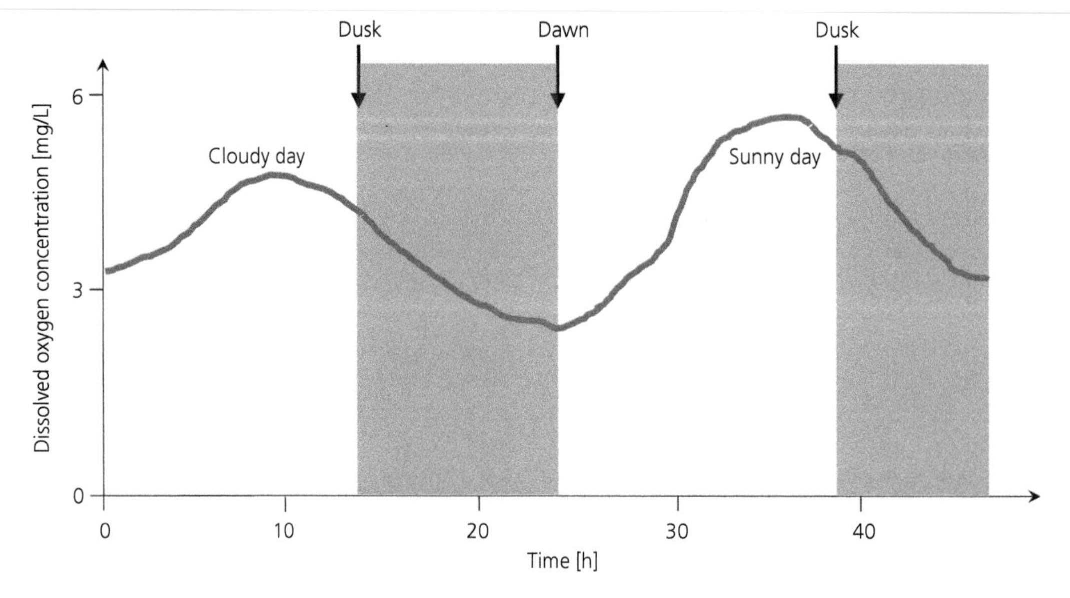

Figure 16.3 Daily fluctuations in dissolved oxygen (DO) concentration in an aquaculture pond that is managed semi-intensively by fertilization and supplemental feeding. DO decreases during the night when photosynthesis ceases but respiration continues. Moreover, DO can vary significantly between sunny and cloudy days and seasons in temperate regions as cloud cover and shorter winter day length reduce light input and photosynthesis.

sludge. If disposed of improperly, such waste may cause salinization of aquifers or surface waters in the disposal area.

16.4 Salinity

Salinity is a measure of the proportion of dissolved inorganic ions in water. It excludes dissolved organic ions and undissolved compounds. Salinity has been expressed in many units including %, ‰, parts per thousands (ppt = psu), g/L, and g/kg. The latter (g/kg) is the recommended international standard unit and expresses the amount of salt (in g) that is dissolved in 1 kg water.

Salinity is sometimes also expressed as conductivity (in milliSiemens, mS) if conductivity meters are used for measurement or in milliEquivalents per kg water (mEq/kg) if the contribution of specific inorganic ions to total salinity is known. Conversion of salinity units from g/kg to mEq/kg uses a simple equation: mEq/kg = mg/kg/equivalent mass of the specific ion in mg/mEq. The molecular masses of specific inorganic ions can be derived from the molecular masses of the corresponding elements that are indicated in the periodic table and the valency of the ion (Table 16.1).

Table 16.1 Equivalent masses of important ions that contribute to salinity.

Ion	Molecular mass (g/M)	Valency	Equivalent mass (mg/mEq)
Na^+	23	1	23
K^+	39	1	39
Cl^-	35	1	35
Ca^{2+}	40	2	20
Mg^{2+}	24	2	12
HCO_3^-	61	1	61
NH_4^+	18	1	18

Values are rounded to the nearest integer.

16.4.1 Salinity scale of environmental waters

The Venice salinity scale classifies waters depending on their salinity into five classes:

1. Freshwater (< 0.5 g/kg);
2. Oligohaline (0.5–5 g/kg);
3. Mesohaline (5 –18 g/kg);

4. Polyhaline (18–30 g/kg); and
5. Euhaline (> 30 g/kg) (Anonymous, 1958).

In praxis, oligo-, meso-, and polyhaline waters are referred to as brackish (0.5–30 g/kg) while euhaline waters are divided into marine (30–40 g/kg) and hyperhaline (or hypersaline, >40 g/kg) waters.

Ocean water is marine and has an average salinity of 35 g/kg, that is, 35 g of inorganic salt, mostly NaCl dissolved in 1 kg of water. One litre of distilled, deionized water weighs 1.0 kg at 0 °C, 0.999975 kg at 4 °C, and 0.997 kg at 25 °C. However, if dissolved ions are added then the weight of water increases slightly, and its density increases accordingly (e.g. above 0.997 g/mL at 25 °C). The increase depends on how many dissolved ions and which type of ions are added, and the measurement temperature. Therefore, salinity expressed in g/kg is slightly different from that expressed in g/L. The same difference applies to osmolality, which is measured in mOsmol/kg, and osmolarity, which is measured in mOsmol/L. Osmolality and osmolarity are defined as the sum of all osmotically active organic and inorganic ions that are dissolved in water.

For seawater (SW) and brackish water, the difference between osmolality and salinity is very small since the amount of dissolved inorganic ions in environmental waters is far greater than that of dissolved organic ions. However, for biological fluids such as blood plasma, haemolymph, and intracellular fluid, the contribution of dissolved organic ions to overall osmotic concentration represents a significant fraction. For example, urea and trimethylamine N-oxide (TMAO) concentrations are very high in shark blood (Yancey *et al.*, 1982). For this reason, osmotic concentration of environmental waters is commonly expressed as salinity while that of biological fluids is expressed as osmolality.

Nevertheless, salinity can also be expressed as osmolality. At 20 °C, SW having a salinity of 35 g/kg has an osmolality of approximately 1120 mOsmol/kg. Coldwater polar and temperate oceans generally have a lower surface water salinity of 28 – 34 g/kg because evaporation rates are lower, and they are closer to freshwater (FW) input from melting polar ice caps. Tropical and subtropical warm water oceans have greater evaporation rates and surface salinities are typically higher (34 – 37 g/kg).

Table 16.2 Salinity of select waters that span a wide range of natural environments inhabited by different aquatic species

Habitat	Salinity (g/kg)	Classification
Dead Sea	226	Hyperhaline
Great Salt Lake	203	Hyperhaline
Salton Sea	49	Hyperhaline
Tropical Ocean surface	36	Euhaline marine
Temperate Ocean surface	33	Euhaline marine
Salt Creek (Death Valley)	27	Polyhaline brackish
River Jordan, Jericho	8	Mesohaline brackish
Baltic Sea, Gulf of Finland	4	Oligohaline brackish
River water	0.1	Freshwater
Rainwater	0.01	Freshwater

Some desert lakes such as the Dead Sea and the Great Salt Lake are extremely hyperhaline, that is, their salinity is severalfold higher than that of the oceans leading to formation of brine (Table 16.2). Although NaCl is also the most abundant salt in most hyperhaline inland waters, the ionic composition of these hyperhaline waters is often very different from that of marine waters. Aquatic organisms inhabiting these salt lakes have adapted to perform well under these conditions.

16.4.2 Salinity tolerance of aquaculture organisms

Marine organisms have adapted to perform best at the specific ionic composition of SW. SW contains virtually all elements, although most of them are only present in trace amounts. Their presence in SW is in the form of ionized (dissolved) salts. Table 16.3 lists the eight most abundant inorganic (salt) ions and their contribution to overall salinity. They include four negatively charged anions and four positively charges cations. NaCl (Na^+ + Cl^-) represents by far the most abundant dissolved salt, accounting for 91% of the overall salinity of SW. Over 99% of the salinity of SW is due to ions comprised of six elements (Na, Cl, Mg, K, Ca, S).

Most FW and marine organisms have a narrow salinity tolerance range, i.e., they are stenohaline. Carps are examples of stenohaline FW fish that do not tolerate salinity exceeding 10 g/kg. Many

Table 16.3 Representation of the most abundant elements in seawater as dissolved anions and cations and their contribution to the overall salinity of seawater (35 g/kg)

Ion	Salinity contribution (g/kg)	Ion type
Na^+	15.2	Cation
Mg^{2+}	1.7	Cation
Ca^{2+}	0.3	Cation
K^+	0.3	Cation
Cl^-	16.6	Anion
SO^{2-}	0.9	Anion
Br^-	<0.1	Anion
F^-	<0.1	Anion

ornamental marine reef fish are also stenohaline, that is, they tolerate a narrow salinity range of 30–40 g/kg. Some species are euryhaline and have a wide tolerance to salinity (wider than 15 g/kg). Euryhaline species include migratory species (e.g. anadromous and catadromous fish), intertidal organisms, and estuarine organisms. For migratory species, the salinity tolerance may vary substantially with developmental state.

Besides the distinction of stenohaline and euryhaline species, aquatic organisms are categorized based on how they cope with a change in environmental salinity. Teleosts (bony fish) are osmoregulators that buffer environmental salinity fluctuations by adjusting the osmolality of their blood via energy-consuming epithelial ion and water transport mechanisms in the gills, kidney, gastrointestinal (GI) tract, and skin (Box 16.1).

Box 16.1 Osmotic gradients between fish blood and environmental waters

Like humans and other vertebrates, teleost fish maintain a blood osmolality close to 320 mosmol/kg (≈9 g/kg). Although this blood osmolality is slightly higher (350 mosmol/kg) in SW and slightly lower (290 mosmol/kg) in FW there is still a large difference to environmental salinity. This difference is 290 mosml/kg − 0 mosmol/kg = +290 mosmol/kg for FW fish and 350 mosmol/kg − 1120 mosmol/kg = −770 mosmol/kg for marine fish. Two observations can be made from this calculation. First, the osmotic gradient is steeper for marine fish (−770 mosmol/kg) than for FW fish (+290 mosmol/kg). Second, the osmotic gradient is opposite in the two environments (negative in SW and positive in FW). Therefore, the direction of active ion transport is also opposite in FW and marine fish. The metabolic energy needed to offset the osmotic gradient is generally higher for marine than FW fish; it not only depends on the osmotic gradient alone but also on the concentration of specific ions and epithelial permeability. Euryhaline fish that perform well over a wide range of salinity often grow fastest in brackish water of 9 g/kg (≈320 mosmol/kg) because the osmotic gradient is negligible at this salinity and no metabolic energy is needed to offset it (Kültz, 2015).

In contrast to bony fishes, most marine invertebrates are osmoconformers, that is, their haemolymph osmolality equals that of the surrounding water. Exceptions do exist-some crabs are partial osmoregulators, and bivalves may close their shells to maintain a different internal osmolality than that of the surrounding environment. Moreover, the hemolymph of FW invertebrates is at least somewhat hyperosmotic relative to the environment although the osmotic gradient is generally much smaller than in osmoregulating fish.

16.4.3 Salinity monitoring

Salinity can be measured in a variety of ways. In general, there are two approaches for measuring salinity: a low-cost, medium-precision approach and a higher-cost, high-precision approach. The first approach uses inexpensive equipment such as a refractometer, a hydrometer, or a conductivity meter. These instruments have a precision of ±0.2 g/kg, which is sufficient for most aquaculture systems. As a reference, human taste buds begin to detect salinity (saltiness) at 2 g/kg.

A refractometer measures the refractive index of water, which is directly proportional to its salinity (Figure 16.4). A hydrometer uses water density, which is proportional to gravity and water salinity as the measurement parameter. Conductivity, which is a function of the concentration of dissolved anions and cations in the water, is measured by a conductivity meter. All these devices are dependent on temperature, which means that salinity measurements must be calibrated for water temperature.

The second approach uses more expensive equipment (osmometers) to accurately determine salinity and osmolality with a precision of ±0.01 g/kg. Osmometers require only very small water samples

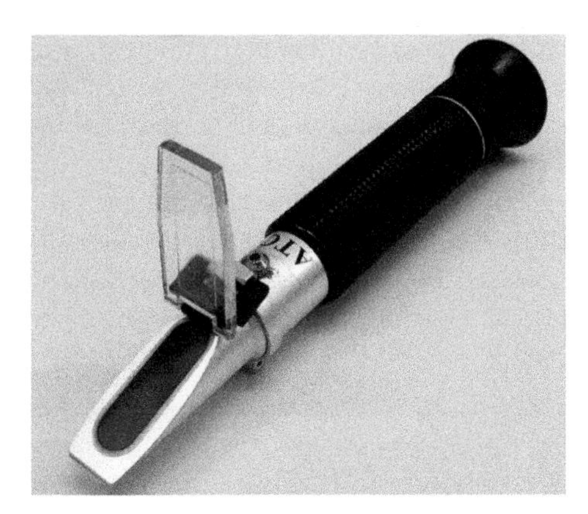

Figure 16.4 A refractometer used to measure salinity in the author's research laboratory. A drop of water is placed on the blue area in the lower left and the plastic lid is then lowered. The salinity can be read directly by viewing the marker line on the graded salinity scale through the eyepiece on the other end (upper right). Photo by Elizabeth Mojica.

(10–20 μL) for measurements and can also be used to determine blood plasma or serum osmolality of aquaculture fish. Two measurement principles are utilized. Vapor pressure osmometers determine the boiling point of water, which increases in proportion to salinity. Freezing point osmometers determine the freezing point of water, which decreases in proportion to salinity (Figure 16.5).

Salinity monitoring and management is particularly critical for hatchery and indoor mariculture because evaporation can cause significant salinity fluctuations that may negatively affect the health and performance of aquaculture organisms and, thus, should be avoided. Salinity can be adjusted by either adding salt (increase) or by adding FW (decrease). In closed circulation systems such as hatchery RAS, salinity can be effectively managed by adding sea salt, SW, or FW in appropriate amounts. The management of salinity in open pond systems depends on flow rate, the volume of pond water, and nearby availability of SW and FW. Salinity, temperature, and other water quality parameters cannot be controlled well in flow-through tank, pond, and raceway systems with high flow rates and in cage and net pen systems.

Salinity is a major management concern for intensive recirculating mariculture systems. Besides salinity, the concentration of specific ions also needs

Figure 16.5 A freezing point osmometer in the author's research laboratory. An exact volume of a 20-μL sample is applied using a special pipette to enable highly precise measurements. This instrument can measure both salinity and osmolality. Data are shown on a digital display after each measurement and can be stored on digital media. Photo by Jens Hamar.

to be managed, for example, Ca^{2+} concentration is critical for proper skeletal formation of many organisms. This additional management burden is much lower for FW aquaculture systems. Moreover, maintenance of tank-, pond-, or raceway-based mariculture systems is complicated by the high corrosiveness of SW, higher technical demands for waste treatment from SW relative to FW, and the added burden of having to manage the disposal of large amounts of salt. For these reasons, FW RAS may represent a more promising solution for the future even though FW as a resource is much more limited than SW.

16.5 Dissolved oxygen

The oxygen concentration in the earth's atmosphere is approximately 20% (by volume). This large amount of atmospheric oxygen has accumulated since approximately 2 billion years ago during an era termed the Great Oxygenation Event

(GOE). The difference between atmospheric oxygen present in air and the maximum dissolved oxygen (DO) concentration in water is tremendous.

This huge difference can be illustrated by converting 20% atmospheric oxygen into mg/L [20% = 200 parts per thousand = 200,000 parts per million = 200,000 mg/L]. At atmospheric pressure, the concentration of DO in water is approximately 5–15 mg/L, depending on the temperature and salinity. Therefore, the oxygen concentration in water is about 20,000-fold lower than that in air!

Consequently, aquatic animals require DO levels that are always near or at saturation levels. Oxygen is required for all aquatic animals but the extent to which different aquaculture species tolerate temporary decreases in oxygen concentration (hypoxia), or even the temporary absence of oxygen (anoxia), differs substantially. Hypoxia occurs when DO concentration decreases below a threshold that varies depending on species and environment; <2 mg/L is often used for defining hypoxia in environmental waters.

16.5.1 Hypoxia

Hypoxia is a condition that causes significant stress in aquaculture species. Hypoxic environmental waters eventually lead to hypoxia in biological fluids such as blood and haemolymph. The threshold for hypoxia in biological fluids is generally much lower than for environmental waters. Hypoxia has wide-ranging biomedical implications and received much scientific and public attention because it also occurs in biological fluids of terrestrial animals and humans (Richalet, 2021).

Species differences in tolerating low DO concentrations and corresponding aquaculture management implications are illustrated by comparing oxygen requirements of common carp with that of salmonids. Common carp lower their metabolic rate and switch to anaerobic metabolism for several months during winter when lakes are covered with ice and oxygen is limited. In contrast, salmonids require saturated or near saturated levels of DO for optimal growth and development and only tolerate short (hours to a few days) drops below those levels.

Many fish species respond to hypoxia by gulping at the water surface. In most cases (except lungfish and other fishes with air-breathing organs) the gulping behaviour does not promote oxygen uptake from air but rather increases the area of the microlayer at the water/air interface by creating ripples. This microlayer is more highly oxygenated due to the diffusional gradient between air and water and increasing its area helps in enriching the water with oxygen. The same principle is used by aeration devices.

16.5.2 Aeration devices

Aeration devices are used for semi-intensive and intensive aquaculture to maintain DO concentration near saturation levels. These devices operate on the principle of increasing the surface area for diffusion of oxygen from air to water along the huge concentration gradient outlined earlier. Because of daily fluctuations in DO levels, it is common practice on many aquaculture farms to perform daily measurements of DO concentration in samples collected at dawn and in the afternoon. DO readings should be taken immediately when samples are collected to prevent changes in DO content while handling them.

Aeration devices can be divided into three broad categories: fountain, propeller (paddle wheel), and bubble aerators. They serve a dual purpose: first, to increase the surface area of the water to air interface to facilitate gas exchange by diffusion, and second, to create a water current to mix oxygen-rich surface water with oxygen-depleted bottom water and counteract temperature and/ or salinity stratification.

Bubble aerators are simple air blowers or air stones that are connected to tubing supplied by high pressure air or liquid oxygen tanks. The latter are sometimes used for aeration during transport of animals by truck or railway. Blowers or air stones are installed at the bottom of the pond or tank and oxygen contained in the bubbles diffuses into the surrounding water. In addition, bubbles generate water movement for mixing and increasing the water surface area through which oxygen can be taken from the air up by diffusion.

Propeller aerators are often used for ponds as they are readily mobile and can be repositioned as needed. The small water droplets generated by these devices have the same purpose as bubbles, that is, they greatly increase the area through which diffusional gas exchange occurs.

Aerators are often operated only for an hour to a few hours daily to minimize energy consumption. They are commonly turned on during early morning hours before dawn when hypoxia or anoxia is most likely to occur because of night-time respiration and lack of photosynthesis (Figure 16.3).

16.5.3 Measurement of DO concentration

Permanent oxygen monitoring equipment consisting of sensors that are interfaced to computer-controlled alarm systems and aerators can be installed to turn on aerators automatically when DO levels drop below a defined limit. However, the initial cost of such equipment and maintenance to prevent instrument failure represent an economic investment. Thus, most aquaculture farms use hand-held DO meters when feeding or performing routine daily monitoring of animals. Modern DO meters automatically compensate for water temperature and salinity.

Salinity and temperature greatly influence oxygen solubility, which has practical implications for aquaculture besides calibrating DO measurements. For example, if stratification of aquaculture pond or lake water due to thermal or saline gradients occurs, then the maximum oxygen availability in different strata will vary. Such variation must be considered when monitoring oxygen levels by taking measurements at multiple depths, in particular when bottom-dwelling animals are being cultured.

The effect of temperature on oxygen saturation limits is quite large. In FW, DO is saturated at 14.8 mg/L at 5 °C. Oxygen concentrations at saturation decrease to 13.0 mg/L at 10 °C, 9.4 mg/L at 20 °C, and 7.8 mg/L at 30 °C. Therefore, warmwater organisms are more oxygen-limited than coldwater organisms.

An increase in salinity also decreases the concentration of DO at saturation. This effect of salinity is a result of the limited solvent capacity of water. Water is a universal solvent for both ions and gases and its solvent capacity for gases is reduced if more ions are dissolved in it (i.e. as salinity increases). At 20 °C, the oxygen concentration at saturation decreases from 9.4 mg/L in FW (0 g/kg) to 8.8 mg/L at 10 g/kg salinity, 8.1 mg/L at 25 g/kg salinity, and 7.6 mg/L at 35 g/kg (= SW) salinity.

Hydrostatic pressure is another abiotic parameter that influences oxygen concentration at saturation. In contrast to temperature and salinity, an increase in hydrostatic pressure increases the concentration of oxygen that can be dissolved in a constant volume of water. However, unlike temperature and salinity, hydrostatic pressure is not a practical concern for most aquaculture farms.

Although DO concentration should be maintained near or at saturation in all aquaculture settings, it should not exceed saturation levels to avoid gas bubble disease, which causes bone and skeletal deformations during development. Supersaturation occurs if cold water that is saturated with oxygen is heated. Heating will lead to gassing out of oxygen, which causes the formation of bubbles that can be trapped in developing tissues and interfere with proper development of fish. Supersaturation can also occur by air being entrained in high-pressure water circulation pumps.

16.6 The aquatic carbonate system (CO_2, hardness, pH)

All animals generate carbon dioxide (CO_2) while consuming oxygen during aerobic metabolism. Whereas DO in water decreases if respiration rates are higher than can be compensated by diffusional uptake from air and photosynthetic production, CO_2 levels will rise. Since CO_2 is a volatile gas, it will escape the water when its concentration exceeds the solubility threshold, which is temperature and salinity dependent. At 10 °C, CO_2 solubility is twice as high (2.5 g CO_2 per kg water) as at 30 °C (1.25 g CO_2 per kg water). Thus, CO_2 can escape more readily in warmwater than in coldwater aquaculture.

However, CO_2 is highly reactive with water and can be rapidly converted to dissociated bicarbonate (HCO_3^-) and carbonate (CO_3^{2-}) anions in aqueous solutions. This chemical conversion process generates hydronium (H_3O^+) or, more simply, hydrogen (H^+) ions, which decreases the pH of water. Moreover, carbonate can react with divalent calcium (Ca^{2+}) and magnesium (Mg^{2+}) cations to form salts that influence water hardness (Box 16.2).

Box 16.2 The aquatic carbonate system expressed as chemical equations

$CO_2 + H_2O \rightleftharpoons H^+ + HCO_3^- \rightleftharpoons 2\,H^+ + CO_3^{2-}$
$CaO + H_2O \rightleftharpoons Ca(OH)_2$
$2\,H^+ + CO_3^{2-} + Ca(OH)_2 \rightleftharpoons CaCO_3 + 2\,H_2O$

This tight chemical interrelationship between the concentration of an important dissolved gas (CO_2), water hardness, and pH is referred to as the aquatic carbonate system.

The aquatic carbonate system is critically important for the operation of all aquaculture farms. A solid understanding of this system is essential for proper management of water quality. The three variables of the aquatic carbonate system—CO_2, pH, and water hardness—and how they relate to each other, how they are monitored, and how they can be managed, are discussed in the remainder of this chapter.

CO_2 generated by respiration of aquatic animals is a primary component of the aquatic carbonate system. However, on aquaculture farms it is not monitored directly but rather indirectly by assessing rates of respiration and photosynthesis via DO measurements and by measuring pH and water hardness. Volatilization of CO_2 is promoted by aeration devices. Electrodes for direct measurement of CO_2 are available but they are less robust than oxygen electrodes.

16.6.1 Water acidification due to climate change

pH is defined as the negative decadic logarithm of the hydrogen ion concentration in aqueous solutions, $-\log_{10}(H^+)$. It is expressed on a scale of 0 to 14, with 0 being highly acidic (very high concentration of H^+), 14 being highly basic/alkaline (very low concentration of H^+), and 7 being neutral. The pH of rainwater is slightly acidic (6) while that of SW is slightly basic (8) and FW pH is often close to neutral (7). For comparison, most beverages that we drink are also slightly acidic.

The intensification of burning fossil fuels, deforestation leading to decreased CO_2 assimilation, and other anthropogenic causes have led to a steady rise of atmospheric CO_2 in previous decades. Thus, the diffusional gradient for CO_2 uptake into oceans and FW bodies has increased, which has shifted the chemical equilibrium in the direction of increased carbonic acid ($2\,H^+ + CO_3^{2-}$) generation (Box 16.2).

This trend of ocean and FW acidification is stressful for aquatic organisms, which are forced to adapt their optimal function to a lower pH range. However, the evolutionary timescale required for adaptation of many aquatic species may be outpaced significantly by the speed at which anthropogenic acidification of their environment occurs. Water acidification is particularly stressful for organisms that rely on calcium carbonate for exoskeleton formation (e.g. molluscs and corals) because bioavailability of this salt decreases with increasing pH.

Moreover, acid rain has greatly decreased the pH of boreal lakes and rivers in Europe and North America during previous decades. Mountain lakes on granite bedrock are particularly susceptible to acid rain because they have a thin sediment layer and low pH buffering capacity. In some lakes pH decreased by several pH units, resulting in massive fish kills (Schofield, 1976; Hopkin, 2005). Acid rain is generated when anthropogenic pollutants produced by industrial processes escape into the atmosphere and react with oxygen and cloud water. For example, volatile pollutants such as nitrous and sulfur oxides emitted into the atmosphere by industrial pollution form sulfuric and nitric acids that greatly decrease the pH of rainwater.

Since most aquaculture systems are open pond, cage, or raceway systems, they are subject to the same acidification trends as natural water systems. In addition, acidification of water results from oxidation of fertilizer that is applied in semi-intensive aquaculture systems. Some fish species are naturally adapted to very high pH, which renders them particularly susceptible to acidification (Brauner, Gonzalez, and Wilson, 2012).

16.6.2 Water alkalinization by liming

Liming is a common procedure used to counteract water acidification in semi-intensive aquaculture, especially for open pond systems. It uses quicklime or limestone. Quicklime is calcium oxide (CaO) while limestone is a naturally occurring mineral consisting of calcite (i.e. calcium carbonate, $CaCO_3$) and dolomite. Dolomite is an equimolar mixture of $CaCO_3$ and magnesium carbonate ($MgCO_3$). Both carbonates react with water to form bicarbonate (HCO_3^-) and the corresponding oxides. Bicarbonate can sequester hydrogen (H^+) ions by forming water and volatile CO_2 and, therefore, increase pH. Calcium and magnesium oxides (CaO and MgO) can react with water to form the corresponding hydroxides, $Ca(OH)_2$ and $Mg(OH)_2$, which can dissociate to form hydroxyl ions (OH^-) that sequester hydrogen (H^+) ions to form water and, thus, increase pH.

Box 16.3 shows the corresponding chemical equations for these reversible reactions for calcium, and they apply equally to magnesium. The direction of these chemical reactions depends on the relative concentration of compounds on each side of an equation. Applying quicklime or limestone shifts the equilibrium toward the direction of hydrogen ion sequestration, which increases pH.

Liming increases pH, water hardness, CO_2 volatilization, and the bioavailability of calcium ions for calcification of bones, mollusc shells, and crustacean exoskeletons. However, the immediate effect of liming on pH and CO_2 volatilization is transient and lasts only for a few days until a new equilibrium has been achieved.

Massive liming often precedes the stocking of aquaculture ponds because it temporarily raises the alkalinity of water to pH 11 and above. Such a high pH is toxic to most organisms, including predators and pathogens. Fertilization of ponds is applied subsequent to liming after pH has decreased to values near neutral to permit the growth of phytoplankton and zooplankton.

Box 16.3 The effect of liming on H^+ sequestration (pH increase) and water hardness

$$CaCO_3 + H_2O \rightleftharpoons CaO + HCO_3^{-} \; {}^{+H+}$$
$$HCO_3^- + H^+ \rightleftharpoons H_2O + CO_2$$
$$CaO + H_2O \rightleftharpoons Ca(OH)_2$$
$$Ca(OH)_2 \rightleftharpoons Ca^{2+} + 2\,OH^-$$
$$OH^- + H^+ \rightleftharpoons H_2O$$

16.6.3 Water hardness

Although liming only transiently elevates pH, it permanently increases pH buffering capacity and water hardness. Such increases are often desirable for aquaculture since acidification of aquaculture waters due to anthropogenic dilution that causes acid rain and increased CO_2 in the atmosphere can be more effectively counteracted. However, if species that require soft water for optimal growth and reproduction are cultured (e.g. ornamental fishes from the Amazon River basin) then liming should be avoided.

Hard water has a high pH buffering capacity because CO_2 conversion to carbonic acid can be balanced by conversion of carbonic acid to calcium carbonate or magnesium carbonate (Box 16.2). Soft water such as rainwater has low pH buffering capacity and tends to be acidic because carbonic acid cannot be converted to calcium carbonate or magnesium carbonate.

Water hardness is a measure of the amount of divalent cations (calcium and magnesium) in FW. The unit of water hardness is milligrams of $[Ca^{2+} + Mg^{2+}]$ per litre water, which is equivalent to parts per million (ppm). An alternative unit that is often used is degrees of German Hardness (dGH). The relationship between these units differs depending on the chemical form of calcium and magnesium. For oxides (CaO) 1 ppm = 1 mg/L = 0.1 dGH while for carbonates ($CaCO_3$) 1 ppm = 1 mg/L = 0.056 dGH.

Table 16.4 summarizes the classification of FWs based on water hardness. Because of liming, most FW aquaculture systems use moderately hard to very hard water with a high pH buffering capacity (Boyd, Tucker, and Somridhivej, 2016).

Table 16.4 Classification of water hardness based on the sum of calcium and magnesium oxides and carbonates in freshwater

Water hardness	mg/L (= ppm)	dGH
Very soft	0–30	0–4
Soft	30–60	4–8
Slightly hard	60–90	8–12
Moderately hard	90–120	12–18
Hard	120–180	18–24
Very hard	>180	> 24

dGH = °German hardness = dKH = °Karbonathärte

Key conclusions

- Clean water has become a precious commodity, which mandates regular monitoring and adjustments of key abiotic and biotic water quality parameters in all aquaculture systems.
- Abiotic water quality parameters that are critical for aquaculture include temperature, photoperiod, turbidity, salinity, concentrations of DO and CO_2, pH, and water hardness.
- Temperature is a key variable for controlling metabolic and growth rates, which is conveyed as the Q_{10} effect. Temperature adjustments are only practical for closed recirculating aquaculture systems (RAS) and heating is easier and more economical than cooling.
- Photoperiod is critical for reproduction and can be manipulated in closed RAS to maximize fecundity and shorten breeding intervals of aquaculture broodstock, enable continuous feeding, and prevent hypoxia.
- Turbidity is a measure of undissolved (suspended) particles in water and a direct indicator of water quality and eutrophication.
- Salinity is a measure of the concentration of dissolved inorganic ions (salts) while osmolality includes the concentration of both inorganic ions and organic osmolytes in aqueous fluids. Osmolality is utilized for biological fluids but is also applicable to environmental waters.
- Aquaculture fish are osmoregulators while most aquaculture invertebrates are osmoconformers.
- Many aquaculture animals have relatively wide salinity and temperature-tolerances, that is, they are euryhaline and eurythermal, have been selected for high environmental resilience during domestication, and perform well in captivity under a wide range of conditions.
- DO concentration must be monitored closely in all aquaculture systems because it is much lower than oxygen levels in air and subject to daily fluctuations. DO should be always maintained at or near saturation levels and hypoxia must be avoided by using proper aeration devices.
- Dissolved CO_2 concentration, pH, and water hardness are closely interrelated abiotic parameters referred to as the aquatic carbonate system; This system is strongly impacted by liming, which is a common procedure before stocking aquaculture ponds.

References

Anonymous. (1958). 'The Venice system for the classification of marine waters according to salinity', *Limnology and Oceanography*, 3(3), pp. 346–347.

Boyd, C. E., Tucker, C. S., and Somridhivej, B. (2016). 'Alkalinity and hardness: critical but elusive concepts in aquaculture', *Journal of the World Aquaculture Society*, 47(1), pp. 6–41. doi:10.1111/jwas.12241

Brauner, C. J., Gonzalez, R. J., and Wilson, J. M. (2012). 'Extreme environments: hypersaline, alkaline, and ion-poor waters', in McCormick, S. D., Farrell, A. P., and Brauner, C. J. (eds.) Fish physiology: euryhaline fishes. Oxford: Academic Press, pp. 435–476. doi:10.1016/B978-0-12-396951-4.00009-8

Carlson, R. E. (1977). 'A trophic state index for lakes', *Limnology and Oceanography*, 22(2), pp. 361–369. doi:10.4319/lo.1977.22.2.0361

Golovanov, V. and Alexey, S. (2007). 'Influence of the water heating rate upon thermal tolerance in common carp (*Cyprinus carpio L.*) during different seasons', *Journal of Ichthyology*, 47, pp. 538–543. doi:10.1134/S0032945207070089

Heine, H. (1844/2019). *Deutschland. Ein Wintermärchen*. Varna: Pretorian Books.

Hopkin, M. (2005). 'Acid rain still hurting Canada', Nature [online]. doi:10.1038/news050808-10

Kültz, D. (2015). 'Physiological mechanisms used by fish to cope with salinity stress', *The Journal of Experimental Biology*, 218(Pt 12), pp. 1907–1914. doi:10.1242/jeb.118695

Réalis-Doyelle, E., Pasquet, A., De Charleroy, D., Fontaine, P., and Teletchea, F. (2016). 'Strong effects of temperature on the early life stages of a cold stenothermal fish species, brown trout (*Salmo trutta L.*)', *Plos One*, 11(5), p. e0155487. doi:10.1371/journal.pone.0155487

Richalet, J.-P. (2021). 'The invention of hypoxia', *Journal of Applied Physiology*, 130(5), pp. 1573–1582. doi:10.1152/japplphysiol.00936.2020

Schofield, C. L. (1976). 'Acid precipitation: effects on fish', *Ambio*, 5(5/6), pp. 228–230.

Somero, G. N., Lockwood, B. L., and Tomanek, B. L. (2017) *Biochemical adaptation: response to environmental challenges from life's origins to the Anthropocene*. Oxford: Oxford University Press.

Yancey, P. H., Clark, M. E., Hand, S. C., Bowlus, R. D., and Somero, G. N. (1982). 'Living with water stress: evolution of osmolyte systems', *Science*, 217(4566), pp. 1214–1222.

CHAPTER 17

Feeds, waste, and stress

'Limited supplies of freshwater in the world mean that technology that can reduce water consumption per kilogram fish produced will be important; this includes reliable, cost-effective re-use technology.'

Lekang, *Aquaculture Engineering* (2019)

Feeds (and fertilizer) represent the main input of energy and matter in aquaculture systems, while waste (in addition to the organisms produced) represents the main output. Balancing these inputs and outputs can best be controlled in recirculating aquaculture systems (RAS). However, most commercial aquaculture operates by using open systems that only permit control of the inputs while discharging most of the outputs into adjacent wild ecosystems.

Aquaculture discharge alters the nutrient balance and succession of natural ecosystems, even if it cannot be noticed in the short term. Ecosystems within the proximity of the aquaculture facilities will be impacted most, but more subtle or indirect effects on distant natural ecosystems are also possible. The extent of alteration depends on the amount of discharge relative to the carrying capacity of an ecosystem. Aquaculture systems can be classified based on their inputs, outputs, reliance on technology versus ecosystem services (closed versus open, extensive versus intensive), use of space, and the type of water body used (Figure 17.1).

17.1 Classification of aquaculture systems based on their inputs and outputs

As pointed out in Chapter 1, I refer to intensive systems only if they are technology-intensive on both ends of the system (input and output). This means that intensive systems utilize technologies to produce and provide feeds and/or fertilizer and increase biomass (e.g. via aeration) on the input end, but they also heavily rely on technologies to treat waste and manage biosecurity on the output end of the system. This definition is different from the more common definition of intensive aquaculture, which only considers the input end of the system and distinguishes intensive from semi-intensive systems based on their exclusive reliance on manufactured feeds.

Aquaculture waste discharge may seem negligible when the vast expanse of the oceans is considered, but the amount of such discharge adds to other anthropogenic pollution and may eventually reach a level that exceeds a critical threshold at which ecosystem succession will accelerate and become noticeable. Such human-induced ecosystem succession represents an adaptive response at a macrocosm scale that alters ecosystems in ways that are less desirable from a human need perspective. Negative impacts of human-induced ecosystem succession on humanity are illustrated by eutrophication of freshwater ecosystems, coral bleaching, and fisheries-induced evolution (Heino *et al.*, 2015; Sully *et al.*, 2019; Wurtsbaugh *et al.*, 2019).

For this reason, the sometimes-popular attitude that semi-intensive cage mariculture should be expanded to vast, supposedly underutilized, areas of the oceans because of limited freshwater resources is incompatible with good environmental stewardship that properly considers the needs of future generations. It represents a popular idea

A Primer of Ecological Aquaculture. Dietmar Kültz, Oxford University Press. © Dietmar Kültz (2022). DOI: 10.1093/oso/9780198850229.003.0017

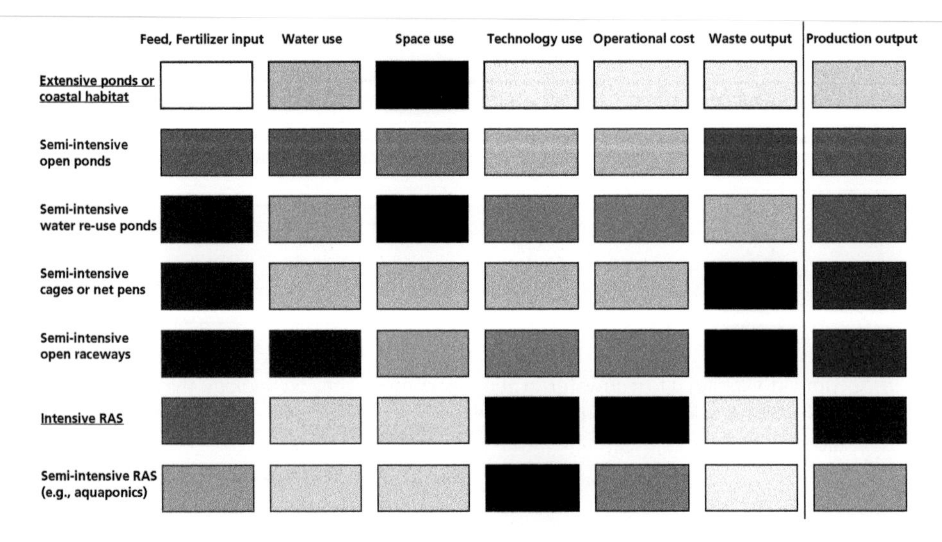

Figure 17.1 Resource footprints (left of vertical line) and production output of the major types of aquaculture systems. Darker squares indicate higher values. Resource footprints are expressed relative to the size of the aquaculture facility while production output is expressed relative to global demand that can be met in an ecologically sustainable manner. Note that true values of water and space use for extensive pond/coastal and cage/net-pen systems are difficult to assess as these forms of aquaculture do not use designated space but fully rely on public waters.

for fixing pressing immediate needs at the expense of future needs because marine cage culture is economically much more attractive in the short-term than applying costly technologies to operate RAS and other water reuse systems. However, the problem of dealing with the waste produced is not solved but rather just deferred into the future and methods of waste removal will be more expensive, if at all possible, because the chance of collecting it in concentrated form has been missed.

If practiced on a large scale, cage-based mariculture will alter marine ecosystems over intermediate to longer timeframes in ways that reduce biodiversity and deteriorate the quality of life for future generations of people. The scientific knowledge accumulated during previous decades about the ecological implications of food production systems teaches us that increased reliance on ecosystem services is not without cost, but rather defers costs into the future.

Cage mariculture solely relies on ecosystem services for processing dispersed waste into a macrocosm scale and does not take advantage of the opportunity to collect and process concentrated waste by applying available technologies for aquaculture mesocosms (Chapter 4). Deferred ecosystem restoration efforts at a macrocosm scale, if possible, will be much more expensive than applying available waste-water treatment technologies at a more confined mesocosm scale.

Aquaculture solutions that are more costly in the short term, that is, RAS and other water reuse systems, fully utilize available knowledge and technologies in lieu of reliance on ecosystem services. These systems minimize the use of clean freshwater, space, and waste discharge. They are much more cost-effective for freshwater aquaculture than mariculture. But RAS cannot compete in terms of short-term profitability with quick-fix approaches that exploit ecosystem services unless economical value that crosses national boundaries and considers long-term outcomes is equated with such services.

Who (if not humanity as a whole) owns ecosystem services and who is eligible to benefit from them? Which of the many benefits they provide should be prioritized? How can ecosystem services be equated with an economical value that offsets the costs of using available knowledge and technologies for recycling waste from aquaculture and other food production systems? How should long-term value be weighed against short-term gain? How can trade imbalances of aquaculture products be addressed in the context of ecological sustainability without putting most of the burden on producer countries? From a scientific perspective, the most important aspect of finding solutions to pressing issues or problems is to ask the right questions. Are we asking the right questions to promote sustainable aquaculture for current and future generations?

17.1.1 Management of aquaculture inputs and outputs

To maintain ecosystem function that best supports human needs in a sustainable manner not only the inputs but also all outputs of aquaculture systems should be controlled and properly managed. Such management can be done by either:

1) collecting waste and processing/ disposing of it in an ecologically sustainable manner, or
2) recycling waste by closing trophic cycles within the system to a maximal extent.

Both these general waste management strategies are resource intensive and costly. Their implementation into commercial aquaculture relies on proper economic incentives that acknowledge their value to maintain the operational feasibility of aquaculture farms. Incentivization of using available technologies and developing new ones for managing aquaculture discharge remain a critical challenge for economists, regulators, and policy makers that must be addressed globally since a large fraction of aquaculture products is traded internationally.

The main inputs into aquaculture systems (feeds, fertilizer, and fertilized eggs or embryos) are highly amenable to strict management. Their exact quantities, composition, and physical properties can be determined, controlled, and tailored to support optimal performance of aquaculture systems. The level of control is much lower for the outputs (waste and biomass of aquaculture animals and other organisms, e.g. bacteria and microalgae). Quantitative capture of outputs is limited to biomass of aquaculture organisms in most aquaculture systems, which are operated as semi-intensive, open systems. Only for RAS and other closed water-reuse systems is it possible to capture most of the outputs. However, even if waste is quantitatively captured, the form and consistency of waste outputs cannot be altered but depend on the properties of the system (i.e. which species are being cultured at what densities under which conditions). This means that waste treatment technologies need to be tailored towards the form and consistency of the waste.

Wastewater discharge from aquaculture and other human activities increases eutrophication, turbidity, sedimentation, and the concentration of potentially toxic chemicals in the receiving public waters, which may adversely impact environmental and human health. For example, eutrophication promotes the spread of conjunctivitis as well as gastrointestinal (GI) and dermatological diseases (Zhidkova *et al.*, 2020), antibiotics present in wastewater may promote the emergence of antibiotic-resistant strains of human pathogens (Silva *et al.*, 2021), and zoonotic diseases (Box 18.1) may be spread via wastewater (Gauthier, 2015). The prevalence of periphyton in the greater vicinity of wastewater discharge sites represents a readily quantifiable indicator of altered ecosystem composition and succession (Section 5.4). Its biomass is often proportional to the distance from the site of waste discharge. Periphyton growth increases assimilation capacity of receiving public waters by changing the structure and dynamics of natural ecosystems.

Whether certification, environmental impact assessment, so-called life cycle analyses, etc., represent useful means to equitably incentivize ecologically sustainable aquaculture production is debatable. These are worthy efforts and often help educate aquaculture producers, policy makers, and the public about potential ecological implications of production systems. However, they are often subjective, limited to a case-by-case basis, and dependent on which criteria are being applied by whom. The cost/benefit ratio, susceptibility to bias and lobbying, and enforcement issues are challenging aspects of such efforts. Improving the ecological sustainability of production systems in general (not just aquaculture or food production) critically depends on the development of better, less biased, more comparable, and more global assessment and incentivization practices that transcend national boundaries equally as the ecosystems affected by them.

17.2 Aquaculture feeds

The output from aquaculture systems in the form of waste can be minimized by improving feed efficiency. When treating an aquaculture system as a black box, then the flow of matter and energy through the system can be summarized in a simplified form (Box 17.1).

Energy conversion ratio represents the ratio of chemical energy in the form of food (feeds) versus chemical energy contained in organismal biomass. For natural ecosystems, this ratio is referred to as the trophic level transfer efficiency (TLTE; see Chapter 4). TLTE is usually 10–20% but can be considerably lower or higher for some food sources and species. A goal of commercial aquaculture is to maximize TLTE to make the most efficient use of available resources. Aquaculture is thought to benefit from generally higher TLTE of aquatic animals compared to chickens, cows, pigs, and other terrestrial birds and mammals. A greater percentage of energy intake can be converted into the biomass of aquatic animals because they are ammoniotelic and poikilothermic, that is, they do not need to allocate energy for temperature regulation and uric acid or urea synthesis. Moreover, unlike terrestrial animals, fish are buoyant and need to expend less energy to oppose gravity, which is reflected in a lower percentage of biomass devoted towards skeletal structures.

Feed efficiency can be regarded as the ratio of produced organismal biomass versus waste. The goal of aquaculture is to maximize this ratio and increase both profitability and ecological sustainability of aquaculture farms. Feed efficiency can only be determined if produced waste is accurately quantified, which is often not the case. In contrast to feed efficiency, it is easier to determine feed conversion ratio (FCR), which is defined as the ratio of feed provided versus the biomass of aquaculture organisms harvested. The FCR is typically between 1 and 3 for intensive or semi-intensive aquaculture systems.

Box 17.1 Main inputs and outputs of matter and energy in an aquaculture system

Clean water + Feeds + Fertilizer + Operational energy input (light, electricity) → **Aquaculture system** → Organismal biomass (product) + Waste + Water

The FCR is somewhat misleading in that modern aquaculture feeds are often dry pellets with very low water content, whereas aquaculture product yields are expressed as wet weight, which contains up to 75% of the weight as water. Therefore, multiplying the FCR by a factor of 3 to 4 produces a more realistic estimate when it is based on the weight of dry feeds. The most accurate way to calculate FCR is to base it on dry feed and dry weight gain. A shift from moist feeds to dry formulated pellets has decreased FCR substantially over the past decades. However, as mentioned, part of this decrease is due to the reduction in water content. Effects of fertilization and natural feeds are difficult to account for in FCR estimates but can be significant for most aquaculture systems currently used, that is, semi-intensive pond systems. While dry formulated feeds are now routinely used in fish and shrimp aquaculture, they are not common as food sources for filter feeders (e.g. molluscs) or detritivores (e.g. crayfish).

Feed efficiency is influenced by many factors, including the contribution of natural versus formulated feeds (Section 17.2.2), the percentage of feed consumed, the percentage of nutrients absorbed in the GI tract of aquaculture animals, and the metabolism of aquaculture animals to convert nutrients into animal biomass. As pointed out in Chapter 16, animal metabolism greatly depends on temperature (e.g. Q_{10} effect) and other abiotic parameters. In addition, metabolism is affected by stress (Section 17.4.1) and disease (Chapter 18). Therefore, managing these parameters to be optimal for the culture of organisms maximizes nutrient absorption via the GI tract, nutrient conversion into biomass, and feed efficiency. Feed efficiency can be further maximized by proper management of feeding, that is, the specific ways in which food is provided. The time of feeding is important because some animals are nocturnal feeders, while others are most active feeding in the morning or afternoon.

The frequency of feeding is another important factor and the needs in that regard vary greatly depending on species and developmental state. Tailoring feeding frequency to those needs maximizes feed efficiency. Feed consistency and the method of application (manually or using automatic feeders) also represent important considerations. Automatic feeding facilitates frequent application of feeds while manual feeding enables better monitoring of feeding behaviour (Suresh and Bhujel, 2019). Automatic feeding is most amenable to dry pellets and less practical for moist or wet feeds. Different types of feeding equipment used for aquaculture include feed blowers, feed dispensers, demand feeders, automatic feeding machines and robots, and computerized feeding systems with central control units. These different feeding systems are described in more detail elsewhere (Lekang, 2019).

Feed consistency also has large impacts on feed efficiency. The size, shape, density, and moisture content of formulated feeds must be tailored to the species and developmental state that are being cultured. Whether live or formulated food is provided depends on the developmental state of the cultured organisms. Many species require live food during early developmental stages in hatchery cultures, but the best developmental window for weaning juveniles off live food is species dependent (Box 17.2).

Box 17.2 Hatchery culture of microalgae

Microalgae cultured in hatcheries as a source of live larval food include flagellates, diatoms, and heterokonts. Species that are rich in protein, for example, *Isochrysis galbana* and *Chaetoceros calcitrans*, are preferred. When pigmentation of aquaculture animals is important, for example, for salmon aquaculture, then microalgae that are rich in astaxanthin such as *Haematococcus spec.* can be included as live food in larval diets. Microalgae range between 5 and 50 μm (0.005 to 0.05 mm) in size and are the preferred live food for very small fish and invertebrate larvae. Microalgae cultures are managed either as discontinuous batch cultures or continuous cultures. For batch cultures, periodic harvest of all but a small fraction of the culture is performed when it reaches stationary phase. The small fraction that is remaining is used to start culture of the next batch. For continuous cultures, the microalgae population is always kept in the exponential phase, and a small volume is harvested continuously. Microalgae cultures are grown under optimal lighting, temperature, salinity, and pH to maximize proliferation of the culture organisms. The nutritional value of microalgae as live larval food depends on the nutrient composition of culture media. Therefore, much research has focused on optimizing culture media for key species of microalgae that are used as live food organisms for aquaculture.

Live food and floating food pellets that do not easily disintegrate are advantageous for aquaculture because they maximize the time of food availability while minimizing food waste. In conjunction with aeration, floating pellets reduce sediment degradation and contribute to maintaining a healthy sediment redox potential that supports aerobic decomposer communities in pond and cage aquaculture systems. Floating food pellets also facilitate the observation of feeding activity. However, semi-floating or sinking food is required if bottom feeders are cultured that have not been acclimated to feed on the surface. Moreover, floating feed presents other challenges in ponds or marine pens if wind or water current drives it towards pond shores, or beyond the boundaries of cages and pens.

Another critical aspect of maximizing feed efficiency is the optimization of feed composition relative to nutritional requirements of the species cultured. For live food, such optimization is reliant on knowing the nutrient composition of food organisms and choosing species that have an optimal composition. For formulated feeds, nutrient composition can be tailored to match the requirements of the culture species. Much research has been and is currently performed on feed optimization for many aquaculture species to improve feed efficiency (Hardy and Kaushik, 2021).

17.2.1 Live food

Aquaculture fish and invertebrate larvae are commonly fed live food that resembles their preferred food source in natural habitats and can be conveniently produced in hatcheries. Various species of phytoplankton (microalgae) and zooplankton (brine shrimp nauplii and adults, copepods, rotifers) are commonly used as live food organisms.

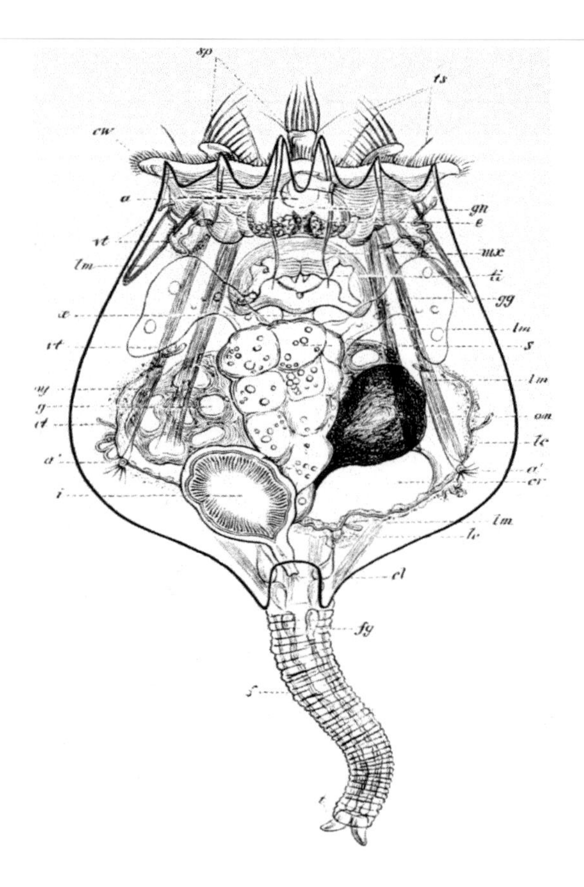

Figure 17.2 Rotifers of the genus *Brachionus* are excellent live food organisms for larval aquaculture. A drawing showing the organization of *Brachionus rubens* is depicted. The size of this organism is 0.3 mm. Drawing reproduced from (Carpenter and Dallinger, 1901) under CC0-1.0.

Phytoplankton cultures are integral components of aquaculture hatcheries. Small ponds that are part of hatcheries can be fertilized to induce blooms of microalgae that serve as live food. However, often closed indoor phytoplankton culture systems such as photobioreactors are used to control the species of microalgae being cultured and exclude larger zooplankton that could prey on or compete with small larvae. A consistent species composition of phytoplankton to be used as live larval food ensures more consistent feed efficiency.

Commercially available concentrates of microalgae are available that can be stored for several months in the refrigerator. They either consist of a single species, for example, *Nannochloropsis spec.*, or a mixture of multiple species, for example, Shellfish Diet 1800® (Reed Mariculture), which contains *Isochrysis spec.*, *Pavlova spec.*, *Tetraselmis spec.*, *Thalassiosira weissflogii*, and *Thalassiosira pseudonana*. These convenient algal concentrates are suitable feeds for many fish and invertebrate larvae. In addition, they can also be used for feeding adult molluscs and other filter feeders as well as zooplankton cultures.

Zooplankton is used as a food source for larger fish and invertebrate larvae. The most common taxa of zooplankters used as live food are rotifers of the genus *Branchionus* (Figure 17.2), brine shrimp of the genus *Artemia* (Figure 11.5), and various copepods (Figure 17.3). These zooplankters are relatively easy to culture, they reproduce and grow rapidly in aquaculture hatcheries, and they have a salinity tolerance that spans freshwater to seawater. Their hardiness and environmental resilience render them suitable live food for both freshwater aquaculture and mariculture.

Branchionus rotifers range between 0.1 and 0.3 mm in size, which is ideal for many fish and invertebrate larvae. In culture, rotifers complete their entire life cycle in just over a week. Culture conditions are kept optimal to encourage rapid asexual reproduction, which results in only females. Sexual reproduction is induced by environmental stress to produce dormant resting eggs, which are metabolically quiescent and encapsulated by a hard shell that prevents dehydration. Resting eggs can be kept dried and out of water for many years and hatched

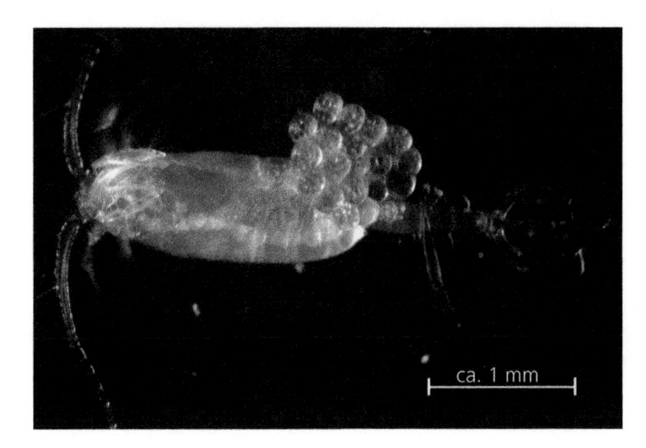

Figure 17.3 Copepod with eggs (blue) attached to the lower thorax/upper abdomen shown on a greatly magnified scale. Photo by Matt Wilson/Jay Clark, NOAA, reproduced under CC0-1.0.

when needed to start a new culture (Clark *et al.*, 2012).

Cultured rotifers are fed either by adding a solution of microalgae from in-house hatchery algal cultures or by adding a small volume of commercially available concentrated microalgae. Batch culture of rotifers in RAS achieves very high densities of 10^4 organisms per mL. In batch culture, rotifers are harvested twice per week. A small volume of the culture is retained as a new starter culture for growing the next batch.

Brine shrimp have a similar biphasic life cycle as rotifers. When environmental conditions are stressful, they also produce resting eggs that can be stored in dried form for many years. Since brine shrimp are crustaceans, their culture is discussed in Section 11.7. Brine shrimp nauplii (0.5 mm) are used as live food for small larvae, while adult brine shrimp (up to 10 mm) are used to feed larger fry.

Copepods are another order of crustaceans with several species that are commonly used as live food in aquaculture hatcheries. The species used for aquaculture range between 0.5 and 2.5 mm in size. In contrast to rotifers and brine shrimp, most copepods do not reproduce asexually and do not have a dormant resting stage. They develop from nauplius to copepodite to adult stages, all of which can be used as live food, depending on the size of the larvae that are being fed.

Large-scale copepod culture under controlled conditions in indoor RAS is still very challenging and most populations consist of a mixture of species in outdoor ponds. Cannibalism represents a problem for many copepod species if different developmental stages are not separated. Despite these challenges, copepods are desirable live food organisms for aquaculture because their nutritional value is higher than that of rotifers and brine shrimp due to higher levels of polyunsaturated fatty acids (PUFAs). Therefore, research on culturing new copepod species and improving culture methods is actively pursued to overcome the challenges pointed out earlier.

Zooplankton used for hatchery production of fish and invertebrate larvae varies in its nutritional value depending on the nutritional value of the organisms used to feed the zooplankton cultures. Ideally, microalgae that are rich in essential amino acids and PUFAs should be used to feed these cultures. When yeast or other food sources that lack PUFAs are used for routine feeding, then emulsion droplets containing high concentrations of essential PUFAs, taurine, *myo*-inositol, and choline are fed for short periods prior to harvesting zooplankton cultures to enrich them with these nutrients.

17.2.2 Formulated feeds

Formulated feeds are produced by mixing all necessary nutrients in the optimal amounts and ratio according to a specific recipe. The mixture is ground into a fine powder, conditioned using steam and pressure, aggregated using binders (mainly starch gelatinization), and then forced through a die under pressure to form pellets of the desired shape and size. The resulting pellets are dried to less than 12% moisture to maximize durability, for example, to prevent mould growth, and facilitate handling. The production of formulated dry feed pellets is ecologically most sustainable if resource use and the associated ecological footprint are minimized. Utilizing organisms that occupy low trophic levels and cannot be used directly for preparing human food are best for minimizing the ecological footprint of formulated feed production. In addition, energy-expensive heating and processing steps for formulated feed processing should be minimized and renewable clean energy sources utilized whenever possible.

The main macronutrient categories in formulated diets are protein, lipid, and carbohydrates. In general, lipids and carbohydrates are considered the main nutrients for providing metabolic energy while proteins are considered the main building blocks for muscle and other tissues (growth). However, this generalization is an oversimplification as lipids are essential components of all cellular and organellar membranes that muscle and other tissues are comprised of and proteins are constituents of non-tissue compounds such as extracellular enzymes, soluble blood components, etc.

Moreover, carbohydrates are attached to proteins yielding glycoproteins that are essential components of cells and the extracellular matrix and, therefore, also represent essential building blocks for growth. Conversely, proteins can be used as a source of metabolic energy, for example, via the biochemical pathway of gluconeogenesis. Half of the twenty common amino acids that comprise tissue proteins cannot be synthesized to the extent needed by aquatic animals. Therefore, they need to be supplied in the diet. The other amino acids can be synthesized *de novo* from precursors through biosynthetic pathways.

Uptake of essential amino acids with food usually takes place in the form of protein that is composed of those amino acids. **Dietary protein** is digested into di- and tripeptides and individual amino acids by a variety of peptidases to facilitate their absorption across the epithelium of the GI tract. Formulated diets enable optimization of the concentrations of each amino acid in addition to total protein content, which contributes to maximizing food use efficiency.

Total dietary protein content is high for most aquatic animals, especially for carnivores. Optimal protein concentrations in the diet are significantly lower in adults than in fry and larvae. Moreover, the protein content of broodstock animals increases by approximately 5% to promote gonad maturation. Carnivorous salmonids require up to 55% protein in their diet for optimal development of fry and 40% as adults.

For omnivores and herbivores, the required protein content is lower than for carnivores (e.g. 45–50% for fry, 30–32% for adults, and 35–37% for broodstock of tilapia and catfish). Although these high protein ratios are optimal in formulated diets, it is equally important to balance the protein requirements of aquaculture animals with their energy requirements, which are met more effectively by lipids and carbohydrates than proteins.

Lipids have the highest chemical energy content per unit mass. Along with other nutrients (carbohydrates and proteins), they can be metabolized to provide chemical energy in the form of adenosine triphosphate (ATP) and reducing equivalents such as NADH and NADPH. Alternatively, lipids are used as building blocks for biological membranes of cells and their organelles. Lipids include three classes: sterols, triglycerides, and phospholipids. Most sterols, such as cholesterol, do not contain fatty acids, while triglycerides and phospholipids are molecules consisting of hydrophobic fatty acids and a hydrophilic head group, which often is either ethanolamine, choline, or inositol. The latter two compounds are essential micronutrients that must be provided in sufficient amounts with the diet. Choline and inositol are also important derivatives of cytoprotective metabolites (CPMs) and compatible osmolytes that protect cells from osmotic and other types of stress (Kültz, 2020b).

Aquaculture animals require essential fatty acids in their diet to synthesize lipids for their biological membranes or derive chemical energy by beta-oxidation. Linoleic acid (18:2ω6) and linolenic acid (18:3ω3) are essential fatty acids that are universally important. Both are polyunsaturated fatty acids (PUFAs, Section 3.3). Some aquaculture species can modify these two essential fatty acids by desaturases and other enzymes to synthesize more highly unsaturated and longer PUFAs in sufficient amounts (De Silva and Wang, 2019). However, many fishes also require eicosapentaenoic acid (EPA, 20:5ω3) and docosahexaenoic acid (DHA, 22:6ω3) in their diet (Figure 3.6).

The optimal amount of PUFAs in the diet ranges between 0.5 and 2% for most aquaculture fishes and invertebrates (D'Abramo, 2019). Overall dietary lipid requirements are higher for carnivores than for herbivores or omnivores because carnivore metabolism has evolved to derive more metabolic energy from lipids and proteins relative to carbohydrates. For example, the optimal lipid concentration in salmonid feeds is 25% compared to only 5–10% in tilapia feeds.

Carbohydrates are commonly present as plant-derived polysaccharides in formulated feeds. These polymers are broken down into their monomeric subunits (glucose, galactose, fructose, and other sugars) and absorbed via the GI tract. Both digestion and absorption of dietary carbohydrates are much more efficient in omnivores and herbivores such as carp, milkfish, and tilapia than in carnivores. The digestibility of plant polysaccharides is a derived trait that has been acquired by only about 15% of all fish species. The majority (85%) of all fishes remain carnivorous and have a limited capacity to utilize plant-derived carbohydrates in their diet. Therefore, the carbohydrate content in formulated feeds of carnivorous species is typically significantly lower (<20%) than that of herbivores and omnivores (up to 45%) (D'Abramo, 2019).

The source of ingredients for formulated aquaculture feeds varies widely and greatly impacts the ecological sustainability of specific feeds. Although the percentage of fish meal and fish oil in formulated feeds has decreased significantly in recent years, these ingredients are still included in most formulated aquaculture feeds (Box 17.3).

Box 17.3 More ecologically sustainable replacements for fish meal and fish oil

Fish meal and fish oil are rich in protein comprised of essential amino acids as well as in lipids and essential polyunsaturated fatty acids (PUFAs) that are required to meet the nutrient needs of aquatic animals, especially high trophic level carnivores. However, the ecological footprint and resource use for producing fish meal and fish oil are high because approximately half of it is currently still derived from capture fisheries, while the other half is contributed by recycling carcasses after filleting. Species used for producing fish meal and fish oil by capture fisheries include menhaden, herring, anchovies, pilchard, capelin, sand eel, and sardines. These species are not captured to provide seafood for direct human consumption but specifically for producing fish meal and fish oil. Even though these species can be consumed directly by humans, efforts over decades to develop products that people will purchase have not increased direct consumption. Therefore, consumer education and preference are important factors for mitigating the unsustainable dependence of aquaculture on fish meal and fish oil. There are four approaches for minimizing the dependence of aquaculture on fish meal and fish oil.

First, the recycling of waste from processing aquaculture and agricultural products and the human food supply chain can be further improved to produce more meals and oils from these wastes as aquaculture food ingredients.

Second, aquaculture of herbivorous and omnivorous animals should be increased to replace the production of carnivores.

Third, plant-based alternatives to animal meals and oils can be further improved. A remaining caveat of using current plant meals and oils is that they lack enough PUFAs and essential amino acids.

Fourth, essential ingredients such as certain amino acids, PUFAs, and vitamins could be produced in cell-based bioreactors.

Caveats that need to be addressed include the cost of purifying specific ingredients from microbial or animal cell cultures and consumer acceptance of feed ingredients derived from genetically modified cells for producing non-GMO organisms.

Besides the three major groups of macronutrients (proteins, lipids, and carbohydrates) formulated aquaculture feeds include a variety of supplements. Common supplements that are often added as premixes are vitamins, minerals, and taurine. These **micronutrients** are required for proper metabolism, enzyme function, respiration, and as antioxidants for relieving stress resulting from the production of reactive oxygen species as a by-product of oxidative metabolism. For example, 0.25 to 1% of a vitamin supplement mixture is used for intensive tilapia aquaculture.

Dietary supplements also include pigments (e.g. astaxanthin, Section 14.1.3), immuno-stimulants, probiotics, and prebiotics. Probiotics are living bacteria that promote nutrient digestion and uptake (e.g. *Bacillus spec.* and *Lactobacillus spec.*). These bacteria can be produced in bioreactors. Their inclusion in formulated feeds requires that no harsh conditions are applied during formulated feed processing, which is also the case when biologically active enzymes are included in feeds to preserve their activity. Prebiotics are chemicals that promote the metabolic activity of probiotic or endosymbiotic bacteria in the GI tract of aquaculture animals. Chemically diverse oligosaccharides are examples of prebiotics that are added to formulated aquaculture feeds.

17.2.3 Nutritional diseases

Nutritional diseases result from the presence of toxic compounds or compounds that interfere with digestion in the diet, from overfeeding, and from deficiencies of essential nutrients. In contrast to infectious diseases, pathogenic organisms (biotic factors) are not causal for nutritional diseases, although malnutrition predisposes aquaculture organisms to secondary infections. Some secondary plant metabolites inhibit metabolism or digestion and must be identified, removed, or inactivated if they are present in plant meals and oils. For example, saponins that are present in soybean meal cause intestinal inflammation of salmon (D'Abramo, 2019). Several other compounds that inhibit digestive enzymes of aquaculture animals are present in sunflower and peanut meals. Identification of these compounds and the development of approaches for their removal or neutralization represents an important area of aquaculture nutrition research.

Secondary plant metabolites, for example, phytic acid, can also cause **deficiencies** in essential minerals. It contains negatively charged (anionic) phosphate groups that sequester divalent cations of essential minerals including calcium, iron, zinc, and magnesium and decrease their bioavailability. Moreover, phytic acid reacts with positively charged amino acids in proteins (lysine, arginine), which results in the formation of insoluble protein complexes and decreased feed efficiency (Wanasundara, 2011). An effective remedy of mineral sequestration by phytic acid has been the inclusion of phytic acid degrading enzyme (phytase) in formulated aquaculture diets.

Overfeeding and too much protein in the diet can cause stress resulting from elevated production of toxic ammonia, which is the main nitrogenous waste produced by catabolism of proteins in ammoniotelic fishes and aquatic invertebrates. In uricotelic aquaculture animals, for example, alligators and turtles, the main nitrogenous waste compound is uric acid. Overfeeding these animals with excess protein leads to gout, which is a painful disease caused by the deposition of excess uric acid crystals in the joints.

To avoid overfeeding, daily monitoring of feeding behaviour is critical in aquaculture. If overfeeding has been diagnosed based on leftover food and a more rapid decrease in water quality, then feeding should be adjusted or cease for a few days. Adjustment of the amounts of feeds provided represents a constant process in aquaculture as food requirements change while animals grow and environmental factors vary in open or semi-closed systems.

In addition to toxic ammonia, reactive oxygen species are generated as toxic by-products of oxidative metabolism. Stress causes an increase in the production of these toxic metabolites, which results in decreased growth rates and impairs the health and welfare of the animals (Section 17.4.1). Excess and deficiencies of any of the major nutrients (protein, carbohydrates, lipids) in food impacts growth and energy metabolism, which are traits that are of central importance for aquaculture productivity. Therefore, much research has focused on optimizing the relative amounts of these macronutrients for aquaculture species and tailoring them to specific developmental stages (Section 17.2.2).

The content of minerals and vitamins has also been optimized in formulated diets to avoid nutrient deficiencies while eliminating excess to maximize feed efficiency. The many syndromes associated with mineral and vitamin deficiencies in aquaculture fish have been summarized previously (Amlacher, 2009). Examples for mineral deficiency syndromes include skeletal abnormalities (lack of calcium, phosphate), muscular dystrophy (lack of selenium), anaemia (lack of iron), cataract formation (lack of zinc), thyroid hyperplasia (lack of iodine), and loss of appetite and renal insufficiency (lack of magnesium). These deficiencies are more common in soft than hard waters because minerals present in hard water can be directly absorbed via the gills and GI tract in addition to ingestion with food.

Common syndromes for vitamin deficiency include anaemia and anorexia independent of which specific vitamin is lacking. However, more specific syndromes that are diagnostic for a deficiency in a specific vitamin have also been described. An example of a vitamin-specific deficiency syndrome is scurvy, which is caused by lack of vitamin C (ascorbic acid). Scorbutic aquaculture salmon have been observed and treated by increasing the amount of ascorbic acid or metabolizable precursors of vitamin C in salmon feeds (Halver and Hardy, 1994). Scurvy was also prevalent in human sailors on long sea voyages during the Middle Ages who did not have access to foods containing sufficient vitamin C.

17.3 Aquaculture waste

Waste in aquaculture systems is produced by 1) unused, decaying food, 2) excreted animal waste, and 3) decaying plant, bacterial, and animal biomass. Decaying plant and bacterial biomass has a significant impact on overall organic waste generation in semi-intensive aquaculture systems that use large amounts of fertilizer. Fertilizer application increases primary productivity of semi-intensive aquaculture systems by stimulating blooms of phytoplankton and subsequently zooplankton at the bottom of trophic pyramids that provide food for aquaculture organisms. However, much of the microbial biomass that is produced by fertilization and responsible for eutrophication is not utilized as food but ends up as decaying organic material.

17.3.1 Fertilizer-induced eutrophication and hypoxia

Common fertilizers used for aquaculture are manure, polyphosphates, and urea. Manure is the least efficient, has high batch variability, and may contain toxins resulting from the metabolization and excretion of pesticides, pharmaceuticals, antibiotics, or other bioactive compounds. Polyphosphates and urea rapidly hydrolyse in water to form orthophosphate (PO_4^{3-}) and ammonium (NH_4^+), respectively. These inorganic ions are taken up as nutrients by plants and bacteria.

Excessive fertilization is common in semi-intensive aquaculture and sometimes necessitates application of algicidal toxins to control algal blooms and species composition (Boyd and Tucker, 2019). Eliminating excess fertilizer and algicide toxin use represents a straight-forward way to increase ecological sustainability while reducing costs and preventing economic losses. In extreme cases, excessive fertilization has caused severe eutrophication and hypoxia that has resulted in massive population crashes of aerobic microbes (bacteria and algae) and fish kills. To prevent hypoxia, fertilization is usually accompanied by aeration in semi-intensive aquaculture. Aeration not only increases dissolved oxygen (DO) concentration but also facilitates escape of carbon dioxide, nitrogen, and toxic ammonia gases into the atmosphere.

17.3.2 Pharmaceuticals and toxins

A variety of other compounds may be present in aquaculture wastewater besides fertilizer, feed remains, faeces, and ammonia. These include hormones, pesticides, antimicrobial compounds, parasiticides, algicides, disinfectant chemicals, anaesthetics, antifouling compounds, and probiotics. Although the amount of these compounds in water discharged from aquaculture facilities is relatively small, their biological effect concentrations are also very low, which means that they can have large effects on organisms even at very low concentration. Some of these chemicals can cause significant global problems, for example, the development of antibiotic resistant strains of bacteria.

Moreover, the illegal application of highly toxic, carcinogenic, or teratogenic chemicals that have the potential to bioaccumulate in edible tissues of aquaculture species is of significant concern. The regulation of using such chemicals and food safety is inconsistent in different countries although in recent years progress has been made towards more uniform regulation across national borders. Nevertheless, enforcement of such regulation is difficult, especially when many smaller scale aquaculture farms need to be monitored.

Contrary to some prevailing opinions, waste discharge regulation should not be based on the dispersal and assimilation capacity of receiving public waters but on the actual amount of waste that is being discharged. Among the many reasons supporting this argument, the dispersal and assimilation capacity of receiving waters is not static but highly dynamic and a snapshot assessment is not suitable for predicting the ecological consequences of waste discharge on ecosystem succession. Another major reason for not basing waste discharge regulation on the waste dispersal and assimilation capacity of receiving waters is that such policies incentivize rather than discourage aquaculture site selection in the most pristine habitats left on earth. Rather than encouraging pollution of these remaining habitats regulatory policies should protect and preserve them.

17.3.3 Waste handling by different aquaculture systems

Aquaculture uses many different approaches that vary greatly in the extent of waste concentration and collection versus dispersal, and whether waste is being recycled within the system or discharged into public waters to rely on ecosystem services. Major aquaculture approaches and their resource footprints are summarized in Figure 17.1. Since many intermediates between these main categories of aquaculture systems exist, comparisons of aquaculture farms are very tedious. The main aquaculture approaches shown in Figure 17.1 are distinguished by waste handling strategies.

Extensive aquaculture uses no supplementary input of energy and matter. Therefore, the ecological balance is maintained even if no waste management is implemented. Essentially all waste generated is part of the natural nutrient cycle of wild ecosystems. For extensive aquaculture, the only variables that alter the amounts of energy and matter in wild ecosystems are the stocking of larvae or juveniles and the harvest of adult organisms. If significant biomass is produced and removed during harvest, then the net nutrient load of natural ecosystems is reduced, which contrasts with a net nutrient increase in public waters receiving waste discharge from all other aquaculture approaches.

Semi-intensive static ponds with minimal flow-through of water are a commonly used system for aquaculture. Waste discharge from these systems occurs via groundwater seepage, rainwater overflow, occasional partial water replacements, and complete pond draining and water discharge during harvest. Sludge and sediment build-up is commonly removed after pond draining. Its disposal into wild habitat can cause significant eutrophication due to high levels of accumulated phosphorus.

Semi-intensive static pond systems afford relatively good opportunities for capturing much of the waste since it is concentrated in a relatively low volume. However, currently most static aquaculture ponds are drained without pre-treatment and accumulated sludge and sediment are discharged into public waters with resulting increases in eutrophication and pollution. Multiple semi-intensive static ponds may be connected in a serial succession and managed by culturing different species or by decreasing stocking densities to account for the gradual decrease in water quality. However, in these static pond systems, water only circulates through each pond once.

Semi-intensive water reuse ponds represent a system of connected ponds that include sedimentation and wastewater treatment ponds. Water is circulated through each pond more than once, ideally continuously if wastewater is treated at capacity to keep water replacement minimal. The size of wastewater treatment ponds in such systems is significantly greater (up to 95%) than that of the other ponds holding the aquaculture animals.

The ecological sustainability of semi-intensive water reuse pond systems is significantly higher than that of static pond systems since most of the waste is being recycled or can be captured and disposed of in a concentrated form. Biofloc technology exemplifies one successful approach for nutrient recycling in these systems (Section 5.5, Box 17.4). Waste discharge from water reuse systems is mainly limited to partial discharge of pre-treated wastewater, groundwater seepage, rainwater overflow, and solids collected in settlement ponds.

Box 17.4 Biofloc and algal turf scrubber technologies

Biofloc technology has been used successfully for shrimp and tilapia aquaculture. The principle underlying this technology is to encourage growth of organic waste decomposers (bacteria) on solid animal waste (feed leftovers and faeces). Bioflocs have a low density, which keeps them suspended in the water column. Decomposing bacteria attach to these solid particles and assimilate both undissolved and dissolved waste into their biomass. Bacterial attachment and growth (proliferation) promotes particle aggregation and increases the size of waste particles by converting them to bioflocs. Thus, bioflocs are aggregates of solid waste and bacteria that function as intrinsic mechanical and biological filters by concentrating solid waste and removing dissolved animal waste to recycle clean water. The bacterial biomass present in bioflocs represents an effective food source for detritivores, for example, shrimp and some omnivorous fishes, which minimizes costs for feeds and fertilizer and maximizes feed efficiency. Moreover, bacterial decomposers present in bioflocs compete with microalgae for certain nutrients present in animal waste (e.g. ammonia) and, thus, counteract water eutrophication. Even if bioflocs are not used as a food source for aquaculture animals, application of biofloc technology offers concentration of solid waste into larger particles and conversion of dissolved organic waste into solid biomass that can be harvested and removed from the water. This principle is also used by Algal Turf Scrubber© (ATS) technology (Figure 3.9). In contrast to freely suspended bioflocs, ATS are microbial mats that consist of both algae and bacteria and are loosely attached to a solid surface, for example, the concrete bottom of wastewater treatment raceways. ATS assimilate dissolved waste into bacterial and algal biomass, which is periodically harvested and used for production of biofuel, agricultural fertilizer, and feed additives. The ecological principle underlying bioflocs and ATS technologies is the same as that of periphyton growth in natural ecosystems (Section 5.4 and Figures 5.1 and 5.2).

Semi-intensive open cage and flow-through systems are the least ecologically sustainable aquaculture systems. Raceways, net pens, and flow-through pond systems rely on large amounts of clean water supply from natural ecosystems. Water exchange rates in raceways and flow-through tanks and ponds are typically three times the volume of the system per day. For trout raceways they are often much higher. Waste that is discharged from open systems is not (or only minimally) concentrated and treated. If gravity cannot be utilized for water movement, then electricity costs for powering water pumps and aerators contribute to resource use.

Cage and net pen cultures are fully immersed in natural waters and all waste is immediately dispersed into the environment. Waste collection methods have been investigated for cages, but none with satisfactory results. Integrated multitrophic aquaculture (IMTA) with bottom-feeding detritivores (e.g. sea cucumbers) and/or seaweed placed under cage areas is the only tangible strategy currently in use to mitigate waste dispersal from cage mariculture (Chapter 5).

High water-intake rates into open flow-through aquaculture systems may impinge or kill aquatic animals present in public waters by drawing them against screens or into water intake pumps, which could alter biodiversity and natural ecosystem succession. Open cage and flow-through systems can achieve very high stocking densities and yields. If this is the case and only manufactured feeds are used, then these systems are commonly classified as intensive systems. However, this book only considers aquaculture systems as intensive if they combine high stocking densities and yields with predominant utilization of wastewater treatment technologies to minimize their spatial and ecological footprints, including the footprint for required ecosystem services.

Intensive recirculating aquaculture systems (RAS) are usually indoor systems that permit very high stocking densities. These systems are technology intensive on the input and output ends of the system and, hence, require investments into infrastructure and maintenance. However, RAS afford full control over all aspects of water quality and offer maximal biosecurity. Clean water use and wastewater discharge are minimal and solid waste is concentrated to facilitate proper disposal.

Resource use of RAS includes electricity for powering pumps, lighting, heating, monitoring, filtration, and sterilization units. It also includes resources needed to build and maintain the required infrastructure. The use of clean, renewable energy sources and ecologically sustainable methods for manufacturing RAS infrastructure minimizes the ecological footprint of this aquaculture approach. RAS are not always operated as intensive systems but can also be operated as semi-intensive systems, for example, aquaponics (Section 5.5) and ornamental aquaculture in home or public aquaria (Chapter 12).

RAS use water treatment systems that are more space efficient than settling ponds and wastewater lagoons. Much research is currently devoted to improving the recycling capacity and space efficiency of RAS filtration systems. One challenge with RAS is accumulation of essential minerals that are toxic at high concentrations in water, for example, copper and zinc. For this reason, feeds used in RAS and wastewater treatment methods must be tailored to account for accumulation of such micronutrients.

Aquaculture wastewater treatment includes five principal steps, which are discussed in the next sections:

1) Removal of solids (undissolved waste) by mechanical filtration;
2) Breakdown of ammonia and organic waste and assimilation into bacterial and plant biomass;
3) Removal of dissolved organic and inorganic ions by chemical filtration;
4) Removal of gases (ammonia, nitrogen, and carbon dioxide) by degassing and, if necessary, adjustment of pH, salinity, and water hardness; and
5) Sterilization by UV or ozonation.

17.3.4 Mechanical removal of particulate, undissolved waste

Animal aquaculture generates similar waste to that in municipal wastewater and waste from animal agriculture. Wastewater contains compounds from unused food, organic fertilizer, animal excrements, microorganisms that metabolize animal waste and fertilizer, bioactive chemicals and pharmaceuticals, inorganic fertilizer and pesticides, and undissolved sediment and sludge. This waste consists of dissolved and undissolved compounds. Undissolved waste is comprised of suspended solid particles that increase turbidity when their density is equal to or less than that of water, in which case they stay in the water column or float on the water surface. These light-weight solids cannot be collected by sedimentation but must be removed by sieves and skimmers. Undissolved waste compounds that are heavier than water will sink and accumulate at the bottom.

Undissolved waste can be removed mechanically if appropriate capture or collection methods are implemented in the aquaculture system. The disposal of this waste in RAS and other water reuse systems can occur in a much more controlled manner than simply pumping it out into natural ecosystems, which is still the predominant method of disposal for open pond aquaculture systems. For cage, net pen, and other open aquaculture systems all solid waste is lost to the environment, which is why sediment under semi-intensive, high-density cage and net pen sites is often overburdened and benthic communities are severely affected.

Promising strategies for facilitating waste capture and treatment for raceways, cage, and net pen systems have been developed. They include in-pond raceway systems (IPRS) and floating 'bag' systems. The latter are similar to net pens where the nets have been replaced with an enclosed bag, which allows for waste to settle at the bottom and removal into collection systems by low-head pumps. Fallowing, a strategy implemented to support recovery of benthic communities at cage and net pen sites, refers to moving aquaculture cages out of an area for certain periods (e.g. one or more years) to alleviate pressure on ecosystem services in that area and permit recovery of water and sediment quality.

Pond-based water reuse systems typically use a sediment settling pond and a subsequent waste-water treatment lagoon or reservoir for water reconditioning. Sedimentation canals can also be used to collect sediment and sludge resulting from faeces, food remains, and other organic waste. Waste-water treatment ponds and canals may take up 95% of the space of water reuse systems with only 5% used to grow aquaculture animals. Seepage and overflow from sedimentation ponds and canals is avoided by periodically draining them to remove the collected solid waste. For RAS, sumps, which are containers that are positioned at the lowest level in the system to collect wastewater by gravity, fulfil a similar role as sediment traps.

Sediment quality is an important indicator of ecosystem health and should be routinely monitored in relevant aquaculture systems. Sampling of sediment at strategic positions within aquaculture systems should be undertaken in regular intervals to analyse its properties. Sediment cores are commonly tested for hydrogen sulfide (H_2S) content, redox potential, and by visual inspection to estimate the proportion of oxidized versus reduced compounds present and the ratio of anaerobic versus aerobic bacteria.

Sludge accumulates as solid waste under hypoxic conditions when dissolved oxygen concentration and redox potential in sediment are low. Redox potential is a measure of the loss of electrons by oxidation in millivolts (mV). Since electrons are negatively charged, a high redox potential is indicative of a highly oxidized (i.e. well oxygenated), sediment. A negative redox potential indicates that chemicals present in sediment are in their reduced state due to poor oxygenation and electron gain.

Organic waste is decomposed by aerobic bacteria that utilize oxidation reactions requiring dissolved oxygen. If aerobic bacterial biomass and metabolism are high due to high concentrations of organic waste, then oxygen will be consumed quickly and hypoxia may result, especially in sediment with poor water circulation and gas exchange. Under these conditions anaerobic bacteria proliferate and decompose organic waste by fermentation reactions. Fermentation does not completely recycle organic waste to inorganic compounds and produces sludge and toxic compounds such as hydrogen sulfide gas (H_2S).

Sediment and sludge collected periodically from aquaculture systems can be disposed of in an ecologically sustainable manner after drying by evaporation of the included water. Disposal methods for concentrated solid aquaculture waste include application on agricultural lands as fertilizer, deposition in landfills, use as earth fills at construction sites, or incineration and processing into construction bricks. The sediment and sludge collected from mariculture systems has a high salt burden and is more difficult to dispose of in an ecologically sustainable manner than solid waste from freshwater systems. This difficulty is another reason for the greater potential of freshwater aquaculture systems regarding the ecologically sustainable development of aquaculture.

17.3.5 Biological recycling and removal of dissolved waste

Dissolved metabolic waste of heterotrophic aquatic animals consists mainly of inorganic ammonia, organic compounds rich in nitrogen and phosphorus, and carbon dioxide. Dissolved waste cannot be removed mechanically but requires biological removal or chemical filtration. Biological removal is based on assimilation of dissolved waste compounds into bacterial and/ or algal biomass. Many secondary metabolites are included in aquaculture waste that are composed of both major and trace elements, including N, C, O, H, S, P, K, Na, Cl, Si, Ca, Mg, F, B, and many others. Fully balancing trophic webs in RAS requires complete recycling of the quantities of all elements into suitable chemical forms for input matter from the entirety of output matter, which is very difficult for artificial (human-managed) systems, even for the simplest model system (Section 4.4).

Figure 17.4 summarizes the major aspects of the nitrogen cycle in aquaculture systems. Nitrogen added with fertilizer and feeds is incorporated into plant, animal, and bacterial biomass in the form of organic nitrogen (protein and other macromolecules). However, much of it is eventually decomposed into ammonia rather than being present in animal biomass harvested from aquaculture. Non-ionic ammonia (NH_3) is a gas that is toxic to aquatic organisms. It can escape from water into the atmosphere, especially under alkaline conditions. However, ammonia present in water also transforms into its less toxic ionic form, ammonium (NH_4^+), which cannot escape into the atmosphere. Dissociation of ammonia into ammonium ion increases pH by producing hydroxyl ions (OH^-) that sequester hydrogen ions (H^+) to form water. Total ammonia nitrogen (TAN) is the sum

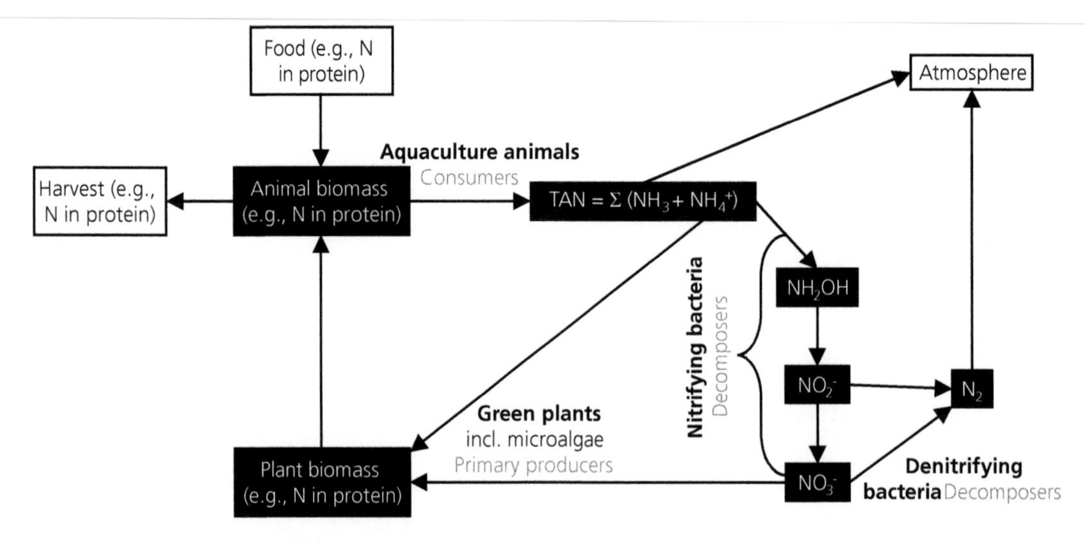

Figure 17.4 Major aspects of nitrogen cycling in aquaculture systems under aerobic conditions. Nitrifying bacteria oxidize ammonium to nitrite (e.g. *Nitrosomonas spec.*) and further to nitrate (e.g. *Nitrobacter spec.*). Denitrifying aerobic bacteria convert nitrite and nitrate to gaseous nitrogen. Black boxes depict components that are recycled within the system. White boxes indicate inputs and outputs. Under anaerobic conditions some ammonia can be oxidized by reacting with nitrite to form gaseous nitrogen. TAN = total ammonia nitrogen.

of NH_3 and NH_4^+, which represents the nitrogenous waste excreted by ammoniotelic aquatic animals.

Depending on its concentration in water, ammonia can:

1) enter fish and some aquatic invertebrates via the gills;
2) be assimilated into plant biomass as an essential inorganic plant nutrient; and
3) be sequentially metabolized into nitrite (NO_2^-), nitrate (NO_3^-), and elemental nitrogen (N_2) by different types of bacteria (Figure 17.4).

Ammonia is toxic to all animals, as it interferes with neuronal function, cell metabolism, and acid-base balance. Since its non-ionic form (NH_3) is more toxic than ionic ammonium (NH_4^+) and the ratio of ionic to non-ionic ammonia depends on pH, ammonia toxicity is pH dependent.

Several test kits for monitoring and adjusting the concentrations of ammonia, nitrite, and nitrate in aquaculture facilities have been developed to prevent ammonia toxicity. Ammonia toxicity varies between aquaculture species and depends on the environmental conditions. In cold freshwater, ammonia toxicity thresholds are generally lowest while they increase in warmer water and at higher salinity. Most test kits are based on colorimetric quantitation methods (Figure 17.5, Box 17.5).

Phosphorus added with fertilizer into water dissociates into its anionic forms—dihydrogen phosphate ($H_2PO_4^-$), hydrogen phosphate (HPO_4^{2-}), and phosphate (PO_4^{3-})—and is sequestered rapidly into sediment and particulate waste. Depuration of aquaculture pond sediments enriched with phosphorus is slow and continues to cause eutrophication even in the absence of any external nutrient input. Organic phosphorus-containing compounds present in food waste and excreted as animal waste are converted to inorganic phosphates by bacteria. Depending on whether aerobic conditions or anaerobic conditions prevail (e.g. in sediment and sludge of aquaculture ponds), different species of bacteria and chemical reactions are involved in the conversion.

Even though complete nutrient recycling may not be feasible with current RAS technologies, very high recycling efficiencies can be achieved with clever systems designs that utilize innovative biological filtration approaches, for example, bioflocs, ATS, bio balls, trickle filters, in conjunction with strains

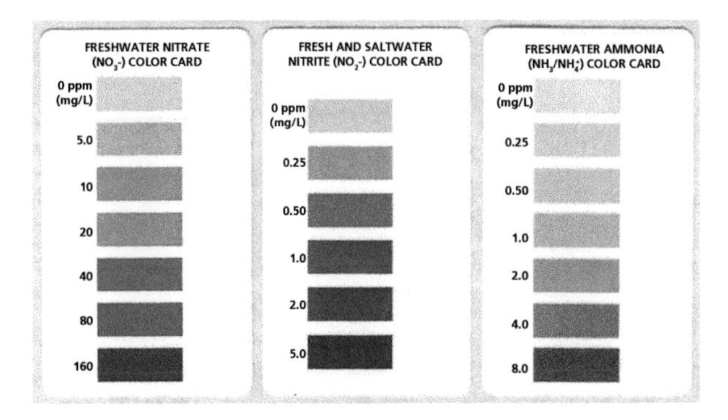

Figure 17.5 Indicator dye-based colour scales for quantifying the concentrations of nitrate (left), nitrite (centre), and total ammonia nitrogen (TAN) in water. Photo by Elizabeth Mojica.

of bacteria and algae that have been optimized for assimilation of aquaculture waste and represent suitable secondary food sources for aquaculture animals. A major challenge with waste assimilation, that is, biological filtration, is that the volume of water and surface area required for quantitative assimilation of waste from semi-intensive and intensive animal aquaculture facilities is enormous. The space required may be a hundredfold or more than that used for the actual production facility. RAS provides opportunities for research and development of biological filter designs that maximize space efficiency to meet this challenge.

Box 17.5 Water quality tests for ammonia, nitrite, and nitrate

Different ammonia test kits are available for freshwater and marine aquaculture systems. Colorimetric ammonia tests and electronic ammonia probes can be used for monitoring total ammonia nitrogen (TAN = $NH_3 + NH_4^+$). The Nesslerization method utilizes a reaction of ammonium with the tetraiodomercurate anion in an alkaline solution (Nessler's reagent) to form a yellow-brown product. The substrate used for this reaction is toxic and handling and disposal of test reagents and solution require extra care. A less toxic alternative to determining ammonia concentration is the Marcelin–Berthelot method. Ammonium ions and phenol react in the presence of an oxidizing agent (hypochlorite, common bleach), at high pH (11–12) to form a vivid blue colour (indophenol blue). For both tests, the colour intensity of the reaction product is proportional to the concentration of ammonia. It can be estimated by checking an intensity scale that is provided with the test kits. Alternatively, colorimetric quantitation can be performed by measuring absorbance at the corresponding wavelength in a spectrophotometer. The Griess test is a colorimetric test used for determining the concentration of nitrite. In this method, a benzene derivative (like phenol or an aromatic amine) is reacted with a diazonium salt forming an N=N bond to produce an intensely coloured (red-violet) azo dye in the presence of nitrite. Again, the intensity of the dye is proportional to the concentration of nitrite, which can be estimated based on colour scale provided with the kit or with a spectrophotometer. Nitrate concentration is also determined using the Griess test except that one additional reaction step is included to first reduce nitrite to nitrate. Cadmium metal ions can be used as the reducing agent. However, since cadmium is highly toxic, alternative reducing agents (e.g. zinc) are now more commonly used even though they may be less potent than cadmium.

An alternative to complete biological recycling of dissolved waste materials is their chemical removal using filter materials that absorb specific chemical waste compounds. Depending on the waste present, a combination of different filter materials may be used. Activated charcoal (carbon) and various ion exchange resins are examples of commonly used filter materials. Dissolved chemical waste is greatly concentrated on filter resins while being removed from the water but the filter resin will eventually be saturated and needs to be reconditioned or disposed of and replaced in regular intervals. Proper disposal of the concentrated chemical waste collected on saturated filter cartridges is much easier than handling waste that has been dispersed into a large volume of public waters.

17.3.6 Wastewater reconditioning by gas exchange and sterilization

Facilitating the exchange of gases between water and air represents an important aspect of wastewater reconditioning in aquaculture. Gas exchange management includes degassing strategies to promote removal of carbon dioxide, ammonia, and supersaturated nitrogen. The other aspect of gas exchange management is oxygenation (aeration, Section 16.5). Devices for enhancing gas exchange minimize the concentrations of carbon dioxide (CO_2), ammonia (NH_3), and nitrogen (N_2) gases, maintain saturated oxygen concentrations, and avoid supersaturation of water with oxygen and nitrogen.

Supersaturation of gases in water must be avoided because it causes gas bubble disease in aquaculture animals. Supersaturation of water with nitrogen can occur when excess nitrogen is produced by denitrifying bacteria (Figure 17.4). Oxygen supersaturation results from high levels of photosynthesis by phytoplankton, for example, in semi-intensive eutrophic ponds. Bubbles start to form as soon as the concentration of these gases exceeds their water solubility. Supersaturation with nitrogen and oxygen also results from rapid changes in water temperature or mixing of water having different temperatures. The concentration of gas dissolved in water is measured using electronic devices and probes that are specific for each type of gas. Modern oxygen meters use oxygen probes that are highly sensitive and automatically correct oxygen concentrations for temperature and salinity (Figure 17.6). The concentration of oxygen and other gases in water is commonly expressed as mg/L, that is, mg of gas dissolved in one litre of water.

When using mg/L as a unit then it is nearly identical to parts per million (ppm) mass units of gas in water because 1 litre of water weighs approximately 1 kg. However, since water used in aquaculture contains dissolved ions (dissolved salt) the weight of 1 L is slightly different from 1kg, depending on whether freshwater, brackish water, or seawater is used.

Another way to express the concentration of a gas in water is its partial pressure (P). The total pressure of an ideal gas mixture is the sum of the partial pressures of the gases in the mixture. The SI unit for partial gas pressure is kilo Pascal (kPa). The partial pressure of a gas in water can be calculated from its concentration (C in mg/L) by using the Bunsen coefficient (β), which is specific for the type of gas, temperature and salinity, and a gas-specific constant (A) (Lekang, 2019). For example, the partial pressure of 5 mg/L oxygen in freshwater at 20 °C would be: $P(O_2) = C(O_2)/\beta(O_2) \cdot A(O_2) = 5/0.031 \cdot 0.071 = 11.5$ kPa.

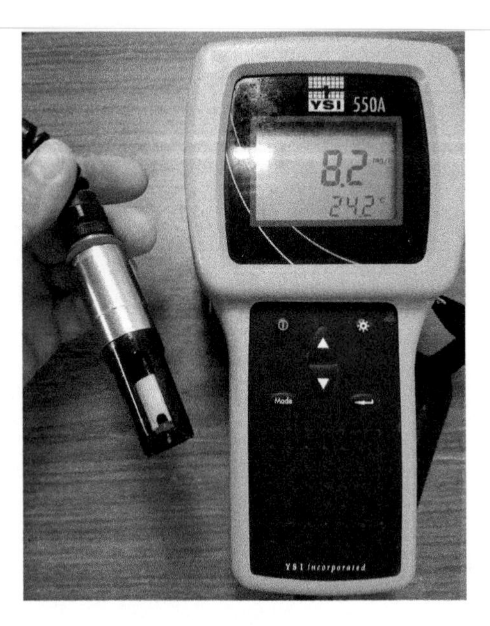

Figure 17.6 Oxygen meter (YSI) with probe used in the author's laboratory for daily measurements of dissolved oxygen concentration (DO) in fish tanks. Photo by Jens Hamar.

Gas transfer across the water/air interface is explained by the two-film theory (Lewis and Whitman, 1924). According to this theory, a microlayer or thin film exists at either side of the air/water interface. These two films inhibit gas exchange between water and air. Turbulence at the air/water interface reduces the thickness of these films and promotes diffusional gas exchange. Therefore, increasing the surface area for diffusion and creating turbulence at the air/water interface are the two main goals of gas exchange devices, including degassing devices and the different types of aerators outlined in Section 16.5.

Degassing devices are sometimes added to RAS units but in many cases biological filters are sufficient for degassing, especially if trickle filters are used. Bubble aerators used for infusion of oxygen or ozone also facilitate removal of supersaturated nitrogen, carbon dioxide, and ammonia by diffusion. Degassers utilize counterflow in space-efficient vertical towers to facilitate diffusional removal of carbon dioxide and other gases. They use packed media that are made of various types of plastics to maximize the area of the air/water interface and turbulence at that interface.

For indoor RAS aquaculture, oxygenation can be fully controlled by infusion of compressed oxygen from liquid oxygen tanks. The rate of oxygen infusion and its dispersion by bubble aeration is carefully adjusted to saturate the water with oxygen while avoiding supersaturation and minimizing loss of infused oxygen to the atmosphere.

Ozone (O_3) is sometimes used for oxygenation instead of oxygen. Ozone imposes severe oxidative stress that kills most organisms present in the water by damaging their DNA and other macromolecules, that is, ozone serves a dual purpose for water oxygenation and sterilization by killing pathogens. Therefore, ozonation of wastewater must be performed in separate treatment tanks and time for ozone degradation must be allowed before reintroducing ozonated water into tanks containing aquaculture animals. The oxidative stress is transient because ozone rapidly decays to oxygen, which is important as water must be purged of ozone before it is supplied to tanks holding aquaculture organisms.

Ozone decays into oxygen by reacting with single oxygen atoms ($O_3 + O \rightleftharpoons 2\ O_2$). The half-life of ozone is temperature dependent and in the range of a few minutes at temperatures that are applicable for aquaculture (Harding *et al.*, 2013). Therefore, the residence time of ozone-treated water in holding tanks must be calculated carefully before supplying it to tanks containing animals.

Irradiation with low-wavelength ultraviolet (UV) light is another common method of sterilization used in aquaculture. Like ozone, low-wavelength UV irradiation causes severe oxidative stress and damages the DNA and other macromolecules of pathogens. The time of exposure to UV can be adjusted to provide effective sterilization while minimizing the residence time of water in sterilization devices.

Ozonation followed by UV radiation represents an effective sterilization strategy because UV light splits O_2 molecules into single oxygen atoms (O). An increase in the concentration of single oxygen atoms accelerates the removal of ozone by its conversion into molecular oxygen (O_2) through the reaction given. The use of ozonation and UV sterilization in RAS units greatly increases the biosecurity of aquaculture farms.

17.4 Minimizing stress and maximizing welfare of fish and invertebrates

Aquaculture organisms, like any others, perform best when they are kept in an optimal environment. Stress and diseases are important factors for management of all aquaculture systems. They strongly affect profitability and environmental sustainability and must be managed properly in semi-intensive and intensive aquaculture systems. Appropriate water quality and diets are key requirements for minimizing stress and maximizing growth and the welfare of aquaculture animals. Effects of the many different types of stress on aquaculture animal performance and welfare are multifaceted and can only be sketched out roughly in the remainder of this chapter. Interested readers will find excellent and more detailed information on these topics in Iwama *et al.*, 2012 and Martos-Sitcha *et al.*, 2020.

17.4.1 Common stresses and their symptoms encountered in aquaculture

Stress is an elusive term and concept that is ubiquitously used but hard to define. However, properly discussing, and analysing, effects of stress on organisms is not possible without first defining what we mean when referring to stress. Stress has been defined as the effect of environmental forces acting on an organism, which causes a strain on the organism (Kültz, 2020a). This definition is equivalent to the stress definition used by physicists for forces acting on an object. However, biologists often substitute the term stressor in lieu of stress (as used in physics) and the term stress in lieu of strain (as used in physics). This inconsistent use of terminology has caused much confusion. Regardless, the ultimate causal threat associated with stress is macromolecular damage leading to dysregulation of physiological systems and cellular and organismal malfunction (Kültz, 2005).

Adverse effects of stress on cellular and organismal function are counteracted by stress response mechanisms that have evolved in all organisms to maintain physiological homeostasis, that is, maintain physiological variables within a narrow range and in global equilibrium. Some species have more potent stress response mechanisms than others. They are referred to as eurytopic species to distinguish them from stenotopic species, which have less potent stress response mechanisms.

For aquaculture management it is critical to know whether the cultured species is eury- or stenotopic, to which types of environmental stress it is most susceptible, how to assess stress response mechanisms as indicators of stress, and how to manipulate stress response mechanisms to alleviate the consequences of stress. The most common types of stress encountered by aquaculture animals are due to four abiotic environmental factors—hypoxia, temperature, salinity, and pH fluctuations—and a suite of logistical and biotic factors, including confinement and crowding, transport, toxic waste, disease organisms (pathogens), predation and cannibalism, intraspecific competition, and handling for disease monitoring, vaccination, sorting, grading, etc.

Stress response mechanisms are energy expensive processes that utilize large amounts of chemical energy in the form of ATP and reducing equivalents. This energy is diverted from other physiological processes, notably digestion, growth, and reproduction, since the total energy budget remains constant or is reduced due to inhibition of feeding activity during stress. Stress-induced reallocation of metabolic energy represents an undesirable outcome for aquaculture.

Aquaculture organisms exposed to stress display characteristic symptoms. These symptoms are manifested at three major levels where stress responses have evolved:

1) the cellular stress response (CSR);
2) the neuroendocrine stress response (NESR); and
3) the psychological and emotional stress response (PESR) (Kültz, 2020a).

For aquaculture organisms, almost all research and consideration has focused on the NESR, although investigations of CSR mechanisms are becoming more commonplace and the existence and potential relevance of PESR mechanisms in fish are being investigated in the context of animal welfare.

NESR stress responses are particularly prevalent in vertebrates, including all aquaculture fish, but they are underdeveloped and incomplete in invertebrates. For this reason, institutional animal care and use committee (IACUC) approval of experiments on animals in academic institutions is not required for invertebrates. Common symptoms associated with the NESR in fish and other vertebrates have been summarized under the term General Adaptation Syndrome (GAS) (Selye, 1936). Three distinct stages are associated with GAS:

1) The alarm stage.
2) The resistance stage.
3) The exhaustion stage.

Table 17.1 summarizes the symptoms associated with each stage that can be used to assess the level of stress in an organism.

17.4.2 Indicators of stress

Symptoms of stress include altered behaviours that can be identified when monitoring animals, for example, during feeding or water quality checks. These behaviours include abnormal feeding activity, erratic or lethargic swimming behaviour, aggressive or fearful behaviours, and gulping at the water surface. In addition, feeding represents an opportunity to inspect animals visually and check for lesions or changes in appearance that are indicative of stress or disease (e.g. frayed fins, changes on body coloration, deformities). It is important to regularly monitor aquaculture animals for signs of stress and disease to diagnose potential problems early and initiate treatments as soon as possible. Therefore, periodic assessments are conducted on many aquaculture farms by randomly sampling a few animals and collecting blood samples.

Table 17.1 Stages of Selye's general adaptation syndrome (GAS) and neuroendocrine stress response (NESR) symptoms associated with each stage

GAS stage (approximate duration)	NESR Symptoms
1. ***Alarm stage*** (seconds to minutes)	• Activation of the sympathetic nervous system • Increased secretion of corticotropin releasing hormone (CRH) from the hypothalamus • Mobilization of chemical energy from readily accessible long-term stores • Inhibition of deposition of chemical energy into long-term stores • Increases in heart rate, blood pressure, and respiration rate • Increased alertness • Increased secretion of catecholamines (adrenaline, noradrenaline) • Increased secretion of analgesics and promotion of local analgesia to increase the pain threshold • Increased levels of blood glucose and oxygen to fuel oxidative metabolism required to counteract consequences of stress
2. ***Resistance stage*** (hours to months)	• Increased secretion of cortisol from the inter-renal gland, which can cause enlargement (hypertrophy) of this gland • Activation of glycogenolysis in the liver • Reduction of glucose uptake in vegetative and storage tissues • Increased lipolysis in adipose tissue • Inhibition of antibody production and number of circulating lymphocytes but activation of other aspects of the immune system (immunomodulation, increased disease susceptibility) • Activation of gluconeogenesis in the liver and, for prolonged stress, also in muscle, which can cause atrophy of these tissues • Reduction of gonadal mass, ceasing of growth and, in severe cases, atrophy causing negative growth (weight loss) • Inhibition of reproduction
3. ***Exhaustion stage*** (minutes to days)	• Resources are being depleted because of faster use of chemical energy than replenishment from feeding • Energy storage is completely inhibited • Rapid fatigue and chronic metabolic syndromes arise due to dysregulation of energy metabolism and hormones (e.g. diabetes) • Sustained increases in blood pressure cause heart disease and strokes • Reproductive disorders develop • Hippocampus neurons are destroyed leading to behavioural and memory disorders • Sever dysregulation of the immune system • Death if severe chronic stress continues

Although many molecular indicators of stress have been identified and characterized in detail, the easiest to use routinely for aquaculture are those that can be measured in blood samples. Blood samples are conveniently collected relatively non-invasively with minimal handling. Common indicators of stress that are being measured in blood plasma samples from aquaculture fish are associated with stage 2 of Selye's GAS (Table 17.1), namely cortisol concentration, glucose concentration, circulating lymphocyte counts, and haematocrit.

Haematocrit is the percentage of solid (red and white blood cells) to the sum of liquid (plasma) and solid blood. This ratio can be easily measured by drawing a small blood sample into a fine capillary glass tube, which is then centrifuged to pellet solid blood cells and separate them from plasma. The interface between solid and liquid blood indicates the haematocrit, which can be read off a graded scale on the capillary glass tube. Haematocrit increases during stress because stress hormones (e.g. cortisol) stimulate haematopoiesis, which is the process of producing red bloods. Haematocrit depends on species, food, and culture conditions but, as a rule of thumb, it is generally less than 50% in unstressed fish and increases to well above 50% in stressed fish.

Unlike erythrocytes (red blood cells) and the neutrophil subpopulation of white blood cells, which account for most blood cells, a specific minor population of white blood cells (lymphocytes) involved in the acquired immune response (adaptive immunity, cell-based immunity) of vertebrates decreases in number during stress. The name lymphocyte derives from the high concentration of these cells in lymph nodes and lymphatic fluid. They include natural killer cells, T lymphocytes, and B lymphocytes (Chapter 18). Both the contribution of lymphocytes to total white blood cell counts and the ratio of lymphocytes to neutrophils decrease during stress. These ratios are sometimes referred to as stress leukograms and can be determined by selective staining of lymphocytes and neutrophils in combination with microscopic analysis using a haemocytometer or automated scanning cytometer.

Lymphocytes are distinguished from neutrophils by staining with Wright–Giemsa stain composed of a mixture of methylene blue, azure A and B dyes, and eosin. After staining, lymphocytes appear purplish blue while neutrophils are more pinkish in colour. Typical ratios of lymphocytes to total white blood cells are between 40% and 80% depending on species, food, state of development, and environmental conditions. Stress-induced decreases of lymphocyte count by 20% have been observed in fish.

Cortisol is an important stress hormone in vertebrates. Unlike adrenocorticotropic hormone (ACTH), corticotropin releasing hormone (CRH), and other stress hormones that are genetically encoded peptides, cortisol is a steroid hormone. Its secretion from the inter-renal gland of fish is stimulated by ACTH, which is produced in the anterior pituitary gland in response to hypothalamic CRH, cortisol is released into the systemic circulation. Its concentration in blood increases robustly during stress and is usually measured by enzyme-linked immunosorbent assay (ELISA).

Another common indicator of stress that is easily quantified in blood samples is the concentration of glucose. Stress releases glucose from long-term polysaccharide stores (e.g. glycogen), promotes its synthesis via gluconeogenesis, and increases its secretion into blood to provide chemical energy that is delivered to tissues with an increased energy demand during stress (heart muscle, brain, haematopoietic tissues, several glands, etc.). Glucose concentration in whole blood or blood plasma samples is quantified using a glucometer, which is a small and relatively inexpensive instrument that requires only a tiny volume for accurate measurement. Glucose concentration in blood is commonly expressed as molarity (mmol/L) or as the fractional mass (mg/dL). Baseline blood glucose levels and the extent of glucose increase during stress differ greatly depending on species and conditions.

17.4.3 Welfare of aquatic animals

Publications related to animal welfare were only available for 84 of the 408 species of aquatic animals used for aquaculture in 2018 and much of the available information has been criticized as providing a 'a production-oriented lens to welfare' (Franks *et al.*, 2021). Thus, opportunities for research on the effects of captive culture stress on the welfare of aquaculture species abound. Animal welfare efforts often focus on a concept labelled the five freedoms:

1) from hunger and thirst;
2) from discomfort;
3) from pain, injury, or disease;
4) from fear and distress; and
5) to express normal behaviour.

This concept is meritorious but, apart from the first aspect (hunger and thirst) far from straight-forward to apply to aquatic animals. Species-specific research on how to define and detect discomfort, pain, fear, distress, and normal behaviour of aquaculture animals is essential for properly applying the five freedoms concept to each aquaculture species.

Animal welfare efforts for aquatic animals have primarily focused on fish because they are vertebrates with a fully functional NESR and predecessors of a PESR (Section 17.4.1). Invertebrates lack a complete NESR although several evolutionarily advanced invertebrates display precursor molecules and other elements of NESR predecessors. Because a fully developed NESR associated with pain perception and emotional processing is lacking in invertebrates, they are exempt from approval of experimental procedures by Institutional Animal Care and Use Committees (IACUC) of academic institutions. Nevertheless, animal welfare efforts are not limited to fishes but are also increasingly implemented for invertebrates. For example, environmental enrichment strategies have been devised for cephalopod aquaculture (Cooke *et al.*, 2019).

Increased stocking densities represent challenges for maintaining environmental conditions in captivity that resemble those encountered by aquaculture animals in their native habitat. If these conditions are different from the optima that the species of interest has evolved under, then stress and disease susceptibility are increased, and animal welfare is impacted. Unlike many species of domesticated terrestrial animals, most aquaculture species have not been domesticated for significant amounts of time on an evolutionary timescale. Therefore, differences between natural and captive environments of aquatic species have a high potential to significantly impair animal welfare.

The effect of domestication on animal welfare is multifaceted. On the one hand, domestication selects for lower levels of stress during handling, higher disease tolerance, and better physiological performance (growth and reproduction). On the other hand, domestication limits the genetic and behavioural diversity of populations and selects for high performance in more uniform, stable environments. For example, selection for increased growth was associated with decreased salinity tolerance of Nile tilapia (Hasan *et al.*, 2014).

Aquatic animal welfare can be improved by developing and implementing less invasive procedures for routine monitoring of health status. For example, molecular indicators measured in mucus secreted from external epithelia and skin can be used as a non-invasive alternative to analysing blood samples to assess stress and the health status of aquatic animals (Martos-Sitcha *et al.*, 2020).

Efforts for improvement of welfare aspects of animal aquaculture have included the investigation and development of more humane methods for harvest and processing to reduce suffering and pain of aquaculture species. It has been recommended that greater emphasis be placed on selecting aquaculture species with the lowest welfare and environmental risks, for example, aquatic plants and phylogenetically lower invertebrates, in lieu of carnivorous fishes that have the highest risks in this regard (Franks *et al.*, 2021). Fortunately, aquaculture species that are superior from an ecological sustainability perspective based on their low position in trophic cascades are generally also the most suitable species from an animal welfare perspective.

Key conclusions

- The main input of energy and matter (excluding water and oxygen) in aquaculture systems are fertilized eggs, food, and fertilizer, while waste and harvested organisms represent the main output.
- Aquaculture systems can be classified based on stocking density, yield, space efficiency, and how inputs are balanced with outputs, that is, to what extent ecosystem services or RAS technology are used.
- Ecologically sustainable management of aquaculture waste employs two different strategies that may contribute in variable degrees to overall waste treatment: 1) collection, processing, and proper disposal of concentrated waste and 2) trophic recycling of waste within the system.
- Aquaculture sustainability is increased when feed efficiency is maximized and feed conversion ratio minimized by optimizing the composition, consistency, and application of food.
- Live food is particularly important in hatcheries for feeding larval stages of aquaculture animals. It consists of phytoplankton (microalgae species) and zooplankton (rotifers, brine shrimp, and copepods).
- Research on many fish and invertebrate species has resulted in formulated feeds with optimal amounts and ratios of amino acids, proteins, lipids, carbohydrates, minerals, vitamins, and other essential nutrients. The key is to maximize the *available* rather than the *total* nutrient levels and balance while minimizing ingredients that are unfavourable, for example, some secondary plant metabolites.
- An optimally nutrient-balanced diet based on formulated feeds and avoidance of overfeeding minimize the incidence of nutritional diseases and maximize the health of aquaculture animals.
- Aquaculture waste is produced from 1) unused, decaying food, 2) excreted animal waste, and 3) decaying plant, bacterial, and animal biomass, which is exacerbated by application of fertilizers.

- Intensive RAS and extensive aquaculture are the most ecologically sustainable forms while open, cage, raceway, and net pen aquaculture are the least ecologically sustainable forms of aquaculture.

Aquaculture wastewater treatment includes five principal steps:

1. Removal of solids (undissolved waste) by mechanical filtration and sedimentation.
2. Breakdown of ammonia and organic waste and assimilation into microbial biomass.
3. Removal of dissolved organic and inorganic ions by chemical filtration.
4. Removal of gases (ammonia, nitrogen, and carbon dioxide) by degassing
5. Oxygenation and sterilization by ozonation and/ or UV irradiation.

- Stress impairs the welfare and performance of aquaculture organisms; it results from their exposure to different environmental forces (water quality) in captivity compared to their natural environments.
- Universal, non-specific symptoms of stress that are easily monitored are used as indicators of animal health and well-being; they include animal behaviour and appearance, haematocrit, lymphocyte counts, and the concentrations of cortisol and glucose in whole blood or plasma.
- Research on aquaculture animal welfare is still sparse but indicates that species which are superior from an ecological sustainability perspective are also most suitable from an animal welfare perspective.

References

Amlacher, E. (2009). *A textbook of fish diseases*. New Delhi: Narendra Publishing House. Available at: https://www.biblio.com/book/textbook-fish-diseases-erwin-amlacher/d/603884012 (accessed: 18 August 2021).

Boyd, C. E. and Tucker, C. S. (2019). 'Water quality', in Lucas, J. S., Southgate, P. C., and Tucker, C. S. (eds.) *Aquaculture: farming aquatic animals and plants*, 3rd edn. Hoboken: John Wiley & Sons Ltd, pp. 63–92.

Carpenter, C. B. and Dallinger, W. H. (1901). *The microscope and its revelations*, 8th edn. Philadelphia: Blackiston's Sons & Co.

Clark, M. S., Denekamp, N. Y., Thorne, M. A. S., Reinhardt, R., Drungowski, M., Albrecht, M. W., *et al.* (2012). 'Long-term survival of hydrated resting eggs from *Brachionus plicatilis*', *PLoS ONE*, 7(1), p. e29365. doi:10.1371/journal.pone.0029365.

Cooke, G., Tonkins, B., and Mather, J. (2019). 'Care and enrichment for captive cephalopods', in Carere, C. and Mather, J. (eds.) *The welfare of invertebrate animals*. Cham, Springer Nature, pp. 179–208.

D'Abramo, L. (2019). 'Nutrition and feeds', in Lucas, J. S., Southgate, P. C., and Tucker, C. S. (eds.) *Aquaculture: farming aquatic animals and plants*, 3rd edn. Hoboken: John Wiley & Sons Ltd, pp. 157–182.

De Silva, S. and Wang, Q. (2019). 'Carps', in Lucas, J. S., Southgate, P. C., and Tucker, C. S. (eds.) *Aquaculture: farming aquatic animals and plants*, 3rd edn. Hoboken: John Wiley & Sons, Ltd, pp. 339–362. doi:10.1002/9781118687932.ch23.

Franks, B., Ewell, C., and Jacquet, J. (2021). 'Animal welfare risks of global aquaculture', *Science Advances*, 7(14), p. eabg0677. doi:10.1126/sciadv.abg0677.

Gauthier, D. T. (2015). 'Bacterial zoonoses of fishes: a review and appraisal of evidence for linkages between fish and human infections', *Veterinary Journal*, 203(1), pp. 27–35. doi:10.1016/j.tvjl.2014.10.028.

Halver, J. E. and Hardy, R. W. (1994). 'L-ascorbyl-2-sulfate alleviates Atlantic salmon scurvy', *Proceedings of the Society for Experimental Biology and Medicine*, 206(4), pp. 421–424. doi:10.3181/00379727-206-43781.

Harding, K., Ntimbani, R., Mashwama, P., Mokale, R., Mothapo, M., Gina, N., and Gina, D. (2013). 'Decomposition of ozone in water', Chemical Technology, July, pp. 6–10.

Hardy, R. and Kaushik, S. J. (eds.) (2021). *Fish nutrition*, 4th edn. San Diego, CA: Academic Press.

Hasan, M. M., Sarker, B. S., Shahriar Nazrul, K. M., and Tonny, U. S. (2014). 'Salinity tolerance level of GIFU tilapia strain (*Oreochromis niloticus*) at juvenile stage', *International Journal of Agricultural Sciences*, 4, pp. 83–89.

Heino, M., Díaz Pauli, B., and Dieckmann, U. (2015). 'Fisheries-induced evolution', *Annual Review of Ecology, Evolution, and Systematics*, 46(1), pp. 461–480. doi:10.1146/annurev-ecolsys-112414-054339.

Iwama, G .K. Pickering,A. D., Sumpter, J. P., and Schreck, C. B. (eds.) (2012). *Fish stress and health in aquaculture*. Cambridge: Cambridge University Press.

Kültz, D. (2005). 'Molecular and evolutionary basis of the cellular stress response', *Annual Review of Physiology*, 67, pp. 225–257. doi:10.1146/annurev.physiol.67.040403.103635.

Kültz, D. (2020a). 'Defining biological stress and stress responses based on principles of physics', *Journal of Experimental Zoology Part A: Ecological and Integrative Physiology*, 33(6), pp. 350–358.

Kültz, D. (2020b). 'Evolution of cellular stress response mechanisms', *Journal of Experimental Zoology Part A:*

Ecological and Integrative Physiology, 33(6), pp. 359–378. doi:10.1002/jez.2347.

Lekang, O.-I. (2019). *Aquaculture engineering*, 3rd edn. Hoboken: Wiley-Blackwell.

Lewis, W. K. and Whitman, W. G. (1924). 'Principles of gas absorption', *Industrial & Engineering Chemistry*, 16, p. 1215.

Martos-Sitcha, J. A., Mancera, J. M., Prunet, P., and Magnoni, L. J. (2020). 'Editorial: welfare and stressors in fish: challenges facing aquaculture', *Frontiers in Physiology*, 11, p. 162. doi:10.3389/fphys.2020.00162.

Selye, H. (1936). 'A syndrome produced by diverse nocuous agents', *Nature*, 138(3479), p. 32. doi:10.1038/138032a0.

Silva, C. P., Louros, V., Silva, V., Otero, M., and Lima, D. L. D. (2021). 'Antibiotics in aquaculture wastewater: is it feasible to use a photodegradation-based treatment for their removal?', *Toxics*, 9(8), p. 194. doi:10.3390/toxics9080194.

Sully, S., Burkepile, D. E., Donovan, M. K., Hodgson, G., and van Woesik, R. (2019). 'A global analysis of coral bleaching over the past two decades', *Nature Communications*, 10(1), p. 1264. doi:10.1038/s41467-019-09238-2.

Suresh, V. and Bhujel, R. C. (2019). 'Tilapias', in Lucas, J. S., Southgate, P. C., and Tucker, C. S. (eds.) *Aquaculture: farming aquatic animals and plants*, 3rd edn. Hoboken: John Wiley & Sons, Ltd, pp. 391–414. doi:10.1002/9781118687932.ch23.

Wanasundara, J. P. D. (2011). 'Proteins of Brassicaceae oilseeds and their potential as a plant protein source', *Critical Reviews in Food Science and Nutrition*, 51(7), pp. 635–677. doi:10.1080/10408391003749942.

Wurtsbaugh, W. A., Paerl, H. W., and Dodds, W. K. (2019). 'Nutrients, eutrophication and harmful algal blooms along the freshwater to marine continuum', *WIREs Water*, 6(5), p. e1373. doi:10.1002/wat2.1373.

Zhidkova, A. Y., Podberesnij, V. V., Zarubina, R. V., and Kononova, O. A. (2020). 'The effect of eutrophication on human health on the example of the Gulf of Taganrog of the Sea of Azov', *IOP Conference Series: Earth and Environmental Science*, 548, p. 052053. doi:10.1088/1755-1315/548/5/052053.

CHAPTER 18

Infectious diseases

'. . . disease problems were considered to be of only secondary importance when extensive farming was the most common practice. With the adoption of semi-intensive and intensive systems, the occurrence of several forms of diseases and consequent mortalities have significantly increased.'

Pillay and Kutty, *Aquaculture: principles and practices* (2005)

Infectious diseases are caused by pathogens that impair the health of hosts. Pathogens include viruses, bacteria, fungi, protozoans, and multicellular parasites. It has been estimated that losses caused by diseases account for approximately 40% of the cost of all aquaculture production. Disease-related losses cost the Asian shrimp industry about US$ 20 billion over the last decade (Owens, 2019).

Infectious diseases are best mitigated by prophylactic (proactive) measures since treatment (reactive) options are often very limited when an infection of aquaculture stock by a specific pathogen has been diagnosed. Pathogen dispersal depends on host density, which is higher in aquaculture than in natural systems. The higher density of hosts in aquaculture is compounded by higher selection pressure for favouring pathogen mutants with increased virulence.

Most information about disease causing pathogens is known for species that have been cultured for many decades in semi-intensive culture systems, for example, carps, salmonids, and penaeid shrimps. A list of economically important diseases of aquaculture fish, crustaceans, and molluscs is maintained by the World Organization for Animal Health, previously known under the name Office International des Epizooties (OIE). As of September 2021, this list included thirty-six prominent diseases of aquaculture animals (OIE, 2021).

18.1 Biosecurity

Biosecurity in aquaculture farms is the process of preventing the introduction and proliferation of pathogens to minimize losses resulting from the transmission of infectious diseases. The importance of strict biosecurity in aquaculture farms extends beyond protecting the health of the cultured organisms. It also minimizes the risk of spreading pathogens to wild populations of the same or similar species. This risk is highest for open cage and net pen aquaculture systems that are stocked with animals at a high density and lowest for recirculating aquaculture systems (RAS).

Moreover, some pathogens of aquaculture organisms also infect humans (Section 18.2.3). Such pathogens cause zoonotic diseases, that is, infectious diseases that can spread from animals to people. The risk of zoonotic diseases increases during domestication of new species because of more frequent interactions between animals and humans. Though infections of fish with some pathogens may be cryptic, humans that are susceptible to immune-related disorders are still at risk of contracting zoonotic diseases from aquaculture organisms that may be asymptomatic (Box 18.1).

In addition to zoonoses, human disease can also result from pathogens that use aquatic organisms as vectors rather than hosts. For example, coliform bacteria include pathogens that cause food poisoning

A Primer of Ecological Aquaculture. Dietmar Kültz, Oxford University Press. © Dietmar Kültz (2022). DOI: 10.1093/oso/9780198850229.003.0018

Box 18.1 Public health implications of zoonotic diseases

Zoonotic diseases (zoonoses) can be transmitted from animals, which represent the natural host species, to humans. Transmission of a zoonosis can occur either directly or indirectly. Direct transmission is by animal-to-human contact via the skin, blood, secretions, or other biological materials containing the pathogen. Indirect transmission is via biological vectors (e.g. parasites that carry the pathogen but are not affected by it) or by ingestion of food contaminated with the pathogen. The transmission of zoonoses via aquaculture pathogens is relatively rare and mainly threatens immunocompromised people. Nevertheless, the incidence has risen steadily and is attributed to an increased consumption of raw or undercooked seafood such as raw oysters or sushi, and increased contact of humans with aquatic animals and their pathogens due to domestication in commercial aquaculture and as ornamental companion animals (Gauthier, 2015).

Zoonoses have been reported for atypical mycobacteria, including *Mycobacterium marinum*, *M. chelonae*, and *M. fortuitum* (Monticini, 2010). Infection of humans with zoonotic mycobacteria occurs via inhalation, ingestion, or physical contact. Disease symptoms in humans manifest as cutaneous inflammations (granulomas) that require antibiotic treatment in combination with surgical excision and can be difficult to cure. *Streptococcus iniae* is another bacterial pathogen that causes zoonosis. *S. iniae* was first isolated from Amazon River dolphins (*Inia geoffrensis*) in the 1970s (Goh *et al.*, 1998). Since then, the list of verified hosts for this pathogen has steadily grown to include many aquaculture fish and humans. This pathogen is transmitted by handling live fish or processing raw fish products. It causes cellulitis in humans, a condition that results from bacterial infection of the skin and inflammation of the affected area. *Streptococcus agalactiae* is another zoonotic species of this bacterial genus.

Vibrio vulnificus and *V. cholerae* have both been implicated in zoonoses that originated from aquaculture organisms, including eels, ornamental fish, and molluscs (Hutson and Cain, 2019). *V. vulnificus* infection of humans cause gastroenteritis and septicaemia. *V. cholerae* infections cause cholera. Treatments include antibiotics and intravenous rehydration therapy. *Clostridium botulinum* occurs in the gastrointestinal (GI) tract of both marine and FW fish and has been reported as a cause for zoonoses that originated from seafood processing of salmonids and catfish (Hutson and Cain, 2019). In humans, *Clostridium botulinum* infection causes neurological disease (botulism) due to production of botulinum neurotoxin by this pathogen. Botulinum toxin interferes with the action of the neurotransmitter acetylcholine and, in severe cases, causes paralysis. Treatments of botulism utilize antitoxins and antibiotics.

Several important non-bacterial zoonoses are caused by multicellular parasites. They include infections by cnidarian parasites of the subphylum Myxozoa (e.g. *Kuoda spec.*) that cause food poisoning in humans. Furthermore, zoonotic helminth parasites are associated with parasitism in humans, including tape worms (*Diphyllobothrium spec.*), liver flukes (family *Ophiostorchiidae*, e.g. *Clonorchis sinensis*), and nematodes (*Anisakis spec.*).

Integrated aquaculture-agriculture (IAA) of birds (chickens, ducks, geese, etc.) has been criticized because of concerns about food safety and risks for human health resulting from viral zoonoses (e.g. outbreaks of avian flu [H5N1]) (Scholtissek and Naylor, 1988).

Moreover, significant concerns for causing influenza pandemics in humans have been raised about mutations of zoonotic viruses in pigs whose manure is used to fertilize IAA aquaculture ponds (Bunting, 2013; Muratori *et al.*, 2000). Such food safety concerns are to be taken very seriously, especially in the post COVID-19 era. They argue against IAA practices unless necessary sterilization procedures and sanitary measures for preventing pandemics caused by zoonotic pathogens are implemented. However, the implementation and monitoring of effective measures for preventing zoonoses in IAA is presently cost-prohibitive.

in humans. They are bioaccumulated by filter feeders such as oysters, when present in surrounding water, for example, because of urban wastewater disposal into natural waters. Furthermore, public health impacts of pond aquaculture in tropical areas include proliferation of vectors of waterborne infectious diseases (e.g. malaria-transmitting mosquitos) by providing suitable habitat.

Strict biosecurity for preventing zoonoses and diseases caused by pathogens that use aquaculture animals as vectors is most feasible for RAS designs that include water sterilization units and good management practices for pathogen screening of aquaculture animals. This feasibility is another reason for RAS being the most ecologically sustainable solution for future aquaculture development.

18.1.1 Infectious disease susceptibility and transmission

Infectious diseases develop because of gene × environment interactions. The genetic makeup of the aquaculture species and genetic variation between individual organisms of a population represents the genetic component of these interactions. This genetic component represents an important determinant of host susceptibility to a particular disease. The environmental component of gene × environment interactions regarding infectious diseases is twofold and comprised of:

1. the virulence of a particular pathogen (i.e. the ability of a pathogenic organism to cause disease in the host) and
2. the sum of all environmental conditions, which indirectly affect host susceptibility (e.g. by causing stress) and virulence.

Infection of a host with a pathogen principally has three possible outcomes. First, the defence system of the host eliminates the pathogen. Second, the pathogen overwhelms the host and proliferates to eventually cause mortality. Third, proliferation of the pathogen is suppressed by the host leading to asymptomatic infection, which can become symptomatic and cause disease at a later stage when environmental conditions change. Asymptomatic hosts that carry pathogens remain infectious during the incubation time, which is the time it takes from host infection (pathogen invasion) to the development of the first symptoms. Under some circumstances, symptoms may never develop even though pathogens subsist within the host at subclinical levels.

Infectious diseases can be transmitted by direct contact of the host with the pathogen present in the environment (water), by another organism (i.e. a vector) that contains the pathogen, or from parent to offspring via gametes or germinal fluid. The first two modes of disease transmission are considered horizontal transmission (shedding from skin, gill, mucus cells, faeces, urine, via food, or via vectors) while the latter is considered vertical transmission between different generations of hosts.

Host specificity is a critical property of pathogens. As is the case of all traits of evolving populations of organisms, host specificity is not static but evolvable and subject to change. Pathogens constantly mutate, and mutation rates are high in microbial pathogens. High mutation rates facilitate exploration of novel host species. Nevertheless, the evolutionary barrier for a pathogen species to conquer a new host species is very high as multiple mutations that support a change in host specificity are generally required. The statistical likelihood for such a specific set of mutations to occur simultaneously in a pathogen is low even in microbial populations that consist of billions of individuals.

18.2 Animal defence mechanisms against infectious diseases

Many pathogens are present in small concentrations in water or other aspects of the environment (e.g. soil in ponds) used for aquaculture, but their propagation in potential hosts is suppressed by host defence mechanisms. Stress impairs these defence mechanisms, which increases disease susceptibility. Infectious diseases are promoted by conditions in aquaculture that impose increased stress on organisms. Aquaculture farms are susceptible to spreading infectious diseases because of high stocking densities of organisms and deterioration of water quality by animal waste (self-pollution). Crowding, handling, and sub-optimal water quality represent stresses in semi-intensive and intensive aquaculture systems that facilitate pathogen propagation and disease outbreaks.

Thus, much research has been devoted to reducing the level of stress imposed on aquaculture organisms. This can be achieved by domestication of specific aquaculture strains via selection or genetic engineering and by tailoring management practices on aquaculture farms to minimize the stresses mentioned earlier. Minimizing stresses that impair defence mechanisms against pathogens (immunity) of aquaculture organisms not only prevents disease outbreaks but also maximizes the health and welfare of animals as well as their growth, which increases aquaculture productivity. The following paragraphs briefly outline the major defence mechanisms of invertebrates and fish against pathogens.

18.2.1 Physical defences of aquaculture animals

Pathogen invasion is the process of organisms that cause infectious disease entering or attaching to hosts. Common sites of pathogen entry into fish and crustaceans, that is, the most susceptible regions, are the gill epithelium, skin, gastrointestinal tract, eyes, wounds, and lesions in skin and areas where scales are lost, or where exoskeleton has been compromised by injury.

Aquatic organisms protect themselves from infection with pathogens by physical defences. The main physical defences are features that protect the skin from lesions, that is, scales of fish, exoskeleton of crustaceans, and shells of molluscs. In addition, susceptible epithelia such as gills and intestinal barriers are physically protected from pathogen entry by a mucus cover. Mucus consists mainly of carbohydrates (sugars) and glycoproteins (proteins with sugar moieties attached) and contains antimicrobial peptides. Therefore, mucus serves a dual defence function as a physical and chemical barrier for in pathogens.

Pathogens attack physical defences of hosts by evolving mechanisms that target physical and chemical host defences in a process called host–pathogen coevolution. This process of reciprocal evolution of interacting hosts and pathogens can be compared to a perpetuating arms race. For example, enzymes of bacterial pathogens have evolved to attack physical defences of their hosts, including chitinases that chemically degrade the exoskeleton of shrimps.

18.2.2 Innate immunity of fish and invertebrates

Both invertebrates (including molluscs and crustaceans) and vertebrates (including fish) have innate immunity. Innate immunity represents a chemical defence system against pathogens. It is also referred to as non-adaptive immunity without cellular memory. Innate immunity is based on the production of chemicals and receptors that recognize and destroy non-self molecular patterns based on the different chemical composition of the surfaces of pathogen cells versus that of host cells. Innate immunity is based on cellular receptors that are expressed by host cells and bind pathogens by distinguishing self from non-self. Specialized cells of aquatic animals express these receptors at high levels, including amoebocytes, haemocytes, coelomocytes, and phagocytes in invertebrates.

The cellular receptors expressed and presented on the surface of these host cells are proteins that recognize foreign patterns on pathogens by having complementary structures that mediate pathogen binding via recognition of conserved, pathogen-specific molecular patterns. Examples for host receptor proteins with pathogen-specific pattern recognition functions include diverse scavenger receptors, lectins, toll-like receptors (TLRs), and pentraxins. Upon binding of their ligands, that is, pathogen cell surface molecules, pattern recognition receptors involved in innate immunity activate intracellular signalling mechanisms in corresponding host cells. These molecular networks signal the presence of pathogens to trigger effector mechanisms that are geared towards pathogen elimination.

Major effector mechanisms of innate immune responses are processes that increase the production of reactive oxygen and nitrogen species (i.e. hydrogen peroxide) that kill pathogens via imposing oxidative stress. Other important effector mechanisms include the activation of phagocytosis (engulfing) and chemical degradation of pathogens, the production of antimicrobial peptides, complement system proteins, and other protein classes that attack and destroy pathogens.

18.2.3 Adaptive immunity of vertebrates

Fishes are the only animals used for commercial aquaculture that have an adaptive immune system in addition to the physical defences and innate immunity mentioned. Adaptive immunity is also termed acquired immunity and based on cellular memory, which enables rapid production of pattern-recognition molecules (antibodies) that recognize specific pathogens and trigger their destruction.

Antibodies (immunoglobulins, Igs) are proteins encoded by genes. They generally consist of multiple subunits, including variable and non-variable components, that is, they are usually encoded by more than just one gene. However, primitive vertebrates like sharks have antibodies that consist of only a single protein encoded by a single gene. These primitive antibodies are called nanobodies. Current genetic engineering research focuses on producing nanobodies that recognize specific pathogens. The goal of this research is to introduce the corresponding genes that encode these nanobodies into animal genomes by transposon-mediated transgenesis (Section 6.3.1) to generate pathogen-resistant strains of commercially important animals as an alternative to vaccination.

Many different cell types and molecules are involved in adaptive immunity. The lymphatic system includes diverse tissues such as the spleen, head kidney, and lymph nodes. A large variety of different white (haemoglobin-free) blood cells (leukocytes) contribute to adaptive immunity. Leukocytes include antigen-presenting cells (APCs) such as B lymphocytes, macrophages, and dendritic cells that express Igs and major histocompatibility complex (MHC) proteins as pattern recognition receptors that recognize cell surface molecules of pathogens.

APCs bind pathogens with high affinity and present them to other leukocytes, including various types of cytotoxic T lymphocytes and natural killer (NK) cells that attack and destroy pathogens and pathogen-compromised host cells. Pathogen destruction is promoted by inflammatory responses that attract immune cells to sites of infection via chemical messengers, such as chemokines and cytokines. As a negative side effect of battling pathogens, severe inflammation can cause significant destruction of host tissue because pathogen-compromised host cells are destroyed and non-selective pathogen defence mechanisms such as production of reactive oxygen species (ROS) by immune cells also imposes oxidative stress on cells in inflamed host tissue.

The concept of vaccination is based on the principle of adaptive immunity, that is, cellular memory of specific lineages of B and T lymphocytes to prior exposure to a specific pathogen. It is therefore limited to fishes (and other vertebrates) and not applicable to invertebrates such as molluscs and crustaceans. Vaccination is commonly performed by presenting molecular patterns of pathogens to the adaptive immune system of hosts to trigger cellular memory of these specific molecular patterns without causing infection with live pathogens. This is typically achieved by intramuscular injection of dead pathogens or purified pathogen surface proteins. Immersion vaccines (i.e. vaccines that are not injected but applied as a bath solution) are also used in aquaculture, but they are much less effective than injection.

Recently, intramuscular injection of plasmid and viral vectors in combination with chemicals that deliver them into host cells is also used for vaccination. Such vaccines are based on the principle of recombinantly expressing genes that encode pathogen surface proteins by utilizing the gene expression, protein translation, and protein processing machineries of host cells to present these pathogen proteins on the surface of host cells or secrete them from host cells such that they can be recognized by the adaptive immune system.

The adaptive immune system is based on the evolution of a region in vertebrate genomes that confers extremely rapid mutation of specific somatic cells, which are called B and T lymphocytes. This hypervariable region (HVR) of vertebrate genomes is altered in response to antigen presentation by a process called variable diversity joining (VDJ) recombination, which randomly mutates the antigen-specific, variable part of vertebrate genomes that encodes the variable region of antibodies (Igs). Lymphocyte lineages that express antibodies with the highest affinity towards the antigenic molecular pathogen pattern that is encountered are then selected by a feedback mechanism based on antigen binding. Once modified by VDJ recombination, the HVR is fixed in specific lymphocyte lineages that reside in lymphatic tissues (i.e. it represents a form of cellular memory). If the same antigen is encountered again, then the corresponding lymphocyte lineage proliferates rapidly to trigger destruction of the pathogen.

18.3 Viral diseases (viroses)

Viruses are different from conventional forms of life in that they lack some genetic material for self-replication. They require elements of the host genome for their propagation. Moreover, most viruses are not enclosed by a plasma membrane (i.e. they are not cellular). For these reasons, there is a debate about whether viruses should be considered a form of life.

Generally, viruses, including those that infect aquaculture organisms, are very small particles (virions) that have tiny genomes comprised of either RNA or DNA. Viral genomes may encode fewer than five proteins, for example, that of the bacteriophage MS2, which is a single-stranded RNA virus that infects bacterial host cells. However, not all viruses are small. Giant viruses are larger than typical bacteria and they contain very large genomes in the megabase size range. Their genomes include many unique genes that are not found in archaea, bacteria, or eukaryotes and genes that encode proteins associated with functions normally only seen in cellular organisms, including elements of the translation machinery (Brandes and Linial, 2019).

18.3.1 Viroses of aquaculture animals

Viral diseases are common in aquaculture species of fish and crustaceans. Many different viruses have been identified in these taxa, including aquabirna, rhabdo, herpes, irido, amnoon, orthomyxo, noda, nima, picorna, baculo, roni, and parvo viruses. Fewer viral pathogens have so far been detected in molluscs used for aquaculture—perhaps because many molluscs are cultured by extensive grow-out. Moreover, even in intensive hatcheries the culture density may be more similar to that in the native habitat for larval molluscs than for fish and crustaceans (Table 18.1).

Many viral diseases of aquaculture fish and crustaceans are transmitted both horizontally and vertically, and asymptomatic carriers that have acquired immunity after an infection are common. Mortality from viral diseases can be as high as 100%, for example, in salmonid hatcheries infested with infectious pancreatic necrosis virus (IPNV). Viral diseases of fish, including IPNV disease (IPNVD), koi herpes virus disease (KHVD), and tilapia lake virus disease (TiLVD, Figure 18.1) in salmonids, carps, and tilapia have caused large economic losses to commercial aquaculture on a global scale (Pierezan et al., 2020). Many viral infections of fish are more prominent in juveniles than adults and lead to antibody production and immunity, which alleviates disease symptoms. However, fish that successfully overcome a viral infection are often asymptomatic carriers of the virus (i.e. they remain infectious throughout their life).

Multiple pathogenic shrimp viruses have been identified, most in the last three decades because of intensification of shrimp aquaculture. Viral diseases had catastrophic impacts on commercial shrimp aquaculture, leading to large global declines in production during the previous two decades, which were particularly devastating in some Central American and Asian countries and have been estimated to have caused economic losses of several billion dollars. For example, white spot syndrome virus disease (WSSVD) was first detected in Chinese white shrimp in China in 1993. It spread to Kuruma shrimp in Korea and Japan, and from there, to Whiteleg shrimp in other Asian countries, the Americas, and Australia.

Another example is Taura syndrome virus (TSV) disease (TSVD), which is named after the Taura River in Ecuador, where it was first detected in 1992. By 1997 this disease had spread from South America to Central and North America and infected the major aquaculture shrimp species (Whiteleg shrimp). In 1998, the first Asian TSVD outbreak was recorded in Taiwan and in 2003 TSV was detected in Thailand. TSV is now present in all major shrimp-producing countries in the Americas and Asia. Juvenile shrimps are often more severely affected by viruses than adults, the latter often being asymptomatic carriers of the virus. Therefore, it is critical to apply diagnostic tests for prevalent infectious diseases before introducing broodstock from an external source to an aquaculture farm in a different country or area.

18.3.2 Diagnosis and treatment of viroses

Viral disease symptoms are often non-specific, and diagnosis requires the use of molecular techniques (PCR and analysis of immortalized cell lines). Only some viral diseases are associated with behavioural or morphological symptoms that are somewhat specific (Table 18.1). However, even if such symptoms are observed, a proper diagnosis of the pathogens still requires follow-up using molecular tools.

Viral pathogens are unambiguously diagnosed by taking a biopsy from an aquaculture organism and analysing it in a diagnostic laboratory. Two types of biopsy analyses are common. Biopsy tissue can

Table 18.1 Common viral diseases of aquaculture animals

Viral pathogen	Host	Disease	Specific symptoms
	Fish		
IPNV Aquabirnavirus (*Birnaviridae*)	Salmonids, carp, eels, tilapia	Infectious pancreatic necrosis virus disease (IPNVD)	Accumulation of milky mucus in the GI tract
YAV Aquabirnavirus (*Birnaviridae*)	Yellowtail and other marine fishes	Yellowtail ascites virus disease (YAVD)	
IHNV (*Rhabdoviridae*)	Salmonids	Infectious haematopoietic necrosis (IHN)	
VHSV (*Rhabdoviridae*)	Salmonids, many other FW and marine fishes	Viral haemorrhagic septicaemia (VHS)	
SVCV (*Rhabdoviridae*)	Common carp	Spring viremia of carp (SVC)	
RHV (*Rhabdoviridae*, *Herpesviridae*)	Eels	Red head virus	Initial haemorrhages in the head region
CCV (*Herpesviridae*)	Channel catfish and blue catfish	Channel catfish virus disease (CCVD)	Vertical, head-up body positioning
KHV (*Herpesviridae*)	Carp	Koi herpes virus disease (KHVD)	Plaque-warty epidermal hyperplasia causing skin lesions and lifting of scales
Lymphocystivirus (*Iridoviridae*)	Many marine species, some FW fishes, e.g. cichlids	Lymphocystis virus disease (LVD)	Small white nodules on body and fins, cellular hypertrophy
TiLV Tilapinevirus (*Amnoonviridae*)	Tilapia	Tilapia lake virus disease (TiLVD)	
ISAV (*Orthomyxoviridae*)	Salmonids	Infectious salmon anaemia virus disease (ISAVD)	
Betanovirus (*Nodaviridae*)	Many marine and FW species	Viral nervous necrosis (VNN)	Blindness
	Molluscs		
Ostreavirus (*Malacoherpesviridae*)	Oysters	Oyster velum viral disease (OVVD)	
Aquabirnavirus (*Birnaviridae*)	Tellina clam	Tellina virus (TV-1)	
	Crustaceans		
WSSV Whispovirus (*Nimaviridae*)	Shrimps	Whitespot syndrome virus disease (WSSVD)	Small white spots on carapax
TSV (*Picornaviridae*)	Shrimps	Taura syndrome virus disease (TSVD)	
MBV (*Baculoviridae*)	Shrimps	Monodon-type baculovirus disease (MBVD)	
YHV (*Roniviridae*)	Shrimps	Yellowhead virus disease (YHVD)	Yellowish carapax
IHHNV (*Parvoviridae*)	Shrimps	Infectious hypodermal and haematopoietic necrosis virus disease (IHHNVD)	Bent rostrum

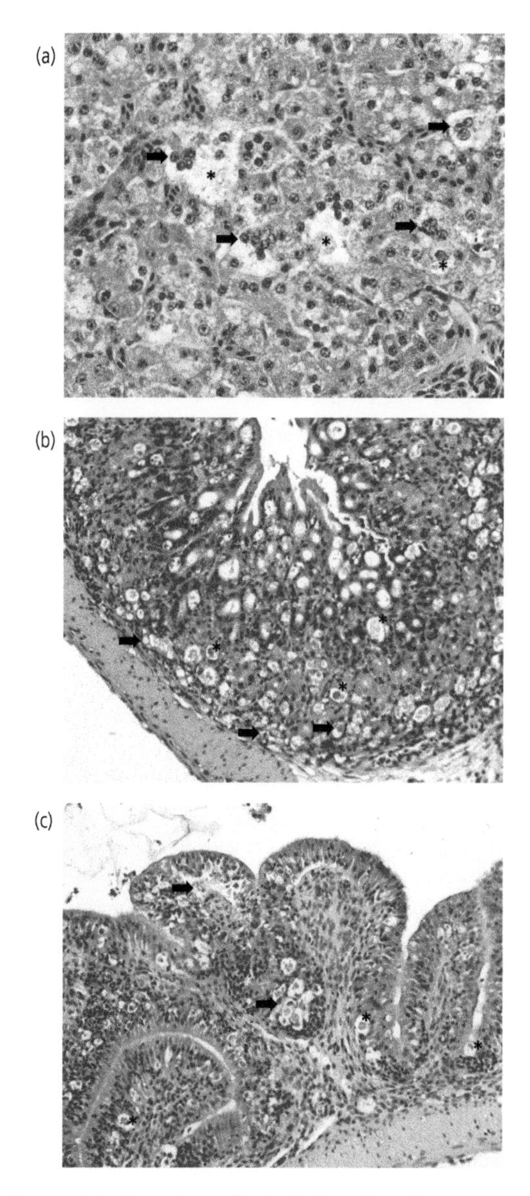

Figure 18.1 Photomicrographs of Nile tilapia (*Oreochromis niloticus*) tissues infected with tilapia lake virus (TiLV). Sections were stained with haematoxylin and eosin. **A)** Liver: hepatocellular necrosis (asterisk) characterized by isolated, fragmented, and shrunken hepatocytes, formation of clear spaces, and formation of syncytial giant cells (arrow) as prominent lesions. **B)** Stomach: necrosis of epithelial cells of the gastric glands (asterisk) characterized by isolated cells or clusters of shrunken cells with loss of cell-to-cell adhesion, eosinophilic cytoplasm, and pyknotic, karyolitic, or absent nuclei. **C)** Intestine: necrosis of epithelial cells of the intestine (asterisk) characterized by isolated cells or clusters of shrunken cells with loss of cell-to-cell adhesion (arrows), eosinophilic cytoplasm, and pyknotic, karyolitic, or absent nuclei. Photos kindly provided by Professors Esteban Soto and Felipe Pierezan.

be used directly for amplification of virus-specific gene(s) using the polymerase chain reaction (PCR) or by infecting a host cell line via contact with biopsy material and observing the replication and cytopathic (destructive) effect of the virus on the cell line.

PCR utilizes viral gene-specific primers, which are short (fifteen to twenty-five base pair) sequences of DNA that are complementary to a unique sequence in the genome of a particular viral pathogen. If this sequence is present in the biopsy material, then these PCR primers will bind and the corresponding DNA will be amplified several billionfold such that it can be visualized on agarose gels as a bright fluorescent band, which indicates the presence of a specific viral pathogen in the biopsy material. PCR requires knowledge of virus-specific genomic sequences, which can be attained by sequencing the genomes of viral pathogens and comparing them against the genomes of other viruses and hosts. Since viral genomes are generally very small, reference genome sequences for novel viral pathogens can be assembled rapidly at low cost after appropriate methods for isolation of a novel pathogen have been established.

PCR can also be performed after propagating viruses in host cell lines. Several cell lines have been developed for diagnosing viral diseases in fish, including the OmB cell line from tilapia (ATCC # CRL-3481™, Figure 1.3), which has been immortalized from Mozambique tilapia (*Oreochromis mossambicus*) neural tissue in the author's laboratory (Gardell *et al.*, 2014). However, development of invertebrate cell lines (e.g. for molluscs and crustaceans) has proved much more challenging despite enormous research efforts in this field over many decades. Recently, a shrimp hybrid cell line (PmLyO-Sf9 cells) has been developed by fusing tiger shrimp (*Penaeus monodon*) lymphoid cells with an insect cell line (Sf9 cells) derived from a moth (*Spodoptera frugiperda*) (Anoop *et al.*, 2021).

The diagnosis of an infectious disease depends on proper identification of the pathogen that causes it. This is often complicated by secondary infections that follow the primary infection and it is not uncommon to detect multiple pathogens in a single biopsy, especially if the biopsy was taken at an advanced stage of the disease. There are no effective treatments for viral diseases and prevention is the only available strategy for coping with viroses of aquaculture organisms (Section 18.6).

18.4 Bacterial diseases (bacterioses)

Bacteria are unicellular organisms that represent one of the two kingdoms of prokaryotes, the other being archaea (Chapter 7). They are enclosed by a lipid bilayer containing many interspersed proteins (plasma membrane) but lack a nucleus. Bacteria are generally much larger and have more complex genomes than viruses, except for some giant viruses. Bacteria are commonly classified based on a procedure developed by Hans Christian Gram that distinguishes them based on the composition of their cell walls. A violet dye is retained in the cell wall of some (Gram-positive) but lost from that of other (Gram-negative) bacteria after Gram staining.

Most known diseases of aquaculture organisms are caused by Gram-negative bacteria of the genera *Aeromonas, Alteromonas, Edwardsiella, Flavobacterium, Francisella, Myxococcus, Pseudomonas, Vibrio,* and *Yersinia*. However, several Gram-positive bacterial pathogens also cause important diseases of aquaculture animals, including the genera *Mycobacterium* and *Streptococcus* (Table 18.2).

18.4.1 Bacterioses of aquaculture animals

Many economically important bacterial diseases have been identified for aquaculture fish, crustaceans, and molluscs (Table 18.2). Described as early as 1894 in Bavarian salmonids, furunculosis is one of the earliest recognized bacterial fish diseases, which is caused by *Aeromonas salmonicida*, and characterized by large subdermal furuncles (boils). Multiple strains of *A. salmonicida* exist that primarily infect salmonids, but some atypical strains of this bacterial pathogen also infect other host species. Horizontal transmission of this disease is promoted by rupturing of the furuncles and release of large numbers of pathogens into the water. Vaccination has been used as an effective preventative strategy for commercial Atlantic salmon aquaculture. Other species of *Aeromonas* that are highly prevalent as pathogens of aquaculture fish include *A. hydrophila* and *A. punctata*. These pathogens cause red fin disease and motile Aeromonas septicaemia (MAS) in eels and tilapia and septicaemic dermo-visceral syndrome (SDVS) in carps (Amlacher, 2009).

Edwardsiellosis is caused by bacteria of the genus *Edwardsiella* (e.g. *Edwardsiella ictaluri* or *Edwardsiella tarda*) which cause enteric septicaemia in Channel catfish and Japanese eels. This disease has led to significant losses for the US catfish industry in the past. It is still primarily treated using antibiotics delivered with feeds although vaccines have been developed and efforts to increase the efficiency of their delivery are being made.

Pathogens within the genus *Flavobacterium* that have commercial importance for aquaculture include *F. psychrophilum, F. columnare,* and some species that cause bacterial gill disease in aquaculture fish and molluscs. *F. psychrophilum* is the causative agent for bacterial coldwater disease (BCWD) in trout. This disease is also known as peduncle disease because of the eroded caudal peduncles during advanced stages. Because BCWD is prominent in trout hatcheries throughout the world, there is great emphasis on vaccine development.

In contrast to *F. psychrophilum, F. columnare* requires warm water and infects warmwater fish hosts (e.g. channel catfish and ornamental species) to cause columnaris disease. The name columnaris derives from the appearance of bacterial pathogens when visualized under the microscope. Although a vaccine exists, antibiotics and chemical baths (e.g. in potassium permanganate, $KMnO_4$) are still the most prevalent treatments.

Mycobacterioses are important diseases that affect many species of freshwater (FW) fish and are caused by various pathogens of the genus *Mycobacterium*. These diseases are referred to as fish tuberculosis because the formation of necrotic epithelial tubercles or granulomas represents a characteristic disease symptom. *M. piscium* is one of the prevalent *Mycobacterium* species that infect many FW fish, while *M. marinum* can also cause disease in some marine fishes. The congeneric *M. tuberculosis* is the causative agent for tuberculosis in humans.

Rickettsia-like organisms (RLOs) are bacteria that reside inside cells of their hosts. The tiny size and

Table 18.2 Common bacterial diseases of aquaculture animals

Bacterial pathogen	Host	Disease	Specific symptoms	Treatment
	Fish			
Aeromonas salmonicida	Mainly salmonids	Furunculosis	Skin blisters	Oxytetracycline, sulfonamides
Aeromonas punctata	Carp	Carp enteritis	Purplish intestine	Oxytetracycline, sulfonamides
Aeromonas hydrophila	Eels, tilapia	Red fin disease, motile aeromonas septicemia (MAS)	Tail and fin rot, haemorrhagic septicaemia	0.5–0.9% table salt (NaCl)
Flavobacterium columnaris	Salmonids and many other fishes	Columnaris disease		Copper sulfate ($CuSO_4$), potassium permanganate ($KMnO_4$), oxytetracycline
Flavobacterium spec.	Salmonids, tilapia, catfish	Bacterial gill disease (BGD)		Potassium permanganate ($KMnO_4$), quaternary ammonium compounds, benzethonium chloride
Francisella spec.	Tilapia, salmonids	Rickettsia-like organism disease (RLOD)		Florfenicol
Myxococcus spec.	Grass carp, salmonids	Gill rot disease		Quaternary ammonium compounds, benzethoni-um chloride, bleach (sodium hypochlorite, NaOCl)
Yersinia ruckeri	Salmonids	Enteric red mouth disease (ERMD)	Small patches of haemorrhagic spots in mouth region	Oxytetracycline, sulfonamides
Edwardsiella spec.	Carp, catfish, eels, tilapia	Edwardsiellosis		Oxytetracycline, sulfonamides
Vibrio anguillarium	Eels, salmonids, mullets, tilapia	Red eel pest, vibriosis	Red spots ventrally and laterally, dark, ulcerating, swollen lesions	Oxytetracycline, sulfonamides
Streptococcus faecalis	Mullets, tilapia	Streptococcosis		Oxytetracycline, sulfonamides, doxycycline, enteroflaxin
Mycobacterium spec.	Many species	Fish tuberculosis and Mott (mycobacteria other than tuberculosis)	White-grey nodules on internal organs	
	Molluscs			
Vibrio spec., Alteromonas spec., Pseudomonas, Flavobacterium	Oysters, clams	Mollusc bacterioses		
	Crustaceans			
Vibrio spec.	Shrimps	Shrimp vibrioses		

intracellular location of RLOs renders their detection and antibiotic treatment much less effective than for other bacteria. Examples of RLOs that represent significant pathogens for aquaculture fish include *Francisella spec.* and *Piscirickettsia salmonis* (Soto et al., 2018). The former causes Rickettsia-like organism disease (RLOD) in tilapia (Figure 18.2), while the latter is the causative agent for salmonid rickettsial septicaemia.

Streptococcosis is a bacterial disease that is prevalent in tropical warmwater fishes in both FW and marine environments. This disease is caused by multiple species of the genus *Streptococcus*. The species of *Streptococcus* that infect aquaculture animals are generally different from those that cause human disease, although there is potential for zoonoses (e.g. for *Streptococcus agalactiae* (Box 18.1)). Vaccine development represents an active area of research to combat streptococcosis.

The genus *Vibrio* has an extremely wide host range that includes many species of aquaculture molluscs, crustaceans, and fish, as well as humans. Examples for vibrioses are acute hepatopancreatic necrosis disease (AHPND) in shrimps (caused by *V. parahaemolyticus*), red eel pest in fish (caused by *V. anguillarium*), and cholera (caused by *V. cholerae*) in humans. AHPND was first discovered in China aquaculture farms in 2010 and is also known as early mortality syndrome (EMS). It causes up to 100% mortality of aquaculture shrimp. Effective vaccines have been developed to mitigate AHPND and other fish vibrioses that are common in commercial aquaculture.

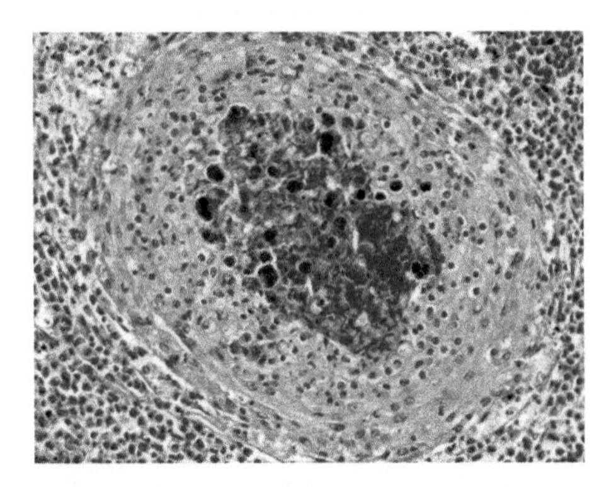

Figure 18.2 Photomicrograph of a tissue section of anterior (head) kidney of Nile tilapia (*Oreochromis niloticus*) infected with *Fransicella orientalis* after staining with haematoxylin and eosin. This image shows a typical granuloma with a necrotic hyperchromatic core surrounded by epithelioid macrophages, lymphocytes, and necrotic cells. Photo kindly provided by Professors Esteban Soto and Felipe Pierezan.

Enteric red mouth disease (ERMD) has resulted in significant losses to the trout aquaculture industry. The causative pathogens for this disease are bacteria of the genus *Yersinia*, mostly *Y. ruckeri*. A different *Yersinia* species (*Y. pestis*) is well known as the pathogen causing the bubonic plague in humans. One of the characteristic symptoms of ERMD in fish disease is haemorrhage around the mouth. Antibiotics have been successfully used for treatment. Effective commercial vaccines are also available to limit the use of antibiotics for treating ERMD.

18.4.2 Diagnosis and treatment of bacterioses

Bacterial diseases result from infection of external and internal tissues. Internal infections with bacteria are referred to as septicaemia (poisoning due to the presence of bacteria in blood or haemolymph). Severe septicaemia causes sepsis (toxic shock syndrome), which is a consequence of acute inflammation of internal tissues and organs, which can be deadly. Ascites, exophthalmia (bulging of the eye), and swollen (inflamed) internal organs are symptoms of septicaemia. External bacterial infections cause syndromes that are also common to other infectious diseases, including lesions, haemorrhage, and changes in gill and skin colour and texture (Section 18.7.3). Many bacterial infections are opportunistic secondary infections that occur when lesions are already present because of prior primary infections with other bacteria, viruses, fungi, or parasites.

Because of the non-specific nature of visible symptoms caused by bacterial infection, it is necessary to collect biopsies from infected specimens to properly diagnose the disease. Biopsies can be tissue samples collected from visible lesions (solid biopsies) or blood samples (liquid biopsies). Liquid biopsies can also be collected routinely in regular (weekly or monthly) intervals from a small number of individuals as part of a preventative disease monitoring programme. Such a strategy can improve treatment outcomes by early detection of pathogens before visible symptoms are recognized as changes in phenotype (appearance and/or behaviour) of the organisms. Biopsies can be analysed on site if diagnostic tools are available or shipped frozen to a diagnostic laboratory, for example, at the University of California Davis School of Veterinary Medicine, which routinely performs tests for a large array of common aquaculture pathogens. The latter option is usually more economical for smaller hatcheries and aquaculture farms.

Diagnosis and treatment options for bacterial diseases are more numerous than for viroses. Major diagnostic methods for bacterioses include PCR and monitoring the cytopathic effect of infecting host cell lines (as described in Section 18.3.2 for viroses). In addition, growth of bacterial pathogens present in biopsies on selective media and pathogen-specific stains in combination with microscopic examination are common diagnostic tools for bacterioses.

Growth on selective media is based on spreading out biopsy material on a Petri dish containing agarose and a medium that includes nutrients needed for bacterial growth, a selective agent (chemical) that permits only certain types of bacteria to grow, and a dye that stains bacteria to facilitate observation of growth on the dish. These tests are rapid and can be completed within twenty-four hours. For example: Rimler–Shotts media specifically promote the growth of *Aeromonas* bacteria while inhibiting that of Gram-positive bacteria and *Vibrio* species by containing the selective agents sodium deoxycholate and novobiocin. Rimler–Shotts media also contain the nutrient maltose and the dye bromothymol blue, which indicates metabolic utilization of maltose by *Aeormonas*.

Media for detecting *Vibrio* species are based on the selective agent thiosulfate citrate in combination with bile salts and the nutrient sucrose (TCBS media). When growth on TCBS media inoculated (spread) with biopsy material is observed, then it indicates the individual from which this biopsy was collected is infected with *Vibrio spec.* Such knowledge informs about available treatment options and best practices for managing a vibriosis outbreak.

An example of a selective stain is the Gram stain, which distinguishes bacteria based on the composition of their cell wall. Most bacterial aquaculture pathogens are Gram-negative and do not retain the violet Gram stain after the staining procedure. However, *Mycobacterium* and *Streptococcus* bacteria are Gram-positive and can be diagnosed using this staining procedure in combination with evaluation of stained biopsy spreads using a light microscope.

Microscopic evaluation of biopsy material stained with selective dyes is a rapid diagnostic procedure that can be completed within a day of receiving the biopsy. Selective stains for diagnosing bacterioses also include fluorescent dyes that are visualized with a fluorescence microscope. An example of a fluorescent dye that selectively stains *Mycobacterium spec.* is auramine, which selectively binds to mycolic acid in the *Mycobacterium* cell wall.

Once diagnosed, treatment and management options for bacterioses caused by specific pathogens can be evaluated and applied. Unlike viroses, many bacterioses can be treated (with varying success) by chemical baths that contain sterilizing agents such as formalin, methylene blue, malachite green, benzalkonium chloride, copper-containing algicides, and strongly oxidizing chemicals such as $KMnO_4$, sodium hydroxide, or bleach (sodium hypochlorite). However, the most effective bactericidal chemicals are also toxic (potentially carcinogenic or teratogenic) to humans. They may be bioaccumulated in muscle tissue of fish and invertebrates and not purged completely even after long culture of the treated organisms in clean water. Therefore, the application of some chemicals in aquaculture is now illegal in many countries.

As with viruses and other pathogens, if the pathogen has a narrower tolerance range to a specific environmental parameter than the host species, then the infected animals can be incubated under conditions that are tolerated by the host but lethal to the pathogen. Baths that utilize altered salinity are common in that regard. However, as with baths containing bactericidal chemicals, such treatments impose significant additional stress on aquaculture organisms.

An alternative treatment to baths is inclusion of bactericidal antimicrobial peptides and/or antibiotics in medicated feeds. One drawback with medicated feeds is that they are inefficient when disease symptoms include anorexia (loss of appetite). Common antibiotics used in aquaculture include sulfonamides and oxytetracycline; the latter is also a potent fungicide. Although antibiotics are very effective and less toxic than the stains and sterilizing agents used in chemical baths, they can be retained in seafood and may cause allergic reactions in some people. Moreover, mass application of antibiotics causes development of antibiotic-resistant strains of pathogens, which limits their usefulness and may pose problems for public health in case of potential zoonoses (Box 18.1).

To provide alternative treatment options to using antibiotics, current research aims at identifying antimicrobial peptides that are most potent in killing pathogens while minimizing harm to symbiotic bacteria that are beneficial for aquaculture organisms (e.g. gut microbiota). Public health implications must also be considered for management of bacterioses since most known zoonotic diseases that are associated with aquaculture organisms are of bacterial origin (Box 18.1).

18.5 Fungal diseases (mycoses)

Fungi are phylogenetically diverse eukaryotes that include primitive unicellular and more advanced multicellular organisms. They function primarily as heterotroph decomposers in ecosystem nutrient cycles by utilizing organic nutrients from either dead, decaying organisms (saprotrophs) or by deriving their nutrients as symbionts or pathogens from living organisms. Like prokaryotes, eukaryotic fungi contribute to mineralization, that is, the breakdown of organic matter into inorganic compounds that represent nutrients for autotroph plants (primary producers).

Many economically important fungal diseases are caused by primitive fungi referred to as water moulds. Water moulds are fungi within the order Saprolegniales in the phylum Oomycota. They are microscopically small filamentous organisms that have a mycelium. A mycelium is a network of long thread-like hyphae that form the vegetative part of these fungi. The mycelium is visible by the naked eye as a fluffy mass of cotton-like tufts, and under the microscope as a meshwork of branching extensions. This typical appearance represents a convenient diagnostic criterion for identifying water moulds (Figure 18.3).

18.5.1 Mycoses of aquaculture animals

Fungal diseases (mycoses) that are economically important for aquaculture are caused by several genera of water moulds. The genera *Aphanomyces* and *Saprolegnia* cause mycoses in crustaceans and fish while the genus *Branchiomyces* has been diagnosed primarily in fish (Table 18.3). Mycoses caused by *Branchiomyces spec.* are associated with a syndrome referred to as gill rot. These fungal pathogens primarily colonize the gills of FW fish, including the dominant species used for aquaculture (carps and tilapia). They derive nutrients from the blood of their hosts and disrupt the branchial circulation, which causes gill deterioration and significant mortality of infected fish.

Saprolegnia species (e.g. *S. parasitica*) are the cause of most mycoses associated with high mortality of eggs and larvae in many fish hatcheries. These water moulds feed on decaying, necrotic tissue and, thus, are opportunistic pathogens causing secondary infections. They are ubiquitously present in FW environments and affect virtually any species of FW fish. However, some marine and brackish water fish are also susceptible to water moulds in cold and warm water. Mycoses caused by *Saprolegnia* commonly start in unfertilized, decaying eggs, from which they can rapidly spread to healthy eggs and larvae resulting in high mortality and often complete loss of offspring in hatcheries. Prevention of primary bacterial infections and lesions due to stress is the best management practice for avoiding secondary *Saprolegnia* infection.

Epizootic ulcerative syndrome (EUS) is a fish disease caused by the fungus *Aphanomyces invadans*. This disease was first identified in ayu in Japan and has been confirmed to infect more than 100 wild and aquaculture species of FW and brackish water fish. Annual economic losses caused by this disease in the 1990s have been estimated around US$ 3 billion (Pillay and Kutty, 2005).

Fungal pathogens of the genus *Aphanomyces* are also important for crustacean aquaculture. The crayfish plague is caused by *Aphanomyces astaci*. This mycosis has caused high mortality and a severe population decline of the Noble crayfish (*Astacus astacus*) in its native distribution range (Europe) after the disease was introduced to Europe with

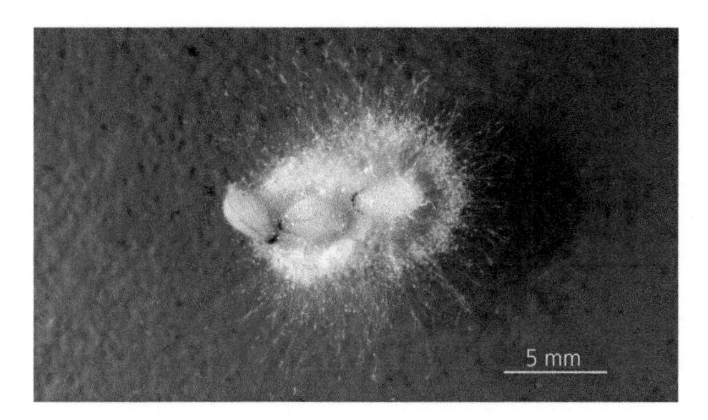

Figure 18.3 *Saprolegnia spec.*, grown on sesame seeds in distilled water one week after inoculation with sample water from Bochumer Wald, Lottental (Germany). The long hyphae (filamentous extensions) are clearly visible. Photo by Oliver Ruiz (Wikimedia Commons, reproduced under CC BY-SA 4.0 licence).

North American red swamp crayfish which is resistant to it.

Some aquaculture molluscs and shrimps are also infected by fungi other than water moulds (order Saprolegniales). For example, some mycoses of oysters and clams are caused by the fungus *Sirolpidium zoophthorum*, which belongs to the order Myzocytiopsidales and the fungus *Ostracoblabe implexa*, which belongs to the order Dothideales, which lacks hyphae (Table 18.3). *Sirolpidium spec.* is also responsible for mycosis in penaeid shrimps by infiltrating soft tissues and causing muscle atrophy and organ failure. In addition, black gill disease of shrimps is caused by melanized lesions resulting from infection with the fungal pathogen *Fusarium solani* from the order Hypocreales (Pillay and Kutty, 2005).

Another shrimp mycosis is hepatopancreatic microsporidiosis or *Enterocytozoon hepatopenaei* parasitic disease (EHP). The pathogen causing EHP and other microsporidians in the order Chytridiopsida are fungal parasites that were originally classified as protozoans. Infection of penaeid shrimps with microsporidian fungal parasites does not generally result in high mortality but decreases nutrient recovery by the digestive system and, therefore, growth. Discolouration of reproductive and muscles tissues and the cuticle are indicative of this disease.

18.5.2 Diagnosis and treatment of mycoses

Most fungal infections can be readily diagnosed by the appearance of a cotton tuft-like hyphal meshwork (mycelium) on affected areas. The hyphae constituting a mycelium can be readily visualized under a light microscope when solid biopsies are analysed. The morphology (appearance) of fungal zoospores under the microscope is used to classify different species of water moulds.

However, since most fungal pathogens of aquaculture organisms have hyphae, and zoospores are sometimes difficult to distinguish, additional diagnostic tools are needed. PCR that utilizes species-specific primer pairs is the most reliable diagnostic tool for identifying a particular pathogen. Growth on selective media and cytopathic effects on host cell lines can also be utilized for diagnosis but these approaches are less commonly used for fungi than for bacteria.

Many fungal pathogens, including *Saprolegnia spec.*, can alternate between asexual and sexual reproduction. Asexual reproduction is via motile zoospores that are produced in a zoosporangium of the parent organism. The zoosporangium is located at the tip of non-septate hyphae. Sexual reproduction is achieved by fusion of two gametes in an oogonium. Fused gametes develop into an oospore that is surrounded by a sturdy cell wall.

Both zoospores and oospores tolerate significant environmental stress, which calls for stringent sterilization procedures for aquaculture equipment exposed to fungal pathogens. Both thick-walled oospores and zoospores are highly environmentally tolerant and endure heat, desiccation, and exposure to disinfecting chemicals for extended periods. Oospores can generally be dormant for longer periods in unfavourable environments than can zoospores. Improvement of environmental conditions, for example, refilling of aquaculture tanks or ponds with water after drying and cleaning,

Table 18.3 Common fungal diseases of aquaculture animals

Fungal pathogen	Host	Disease	Specific symptoms	Treatment
	Fish			
Saprolegniales (multiple species from different families, incl. *Saprolegnia parasitica*)	Many FW species	Saprolegniasis	Cover of white or brownish cottonwood-like meshes of hyphae (mycelium) over skin, gills, and eyes	Table salt (NaCl), copper sulfate ($CuSO_4$), potassium permanganate ($KMnO_4$), formalin, malachite green, hydrogen peroxide
Branchiomyces sanguinis, *B. demigrans*	Carp, catfish, tilapia, other FW fishes	Branchiomycosis	Presence of mycelium	Quicklime, copper sulfate ($CuSO_4$), benzalkonium chloride
Aphanamyces invadans	Marine fishes	Epizootic ulcerative syndrome (EUS)	Skin ulcers, presence of mycelium	
Dermocystidium anguillae	Eels	Fungal eel disease	Presence of mycelium	
	Molluscs			
Ostracoblabe implexa	Oysters, clams	Shell fungus	Dark raised conchiolin 'warts' on inner shell, adductor muscle detachment	
Sirolpidium zoophthorum	Oysters, clams	Larval mycosis	Presence of mycelium	
	Crustaceans			
Saprolegniales (multiple species from different families)	Many FW species	Saprolegniasis	Cottonwood-like meshes of hyphae (mycelium)	Table salt (NaCl), copper sulfate ($CuSO_4$), potassium permanganate ($KMnO_4$), formalin, malachite green, hydrogen peroxide
Aphanomyces astaci	European FW crayfish	FW crayfish disease	Presence of mycelium	
Enterocytozoon hepatopenaei	Shrimps	Hepatopancreatic microsporidiosis (*Enterocytozoon hepatopenaei* parasitic disease, EHP)		

triggers spore germination to give rise to new fungal organisms. Because of their high environmental tolerance, fungal spores can persist in harsh conditions until they encounter a host to infect. For this reason, the treatment of mycoses includes water filtration using mesh sizes that capture zoospores and oospores in addition to baths containing fungicidal chemicals.

Treatment of advanced fungal disease is often unsuccessful and mortality rate is high. However, if mycoses are diagnosed at an early stage and environmental conditions are adjusted to minimize stress and eliminate primary infections by other pathogens, then chemical baths can be effective remedies. The most effective fungicides include carcinogenic chemicals such as malachite green and formaldehyde. However, these chemicals have been banned in many countries. Alternative treatments are based on generation of salinity stress by adding NaCl, and oxidative stress by producing oxygen radicals that kill the fungus, for example, by application of $KMnO_4$, hydrogen peroxide, or ozonated water. More specific fungicides are also used for treating mycoses. The antibiotic oxytetracycline is often used to treat both primary bacterial and secondary fungal infections.

18.6 Diseases caused by protozoans and macroparasites (parasitoses)

Parasitoses represent an infection of a host organism with another organism (the parasite) that causes harm to the host. Parasites rarely cause mortality as they have evolved to depend on a living host. However, parasitism causes significant growth retardation, immunosuppression, and stress, which renders aquaculture organisms susceptible to secondary infection by other pathogens. Parasitoses can be distinguished based on where the parasites reside in the host. Endoparasites are endobiotic eukaryotes that live inside the host while ectoparasites are epibiotic eukaryotes that are attached to exterior surfaces (e.g. skin, gills) of the host. The most prevalent parasites of aquaculture organisms are microparasitic protozoans and larger, multicellular animals (macroparasites). Macroparasites of aquaculture animals that have caused large economic losses on a worldwide scale include cnidarians, helminths, and crustaceans.

18.6.1 Microparasites of aquaculture animals

Many economically important parasitoses that are caused by eukaryotic protozoans have been identified in aquaculture fish and molluscs but they are much less prevalent in aquaculture crustaceans (Table 18.4). Freshwater white spot disease or ichthyophthiriasis (Ich) is a parasitosis caused by the ciliated protozoan *Ichthyophthirius multifilis*. This ubiquitous disease affects many species of FW fish and can cause high mortality. This epibiotic protozoan microparasite is usually transmitted horizontally by an infected carrier fish.

The life cycle of *I. multifilis* requires only fish as a host and consists of four stages (trophont, protomont, tomont, and theront). The trophont represents the ectoparasitic stage that penetrates the host epidermis and derives nutrients from sloughed host skin cells. This causes deterioration of the skin and the formation of characteristic white spots where trophonts develop into protomonts that detach from the host and freely swim in the water for a few hours using cilia. The protomont attaches to a substrate (e.g. the tank walls, soil), loses the cilia, and encysts to form a tomont. The encysted, roundish tomont stage is enclosed within a gelatinous capsule that is highly tolerant of environmental stresses and difficult to eradicate. It divides by binary fission during a period of three to twenty-eight days, ruptures eventually, and releases hundreds of tomites, which transform into free-swimming theronts. Theronts represent the infective stage of this ciliate microparasite. They contain a perforatium, a cilia-covered structure that facilitates infection of a new host fish within the one- to two-day lifespan of the theront.

Marine white spot disease (marine Ich) is caused by another epibiotic species of ciliate protozoan (*Cryptocaryon irritans*) and affects many marine fishes. This microparasite has a four-stage life cycle similar to *I. multifilis*. Ciliate gill disease of FW fish is caused by a variety of epibiotic ciliate protozoans, including the genera *Trichodina*, *Epistylis*, *Apiosoma*, and *Chilodonella*. Infection of fish with ciliate ectoparasites can cover large surfaces of the body, severely impair gill and epidermal barrier functions, and cause high mortality in aquaculture hatcheries, grow-out systems, and ornamental aquaria. Several epibiotic protozoan species (*Epistylis spec.*, *Vorticella spec.*, *Zoothamnium spec.*) also cause ciliate gill disease in shrimps. This disease impairs the vital respiratory and osmoregulatory functions of crustacean gills and can cause mortality resulting from hypoxia and breakdown of body water, electrolyte, and acid-base homeostasis.

Table 18.4 Common parasitoses of aquaculture animals caused by protozoans

Protozoan pathogen	Host	Disease	Specific symptoms	Treatment
	Fish			
Ichthyophthirius multifilis	Carp, Salmonids, many FW species	Ichthyophthiriasis (Ich)	Small white tubercles on the body, corneal lesions	Table salt (NaCl) for ciliated stages, malachite green, formalin
Cryptocaryon irritans	Marine species	Marine white spot disease (marine Ich)	White spots on the body	FW immersion
Trichodina spec., Apiosoma spec., Epistylis spec., Chilodonella spec.	Eels, catfish, other FW fishes	Ciliate gill disease	Fog-like cover of mucus over skin, clamped fins	Formalin
Neoparamoeba perurans	Salmonids, other marine fishes	Amoebic gill disease (AGD)		Freshwater
Ichthyobodo necator (prev. *Costia necatrix*)	Carp, catfish, other FW and marine fishes	Ichthyobodosis, costiasis	Dull spots on the sides of the body	Potassium permanganate ($KMnO_4$), malachite green, formalin
Amylodiunium ocellatum	Marine and brackish water fishes	Dinoflagellate disease		
Cryptobia spec.	marine fishes, cichlids	Flagellate protozoan disease	Intestinal granulomas, parasite presence in blood	Formalin, malachite green, FW baths for brackish and marine species
	Molluscs			
Perkinsus spec.	Oysters, clams, mussels, abalone	Dermo disease		
Haplosporidium spec.	Oysters	Minchinia disease		
Bonamia spec.	Oysters	Bonamiosis		
Polydora spec.	Oysters	Shell-boring polychaete disease		
Mytilicola spec. (copepods)	Oysters	Oyster red worm disease		
Trichodina spec.	Oysters	Ciliate oyster disease		
	Crustaceans			
Epistylis spec., Vorticella spec., Zoothamnium spec.	Shrimps	Ciliate gill disease		

Amoebic gill disease (AGD) of marine fishes is caused by the protozoan amoeba *Neoparamoeba perurans*. This disease has a large impact on salmon aquaculture but also affects other marine aquaculture fishes, echinoderms, and crustaceans. Symptoms of AGD in fish include hyperplasic and proliferative lesions of gill tissue, mucus build-up on gills, and mottled, partially pale gills due to compromised blood circulation.

Ichtyobodosis affects many species of FW and marine fishes and is caused by infection with protozoan flagellate ectoparasites of the genus *Ichthyobodo*, including *Ichthyobodo necator* (prev. *Costia necatrix*). The life cycle of this pathogen consists of a free-swimming stage and a sessile stage that attaches to skin and gill epithelia of host fish. Both life cycle stages of this ectoparasite are susceptible to sterilizing chemicals such as formalin, hydrogen peroxide, and $KMnO_4$.

Dinoflagellate disease is caused by *Amylodiunium ocellatum* and has been responsible for large-scale epidemics in mariculture farms with very high mortality throughout the world. The life cycle of this ectoparasite is similar to *I. multifilis* and the encysted tomont is difficult to eradicate due to its very high environmental stress tolerance.

Another microparasitic disease is trypanosome disease (trypanosomiasis), which is caused by infection with endoparasites of the genus *Cryptobia*. Infection of FW and marine fishes and invertebrates with these endoparasites is commonly mediated via biological vectors, for example, leeches that mediate transfer of pathogens into the blood or haemolymph of hosts.

Dermo disease is an economically important microparasite disease that affects virtually all aquaculture molluscs, and is caused by pathogens of the genus *Perkinsus*, including *Perkinsus marinus*. The name 'dermo disease' derives from the previous name of the pathogen *Perkinsus marinus*, which was *Dermocystidium marinum*. Disease symptoms include degraded internal foot muscle and mantle tissue. Table 18.4 lists several other economically important protozoan diseases of aquaculture molluscs.

18.6.2 Macroparasites of aquaculture animals

Many economically important parasitoses are caused by macroparasites in fish and molluscs but they are less prevalent in aquaculture crustaceans (Table 18.5). Significant macroparasites of aquaculture animals may have multiple hosts during their life cycle, that is, they depend on another host species in addition to the aquaculture species to complete their life cycle. This property is important when considering treatment options, which may include eradication of the intermediate host to break the life cycle of the parasitic pathogen.

Myxozoans are small, multicellular animals of the phylum cnidaria. They include endoparasites that infect aquaculture fish. Most myxozoan parasites require an intermediate host, which is commonly an oligochaete worm in FW and a polychaete worm in SW. Myxospores develop in infected fish and are released to the environment where they find and infect the intermediate invertebrate host and proliferate to produce actinospores in intestinal tissue of the intermediate host. Actinospores are then released into the water and infect new fish hosts to proliferate and produce myxospores in internal host tissues. *Myxobolus cerebralis* causes whirling disease (myxobolosis), which is particularly prevalent in FW hatcheries of trout and other salmonids. The name of the disease reflects the occurrence of erratic tail-chasing behaviour in some infected fish. This pathogen requires tubifex oligochaetes, which are often used as live food for ornamental fishes, as intermediate hosts. Mortality from this disease is very high and survivors are compromised by neurological and skeletal defects.

Proliferative kidney disease (PKD) is another economically important myxozoan disease that causes significant losses in commercial salmonid aquaculture. The main pathogen causing this disease is *Tetracapsuloides bryosalmonae*, which requires a bryozoan intermediate host. This endoparasite infests internal fish tissues, particularly the kidney, and causes severe inflammation and degradation of affected tissues.

Dactylogyrosis is a trematode (helminth) ectoparasite disease that produces characteristic symptoms due to the presence of the causative pathogens described as gill flukes (e.g. *Dactylogyrus spec.*) and skin flukes (e.g. *Gyrodactylus spec.*). These pathogens can infect many species of FW and marine fishes. In contrast to gill and skin flukes, blood flukes are caused by trematode endoparasites of the family *Aporocotylidae* that require a polychaete or gastropod as an intermediate host to infect FW and marine aquaculture fishes. Nematode (roundworm)

Table 18.5 Common macroparasites of aquaculture animals

Parasite	Host	Disease	Specific symptoms	Treatment
	Fish			
Myxozoan cnidarian parasites (*Myxobolus spec.*)	Trout, carp, mullets	Whirling disease, myxobolosis	'Blacktail' syndrome	
Myxozoan cnidarian parasites (*Tetracapsuloides bryosalmonae*, *Henneguya salminicola*)	Salmonids	Proliferative kidney disease (PKD)		Malachite green, fumagillin
Dactylogyrus spec., *Gyrodactylus spec.*, *Thaparocleidus spec.*, *Pseudodactylogyrus anguillae*, *Neobenedenia girellae*	Carp, eels, salmonids, and other FW species, many marine species	Dactylogyrosis, (trematode ectoparasite disease, monogenean gill disease)	Gill and skin flukes	Ammonia, bromex, trichlorphon, formalin potassium permanganate ($KMnO_4$), praziquantel, formalin, mebendazole
Capillaria spec., *Anisakis spec.*, many others	Marine fishes	Nematode disease	larval stage of visible through skin, as coiled or rod-shaped inclusion in muscle tissue	Trichlorfon, niclosamide, levamisole or mebendazole mixed in feed
Argulus spec. (maxillipod crustaceans)	Mostly FW fishes, incl. carp salmonids, eels, breams	Argulosis (Fish lice disease)	Visible parasite attached to body	Potassium permanganate ($KMnO_4$), malathion, trichlorphon, bromex, Lysol
Caligid copepods, e.g. *Lepeophtheirus salmonis*	Salmonids, mullets	Sea lice disease	Lumpy body surface covered with excess mucus	Organophosphates, pyrethroids, FW immersion, hydrogen peroxide
Ergasilus spec. (copepods)	Carp, other FW fishes, marine fishes	Gill lice disease	Visible parasites on gills	
Lernaea spec. *(copepods)*	Carp and other FW species	Lernaeosis (anchor worm disease)	Lesions on body with attached parasites	Table salt (NaCl), potassium permanganate ($KMnO_4$)
	Molluscs			
Cliona spec.	Oysters	Boring sponge shell disease	Holes in shell	
Polydora spec.	Oysters	Mudworm shell disease	Holes in shell and blisters on inner shell	
Snails of the family *Pyramidellidae*, e.g. *Boonea impressa*	Oysters	Pyramidellid disease		

endoparasites of fish include zoonotic species, for example, *Anisakis spec.* and *Capillaria spec.*, and species that infect only a very narrow range of hosts, for example, *Anguillicoloides crassus*, which only infects eels of the genus *Anguilla*. Some parasitic nematode species require intermediate hosts

Figure 18.4 Multiple specimens of the parasitic micromollusc *Otopleura mitralis* (family Pyramidellidae) feeding on a cnidarian host. Photo by author.

while others can be transmitted directly from fish to fish.

Major crustacean ectoparasites of fish include maxillipeds of the genus *Argulus* and many different species of copepods. *Argulus spec.* causes fish lice disease (argulosis) primarily in FW fish, including the major aquaculture species. Caligid copepods (e.g. *Lepeophtheirus salmonis*) are the causative pathogens for sea lice disease, which represents a major challenge for salmon mariculture grow-out and also affects other marine fishes. Gill lice disease of FW and marine aquaculture fish is caused by copepods of the genus *Ergasilus*. Copepod ectoparasites of the genus *Lernaea* cause anchor worm disease (lernaeosis) in carps and other FW fishes.

Macroparasite diseases of aquaculture molluscs include boring sponge shell disease, which is caused by *Cliona* species (phylum Porifera) and mudworm shell disease, which is caused by *Polydora* species (phylum Annelida). Both diseases are characterized by visible holes in mollusc shells and by blisters on the inner surface (nacre) of the host's shell. In addition, ectoparasites of oysters and other aquatic invertebrates include snails of the family *Pyramidellidae*, which cause pyramidellid disease (Figure 18.4).

18.6.3 Diagnosis and treatment of parasitoses

Diagnosis of diseases caused by large ectoparasites can be readily made by visual inspection of aquaculture organisms. Microparasites can be diagnosed by microscopy of biopsy material. For endoparasites it is important to collect biopsies from relevant infected tissues, for example, blood samples for blood flukes and muscle biopsies for some nematodes to maximize the likelihood of early detection of pathogens. Molecular tools such as PCR represent the most reliable method for diagnosing a specific pathogen and determining the species if proper reference genomes are available. Therefore, much effort has been and continues to be devoted to whole genome sequencing (WGS) of aquaculture pathogens. DNA sequence regions that are species-specific, that is, do not occur in other species, are then identified using comparative genomics approaches and used for PCR primer design.

Once a reference genome for a pathogen species is available, tools other than PCR, including transcriptomics and proteomics, also represent powerful diagnostic tools. Transcriptomics is the analysis of all RNA sequences present in biological samples (biopsies) while proteomics is the analysis of most protein sequences in such samples. If these approaches detect the presence of pathogen-specific RNA or protein sequences, then infection of hosts with a particular species of pathogen is confirmed.

Treatments of parasitoses are based on bath immersion and medicated feeds. Some baths utilize exposure to different salinity or temperature if the pathogen tolerance range to these abiotic parameters is narrower than that of the host. Many antiparasitic chemicals are used in baths and medicated feeds but the legality of their use may be limited and may differ depending on country. Common sterilizing baths utilize formalin, copper sulfate ($CuSO_4$), $KMnO_4$, and a large array of more specific parasiticides (anti-parasitic chemicals).

Dimetridazole is used as an anti-parasitic (parasiticidal) chemical to treat infections by protozoan microparasites. Parasiticidal antihelminth compounds used in aquaculture include praziquantel, levamisole, mebendazole, fembendazole, and piperazine. Insecticides such as pyrethroids and organophosphates (e.g. trichlorphon (dipterex)), are used for treatment of parasitism caused by copepod and *Argulus* ectoparasites since these ectoparasites are crustaceans that are phylogenetically related to insects. Crustaceans and insects are both arthropods and share susceptibility to chemical insecticides.

Interestingly, exploration of co-culture of symbiotic cleaner wrasses (family *Labridae*) that feed on ectoparasites of seafood aquaculture species is increasing on a commercial scale to mitigate infestation of aquaculture stocks with ectoparasites and combat the development of drug resistance to chemical parasiticides in pathogen populations.

18.7 Disease prophylaxis

Many diseases of aquaculture organisms are difficult to treat and often treatment options are very limited, or treatment is not possible once a disease outbreak has been diagnosed. Therefore, proper implementation of prophylactic measures is of paramount importance for all aquaculture farms. Plans and facilities for regular monitoring of organisms and rapid quarantine procedures are imperative to diagnose disease outbreaks early and keep them contained to minimize damage and losses.

A biosecurity and risk management plan should contain standard operating procedures (SOPs) that can be followed without delay in case of a disease outbreak. SOPs for managing infectious diseases should include:

1. plans for disease prevention and detection;
2. plans for disease containment and eradication;
3. plans for correction of suboptimal conditions that increase disease susceptibility by causing stress; and
4. steps for following applicable regulatory guidelines and policies.

Common sources of pathogen introduction are water and infected aquaculture organisms. Organisms that are introduced to any aquaculture system from an outside source should first be quarantined and monitored for relevant pathogens before introducing them into the system. This praxis should apply to broodstock, eggs, larvae, and juveniles that are caught from wild habitat or transferred from a different facility. Similar consideration should be given to live food such as *Tubifex* or blood worms, which can be vectors of infectious disease.

18.7.1 Treatment of contamination sources

Water from external sources should be filtered and/or sterilized using filter sizes and sterilization methods that effectively eliminate pathogens. This is most feasible for RAS units and least feasible for open pond and flow-through tank, cage, and raceway systems that are susceptible to continuous exposure to potential pathogen influx. Therefore, of all aquaculture systems used, stringent biosecurity measures can best be implemented in RAS.

Water sterilization methods that are common in RAS units include ozonation, UV irradiation, and chlorination (Section 17.3.6). Chlorination uses 200 ppm chlorine and must be done with great caution since wastewater that is high in organic compounds promotes the formation of chloramines, which are toxic, difficult to purge, and may cause environmental harm or stress to aquatic organisms. Likewise, ozonation must be performed with caution and the residence time of treated water in holding units must be long enough to allow for decay of ozone to avoid oxidative stress on organisms in receiving waters.

Specific pathogen-free (SPF) shrimp and fish can be grown in fully closed RAS, which represents a highly efficient strategy for disease prevention while permitting very high-density culture. Given the devastating outbreaks of viral and bacterial diseases on shrimp and fish farms in many countries in the early 2000s, this preventative aspect represents a tangible incentive for investing funds into transitioning from semi-intensive open pond systems to intensive RAS.

Cage-based aquaculture systems have no option of water filtration. Preventative measures in pond-based aquaculture systems often rely on sterilization methods based on application of quicklime and/or bleach to disinfect aquaculture ponds and tanks prior to filling them with water and stocking. This practice greatly reduces pathogen and predator occurrence in the soil but does not eliminate any potential introduction with water when ponds are filled unless appropriate filters are used.

18.7.2 Regular health monitoring

Regular monitoring of aquaculture organisms represents a critical prophylactic measure for early detection and containment of diseases. Health monitoring must be conducted by trained personnel who are familiar with the behaviour and appearance of the species cultured and the procedures and assays required for health assessment. Daily monitoring should be performed during feeding, and weekly or monthly monitoring programmes can be implemented by examining a small number of representative individuals more closely. Such examination can include the collection of biopsies and routine diagnostic screening for common pathogens, which can be combined with other procedures (e.g. regular thinning, size grading).

Cast-nets are often used during feeding time to sample a small number of animals for routine inspection and monitoring. This can be done on a daily, weekly, or monthly basis depending on species. The monitoring interval represents a trade-off between cost and handling stress on the one hand and the ability for timely detection of potential problems on the other hand.

Routine monitoring and examination of aquaculture animals for common symptoms associated with infectious diseases can identify problems early and trigger a standardized follow-up action/management plan that includes quarantine and more detailed examination of any animals showing abnormal behaviour or other disease symptoms. Routine monitoring can identify whether a potential problem exists and is based on recognizing relatively non-specific visual symptoms of disease.

In most cases, non-specific symptoms have poor diagnostic value for identifying the specific pathogen that is the cause of a disease. Even some of the more specific behavioural and morphological abnormalities mentioned in Tables 18.1–18.5 are often ambiguous and not suitable for identifying the causal pathogen. Therefore, follow-up using the molecular or microscopy-based tools discussed is required for proper diagnosis.

Nevertheless, non-specific symptoms of disease are highly valuable for low-cost, rapid, and early detection of potential disease problems in aquaculture. Their detection can be used to trigger a follow-up action plan for proper diagnosis and containment. Often, non-specific symptoms appear rapidly following acute infection of aquaculture animals with a pathogen and can be recognized within a few days after infection. Such symptoms include morphological and behavioural abnormalities.

18.7.3 Behavioural and morphological symptoms of disease

Behavioural symptoms that indicate a potential disease problem include erratic or sluggish swimming, changes in feeding activity (e.g. anorexia), increased aggressiveness, hyperventilation, impaired opercular movement, altered faecal casts, rubbing of animals against hard surfaces, respiratory distress (e.g. gulping at the water surface), lethargy or disorientation, general debilitation and listlessness, and trembling.

Morphological abnormalities that represent non-specific disease symptoms of aquaculture animals include erosions, ulcers, and necroses (dead tissue) on the external body, changes in gill colour and body pigmentation, haemorrhages, ascites (swollen belly), abdominal distension, exophthalmia or sunken eyes, over-production of mucus, clamped or frayed fins (in fish), softness of exoskeleton and shell lesions (in crustaceans and molluscs), and emaciated appearance.

Any of these symptoms should activate a valid quarantine protocol immediately for a proper duration. Quarantine is most effective if the pathogen is not already prevalent in the area, that is, if an exotic pathogen has been imported from another area and is not already established in the natural environment surrounding the aquaculture system. Most outbreaks of infectious aquaculture diseases are the result of relocating aquaculture animals, either live or frozen, to new areas that were originally free of the corresponding pathogen. Once infection has been diagnosed, the infected stocks must be removed or further quarantined until successful treatment is confirmed (if chemotherapy or other treatments are possible). Reinfection must be prevented, and water quality conditions should be improved to alleviate stress.

18.7.4 Preventative measures

Ideally, infectious diseases should be prevented before they occur. Prevention is better than treatment because it is less costly and avoids side effects such as the destruction of biological filters (decomposers and algae) by drugs used for treatment, as well as the seepage of drugs into natural waters (e.g. when used in earthen ponds). Because disease prevention represents a more effective strategy than treatment it is increasingly emphasized in aquaculture efforts.

Sometimes antibiotics and other drugs are added to prevent disease during transportation when animals are highly susceptible to injury and stress. These preventative practices can promote the rise of drug- and antibiotic-resistant bacteria, which is undesirable in the long-term not only for aquaculture, but also for public health.

Vaccination is a better preventative strategy that relies on adaptive/acquired immunity (i.e. cellular memory), which is limited to vertebrates. Therefore, vaccine development in aquaculture focuses on commercially important fish species. Vaccine development against viral aquaculture diseases represents a particularly active area of research because of lack of effective viroses treatments. However, vaccines have also been developed for several bacterial pathogens that are economically important for aquaculture. Most vaccines are developed by killing the pathogen or purifying select antigens (pathogen surface proteins) that are recognized by the immune system of fish and trigger adaptive immunity in the host by producing corresponding antibodies (Section 18.2.3).

As mentioned, traditional vaccine delivery is by intramuscular injection, but this method is time-consuming, costly, and induces significant handling stress. A more economical and less-invasive vaccine delivery method is bath immersion. Currently intensive research hopes to improve immersion delivery efficiency, which is still significantly lower compared to injection.

Vaccination reduces the need for treatment with toxic and potentially harmful chemotherapeutic drugs and antibiotics and has been shown to be effective. For example, tilapia vaccinated against *Streptococcus* bacteria show greatly reduced mortality over a period of a year compared to unvaccinated tilapia. Besides vaccination, disease prevention is facilitated by aseptic management practices that should be implemented in general SOPs for all aquaculture systems.

18.7.5 Generation of disease-resistant strains

Selective breeding or genetic engineering of specific disease resistant stocks of aquaculture organisms represents a potent preventative strategy. However, when pursuing this strategy, the process of host-pathogen coevolution needs to be considered and genetic improvements of stocks must be updated from time to time. Host–pathogen coevolution is the concerted continuous natural selection of organisms that results from the reciprocal interaction of hosts and pathogens, which enables the pathogen to 'catch-up' with improved host defence mechanisms.

Reducing stress by artificial selection for increased stress resilience and controlling stress factors represent critical strategies of disease prophylaxis (Section 17.4). Stress on organisms can be significantly reduced by selective breeding (domestication) and by eliminating poor aquaculture practices. Preventing injury from overcrowding, malnutrition, overfeeding, suboptimal abiotic water quality parameters (Chapter 16 and Section 17.3), and intraspecific competition reduces the occurrence of lesions and the likelihood of immunomodulation, which minimizes the likelihood of infection with opportunistic pathogens.

18.7.6 Sanitation and hygiene

Disease prophylaxis also includes proper sanitary procedures and good hygiene in aquaculture farms. Proper disinfection procedures should always be applied to commonly used equipment, dip nets, instruments for monitoring water quality, etc., by dipping them into disinfectant solutions. These are concentrated solutions that are used to clean hatchery concrete floors, gloves for handling fish, and solid surfaces such as the walls of tanks and bottom of dried-up ponds. They contain polyvinylpyrrolidone-iodine complexes (PVP iodine or iodophor, 250 ppm), chlorine (2%), NaOH, or quicklime (1%, to temporarily increase pH).

Routine cleaning and drying of equipment, tanks, raceways, or ponds should always be for a period that is sufficient for proper disinfection. Better and coordinated management of genetic, nutritional, stress, and health aspects of aquaculture farms along with the development of more powerful diagnostic procedures for screening organisms introduced into aquaculture systems from external sources improve disease prophylaxis.

Key conclusions

- Infectious diseases have been responsible for the loss of almost half of all aquaculture production.
- Infectious disease incidence can be minimized by strict biosecurity measures, which can best be implemented in closed recirculating aquaculture systems (RAS).
- Infections are caused by pathogens transmitted directly (by physical contact) or indirectly (via biological vectors) and horizontally (within the same generation) or vertically (from parents to offspring).
- Many infectious pathogens are present in the environment, but healthy aquatic animals suppress them by employing pathogen defence mechanisms.
- Immunity is compromised by stresses that are pervasive in aquaculture and promote disease outbreaks, including a much higher density of host organisms in semi-intensive and intensive systems.
- Defence mechanisms of animals against pathogens include physical barriers, innate immunity, and adaptive immunity. The latter is based on cellular memory and limited to vertebrates (aquaculture fish).
- Four types of pathogens cause significant infectious diseases of aquaculture organisms: viruses, bacteria, fungi, and parasites (protozoa and metazoa). Some of them cause zoonoses.
- Viroses are numerous in aquaculture fish and crustaceans and caused by many different types of viruses; They are diagnosed using PCR and cytopathic effects in host cell lines and are best prevented, for example, by vaccination and proper management, since effective treatments are lacking.
- Bacterioses are caused mostly be many types of Gram-negative bacteria but also some Gram-positive bacteria. They are diagnosed by PCR, growth on selective media, and microscopy in combination with specific stains. Treatment options include antibiotics, medicated feeds, and chemotherapeutic baths.
- Mycoses are prevalent in all aquaculture organisms and cause high mortality of eggs and larvae. Diagnosis is based on microscopy and PCR. Treatments include fungicidal baths and medicated feeds.
- Parasitoses are caused by protozoan microparasites and metazoan macroparasites. They are diagnosed using visual inspection of hosts (ectopic macroparasites), PCR, and microscopy, and treated using parasiticidal chemotherapeutics and co-culture with cleaner fish.
- Prophylaxis is the best approach for minimizing disease problems in aquaculture. Disease prevention via proper biosecurity, sanitary and risk management plans include quarantine, water sterilization, disinfection, stress reduction, and regular monitoring procedures to best support animal health.
- Disease prevention is facilitated by using SPF stocks and by selecting or engineering domesticated strains of aquaculture organisms with increased disease resistance.

References

Amlacher, E. (2009). *A textbook of fish diseases*. New Delhi: Narendra Publishing House.

Anoop, B. S., Puthumana, J., Vazhappilly, C. G., Kombiyil, S., Philip, R., Abdulaziz, A., and Bright Singh, I. S. (2021). 'Immortalization of shrimp lymphoid cells by hybridizing with the continuous cell line Sf9 leading to the development of "PmLyO-Sf9"', *Fish & Shellfish Immunology*, 113, pp. 196–207. https://doi.org/10.1016/j.fsi.2021.03.023

Brandes, N. and Linial, M. (2019). 'Giant viruses—big surprises', *Viruses*, 11, p. 404. https://doi.org/10.3390/v11050404

Bunting, S. W. (2013). *Principles of sustainable aquaculture: promoting social, economic and environmental resilience.* Routledge.

Gardell, A. M., Qin, Q., Rice, R. H., Li, J., Kültz, D. (2014). 'Derivation and osmotolerance characterization of three immortalized tilapia (*Oreochromis mossambicus*) cell lines', *Plos One*, 9, pp. e95919. https://doi.org/10.1371/journal.pone.0095919

Gauthier, D.T., 2015. Bacterial zoonoses of fish: a review and appraisal of evidence for linkages between fish and human infections. Vet J 203, pp. 27–35. https://doi.org/10.1016/j.tvjl.2014.10.028

Goh, S. H., Driedger, D., Gillett, S., Low, D. E., Hemmingsen, S. M., Amos, M., Chan, D., Lovgren, M., *et al.* (1998). '*Streptococcus iniae*, a human and animal pathogen: specific identification by the chaperonin 60 gene identification method', *Journal of Clinical Microbiology*, 36, pp. 2164–2166.

Hutson, K. S. and Cain, K. D. (2019). 'Pathogens and parasites', in Lucas, J. S., Southgate, P. C., and Tucker, C. S. (eds.) *Aquaculture: farming aquatic animals and plants.* Oxford: John Wiley & Sons Ltd, pp. 217–247.

Monticini, P. (2010). *The ornamental fish trade - production and commerce of ornamental fish: technical-managerial and legislative aspects.* Rome: United Nations Food and Agriculture Organization.

Muratori, M. C. S., Oliveira, A. L. D., Ribeiro, L. P., Leite, R. C., Costa, A. P. R., Silva, M. C. C. D. (2000). '*Edwardsiella tarda* isolated in integrated fish farming', *Aquaculture Research*, 31, pp. 481–483. https://doi.org/10.1046/j.1365-2109.2000.00448.x

Owens, L. (2019). 'Disease principles', in Lucas, J. S., Southgate, P. C., and Tucker, C. S. (eds.) *Aquaculture: farming aquatic animals and plants.* Oxford: John Wiley & Sons Ltd, pp. 203–216.

Pierezan, F., Yun, S., Piewbang, C., Surachetpong, W., and Soto, E. (2020) 'Pathogenesis and immune response of Nile tilapia (*Oreochromis niloticus*) exposed to Tilapia lake virus by intragastric route', *Fish & Shellfish Immunology*, 107, pp. 289–300. https://doi.org/10.1016/j.fsi.2020.10.019

Pillay, T. V. R. and Kutty, M. N. (2005). *Aquaculture: principles and practices*, 2nd edn. Oxford: Wiley-Blackwell.

Scholtissek, C. and Naylor, E. (1988). 'Fish farming and influenza pandemics', *Nature*, 331, 215. https://doi.org/10.1038/331215a0

Soto, E., Griffin, M. J., Morales, J. A., Calvo, E. B., De Alexandre Sebastião, F., Porras, A. L., Víquez-Rodríguez, X., Reichley, S. R., *et al.* (2018). '*Francisella marina sp. nov.*, etiologic agent of systemic disease in cultured Spotted Rose Snapper (*Lutjanus guttatus*) in Central America', *Applied and Environmental Microbiology*, 84, pp. e00144–18. https://doi.org/10.1128/AEM.00144-18

World Organisation for Animal Health (OIE). (2021). 'Animal diseases', OIE - World Organisation for Animal Health [online]. Available at https://www.oie.int/en/what-we-do/animal-health-and-welfare/animal-diseases/ (Accessed 8 September 2021).

Index